W9-BHU-315

Geodynamics Series

Geodynamics Series

1 Dynamics of Plate Interiors
A. W. Bally, P. L. Bender, T. R. McGetchin, and R. I. Walcott (Editors)

2 Paleoreconstruction of the Continents
M. W. McElhinny and D. A. Valencio (Editors)

3 Zagros, Hindu Kush, Himalaya, Geodynamic Evolution
H. K. Gupta and F. M. Delany (Editors)

4 Anelasticity in the Earth
F. D. Stacey, M. S. Patterson, and A. Nicholas (Editors)

5 Evolution of the Earth
R. J. O'Connell and W. S. Fyfe (Editors)

6 Dynamics of Passive Margins
R. A. Scrutton (Editor)

7 Alpine-Mediterranean Geodynamics
H. Berckhemer and K. Hsü (Editors)

8 Continental and Oceanic Rifts
G. Pálmason, P. Mohr, K. Burke, R. W. Girdler, R. J. Bridwell, and G. E. Sigvaldason (Editors)

9 Geodynamics of the Eastern Pacific Region, Caribbean and Scotia Arcs
Ramón Cabré, S. J. (Editor)

10 Profiles of Oragenic Belts
N. Rast and F. M. Delany (Editors)

11 Geodynamics of the Western Pacific-Indonesian Region
Thomas W. C. Hilde and Seiya Uyeda (Editors)

12 Plate Reconstruction From Paleozoic Paleomagnetism
R. Van der Voo, C. R. Scotese, and N. Bonhommet (Editors)

13 Reflection Seismology: A Global Perspective
Muawia Barazangi and Larry Brown (Editors)

14 Reflection Seismology: The Continental Crust
Muawia Barazangi and Larry Brown (Editors)

Mesozoic and Cenozoic Oceans

Kenneth J. Hsü
Editor

Geodynamics Series
Volume 15

American Geophysical Union
Washington, D.C.

Geological Society of America
Boulder, Colorado
1986

Publication No. 0131 of the International Lithosphere Program

Library of Congress Cataloging-in-Publication Data

Mesozoic and Cenozoic oceans.

(Geodynamics series ; v. 15) (Publication no. 0131 of the International Lithosphere Program)
1. Paleoceanography—Congresses. 2. Geology, Stratigraphic—Mesozoic—Congresses. 3. Geology, Stratigraphic—Cenozoic—Congresses. 4. Paleoclimatology—Congresses. 5. Ocean circulation—Congresses. I. Hsü, Kenneth J. (Kenneth Jinghwa), 1929– II. Series. III. Series: Publication of the International Lithosphere Program ; no. 0131.

QE39.5.P25M47 1986 551.46 86-14062
ISBN 0-87590-515-3
ISSN 0277-6669

Printed in the United States of America.

CONTENTS

CONTENTS

FOREWORD

Raymond A. Price

Past-President, International Lithosphere Program
and
Director General, Geological Survey of Canada, 601 Booth Street, Ottawa, Ontario K1A OE8

The International Lithosphere Program was launched in 1981 as a ten-year project of inter-disciplinary research in the solid earth sciences. It is a natural outgrowth of the Geodynamics Program of the 1970's, and of its predecessor, the Upper Mantle Project. The Program — "Dynamics and Evolution of the Lithosphere: The Framework of Earth Resources and the Reduction of Hazards" — is concerned primarily with the current state, origin and development of the lithosphere, with special attention to the continents and their margins. One special goal of the program is the strengthening of interactions between basic research and the applications of geology, geophysics, geochemistry and geodesy to mineral and energy resource exploration and development, to the mitigation of geological hazards, and to protection of the environment; another special goal is the strengthening of the earth sciences and their effective application in developing countries.

An Inter-Union Commission on the Lithosphere (ICL) established in September 1980, by the International Council of Scientific Unions (ICSU), at the request of the International Union of Geodesy and Geophysics (IUGG) and the International Union of Geological Sciences (IUGS), is responsible for the overall planning, organization and management of the program. The ICL consists of a seven-member Bureau (appointed by the two unions), the leaders of the scientific Working Groups and Coordinating Committees, which implement the international program, the Secretaries-General of ICSU, IUGG and IUGS, and liaison representatives of other interested unions or ICSU scientific committees. National and regional programs are a fundamental part of the International Lithosphere Program and the Chairman of the Coordinating Committee of National Representatives is a member of the ICL.

The Secretariat of the Commission was established in Washington with support from the U.S., the National Academy of Sciences, NASA, and the U.S. Geodynamics Committee.

The International Scientific Program initially was based on nine International Working Groups.

WG-1 Recent Plate Movements and Deformation
WG-2 Phanerozoic Plate Motions and Orogenesis
WG-3 Proterozoic Lithospheric Evolution
WG-4 The Archean Lithosphere
WG-5 Intraplate Phenomena
WG-6 Evolution and Nature of the Oceanic Lithosphere
WG-7 Paleoenvironmental Evolution of the Oceans and Atmosphere
WG-8 Subduction, Collision, and Accretion
WG-9 Process and Properties in the Earth that Govern Lithospheric Evolution

Eight Committees shared responsibility for coordination among the Working Groups and between them and the special goals and regional groups that are of fundamental concern to the project.

CC-1 Environmental Geology and Geophysics
CC-2 Mineral and Energy Resources
CC-3 Geosciences Within Developing Countries
CC-4 Evolution of Magmatic and Metamorphic Processes
CC-5 Structure and Composition of the Lithosphere and Asthenosphere
CC-6 Continental Drilling
CC-7 Data Centers and Data Exchange
CC-8 National Representatives

Both the Bureau and the Commission meet annually, generally in association with one of the sponsoring unions or one of their constituent associations. Financial support for scientific symposia and Commission meetings has been provided by ICSU, IUGG, IUGS, and UNESCO. The constitution of the ICL requires that membership of the Bureau, Commission, Working Groups, and Coordinating Committees change progressively during the life of the project, and that the International Lithosphere Program undergo a mid-term review in 1985. As a result of this review there has been some consolidation and reorganization of the program. The reorganized program is based on six International Working Groups:

WG-1 Recent Plate Movements and Deformation
WG-2 The Nature and Evolution of the Continental Lithosphere
WG-3 Intraplate Phenomena
WG-4 Nature and Evolution of the Oceanic Lithosphere
WG-5 Paleoenvironmental Evolution of the Oceans and the Atmosphere
WG-6 Structure, Physical Properties, Composition and Dynamics of the Lithosphere-Asthenosphere System

and six Coordinating Committees:

CC-1 Environmental Geology and Geophysics
CC-2 Mineral and Energy Resources
CC-3 Geosciences Within Developing Countries
CC-4 Continental Drilling
CC-5 Data Centers and Data Exchanges
CC-6 National Representatives
Sub-Committee 1 - Himalayan Region
Sub-Committee 2 - Arctic Region

This volume is one of a series of progress reports published to mark the completion of the first five years of the International Geodynamics Project. It is based on a symposium held in Moscow on the occasion of the 26th International Geological Congress.

Further information on the International Lithosphere Program and activities of the Commission, Working Groups and Coordinating Committees is available in a series of reports through the Secretariat and available from the President — Prof. K. Fuchs, Geophysical Institute, University of Karlsruhe, Hertzstrasse 16, D–7500 Karlsruhe, Federal Republic of Germany; or the Secretary-General — Prof. Dr. H.J. Zwart, State University Utrecht, Institute of Earth Sciences, P.O. Box 80.021, 3508 TA Utrecht, The Netherlands.

R.A. Price, President
Inter-Union Commission on the Lithosphere, 1981-85

PREFACE

MESOZOIC AND CENOZOIC OCEANS: AN INTERIM REPORT OF THE WORKING GROUP 7 OF THE INTERNATIONAL LITHOSPHERE PROJECT

Marine sedimentary rocks now exposed on land are mainly deposits of shallow shelf seas, although ancient deep-sea sediments do occur in the mountains. The Alpine radiolarites and pelagic lime-stones have been compared to Recent ocean oozes, after the exploration of the H.M.S. Challenger during the second half of the last century, but the ship that revolutionized the earth science was its modern namesake, D/V Glomar Challenger. Thanks largely to the 15-year effort by the Deep Sea Drilling Project of the Joint Oceanographic Institutions Deep Earth Sampling Program, we have made quantum jumps in our understanding of the history of the ocean. Time has been added as the 4th dimension to oceanography, and the birth of a new inter-disciplinary science, paleoceanography, was announced with the recent publication of the publication of the first issue of a journal of that name.

The International Lithosphere Program is a successor to the International Geodynamics Program of the 1970s. The program is coordinated by the Inter-Union Commission on Lithosphere (ICL), a commission of the International Council of Scientific Unions. The Geodynamics Program had emphasized the explorations of the oceans, and the new program, on the recommendation of the Bureau of the ICL, has been aimed to pay more attention to the continental geology. Nevertheless the oceans, which covered 3/4 of the surface of the lithosphere, are not to be ignored. Eugene Seibold, former President of the International Union of Geological Sciences and a member of IUGS"s Marine Geology Commission, came to Zurich in 1979 and discussed with me the need of establishing a working group on ocean history within the framework of ICL. Particularly we saw a need of coordinating the efforts of land geology and those of marine geosciences. My enthusiastic support Seibold's idea was to result in a telephone call from Ted Flinn, the General Secretary of the ICL, asking me to chair the Working Group 7 on Paleoenvironmental Evolution of the Ocean and Atmosphere. After much hesitation, I consented to take up the responsibility for the first five years of the 10-year program.

We recognized that the recent developments in the investigation of the ocean basins and continental margins offer the prospect of inter-disciplinary and quantitative analyses of the Cenozoic and Mesozoic hydrosphere and atmosphere. The integrated applications of stratigraphical, paleomagnetical, sedimentological, geophysical, and geochemical principles and techniques, combined with rapidly evolving technology for deep-sea drilling and other observational and sampling methods have finally provided access to a record of the paleoenvironemtal evolution of the oceans and atmosphere. This record, largely preserved beneath the seafloor, is the key to decipher the ocean history during the last 150 million years. Furthermore, the lessons from the past should permit us to gain insight into the question on the environmental impact of human activities. The principal objectives of our Working Group, as defined by the Bureau of the ICL, were, therefore:

1. Outline the marine record of the Mesozoic and Cenozoic paleoenvironments and paleocirculations in the Earth's oceans and atmosphere.
2. Integrate this with the stratigraphic record that is preserved on the continents.
3. Extend this back through time as far as permitted by the evidence preserved on the continents.
4. Assess the impact of human activities on global environments and climate.

I recognized at the very outset the need for a limitation of the objectives. Geochemists worrying about Precambrian atmosphere have aims, and work with methodologies, quite different from those of sedimentologists studying deepsea drilling cores. In order to produce concrete result in five years, I decided to focus our attention to the first two objectives, so as to come up with an opus on the Mesozoic and Cenozoic Oceans as the interim report of the Working Group. We expect that the newly reconstituted working group, under the chairmanship of Jorn Thiede, will tackle the third objective, while the last is likely to become a main theme for the proposed Global Change Program of the International Council of Scientific Unions.

A history can be divided into time slices. I started out seeking team leaders to work on the evolution of the late Mesozoic, the Paleogene, the Miocene, and the Plio-Quarternary Oceans, as well as on the three more revolutionary changes at the end of Cretaceous, of Eocene, and of Miocene. Thanks to the cooperation of many friends and colleagues, I am able to present in the volume the history as I have planned.

At an 1981 working session of our WG in Los Angeles, Wolf Berger brought up the point that we should emphasize the operations of the ocean processes instead of the history. We needed analyses of ocean dynamics, of ocean climate, of ocean chemistry, and of ocean life. His argument was well received, but the WG members decided during a WG meeting at Bonn, Germany, 1982, that an integrated synthesis of the mutual relations of various processes can be best accomplished through a historical approach. We have, therefore, agreed that the interim report should consist of two parts: I. Processes and II. History.

This collection of contributions was first presented orally in a symposium during the 27th International Geological Congress at Moscow, August, 1984. The first part of the volume includes five contributrions. Eric Barron gives a status report on physical paleoceanography. The goal is to recreate past oceanic conditions, particularly the structure of ocean currents as related to temperature and salinity fields. To complement interpretations of observational data, Barron presents a mathematic model based on laws governing the atmospheric and oceanic circulations. Surface circulation has been reconstructed on the basis of mean atmospheric winds. Examples relating the surface-pressure distribution and the sealevel height, and the topographical relief on land have been given to portray the relation of atmospheric circulation to paleogeography. The importance of thermohaline circulation in forming bottom-watermasses is emphasized and the climatic implications of ocean heat-transport explored.

The second article by D.G.Seidov presents a numerical modelling of the ocean circulation and paleocirculation. His model is first tested in numerical experiments through a simulation of modern circulation, and then applied to reconstruct paleocirculation patterns of the late Mesozoic and Cenozoic Oceans. Using the data on continent-ocean distribution determined by the seafloor spreading theory, the influence of continental displacement on global climate has been quantitatively evaluated. The theoretical modelling results have been found to be in general agreement with the conclusions based upon interpretations of observational data. The modelling shows, for exmple, that warm abyssal waters, postulated for Cretaceous oceans, could have resulted from a global evaporation-rate and/or subpolar temperatures higher than those of today. The next short article by V.H. Enikeev makes use of the Seidov model to reconstruct the paleoceanography of the last glacial stage. Numerical solutions for the ocean and air temperatures, current velocities and fluxes at various depths, have been obtained for the Recent and for the Würm Stage, and the results are presented graphically.

The articles on modelling are followed by L.A. Frakes's review on Mesozoic-Cenozoic climatic history and causes of glaciation. He noted that the global climate recovered after late Permian chill and was anomalously warm during much of the Mesozoic. A punctuated decline in mean global temperature started in Late Cretaceous and continued till the Quaternary when continental glaciation again prevailed. The climatic changes were probably related to the carbon dioxide content of the atmosphere and the global albedo effect. Those are, in turn, related to volcanism and global tectonics.

The final paper of the Part I by E.A. Boyle is an elegant new synthesis, in which the relation of global climate, ocean chemistry and ocean circulation are investigated through the use of a 4-box model. The changes in atmospheric carbon-dioxide content since the last glacial stage have been related quantitatively to intensity of surface upwelling, ocean chemistry, and polar-temperature changes, and to production of bottom-water masses.

The Part II of the volume starts with an article by S.O.Schlanger on the history of the Cretaceous ocean. The author singles out the high frequency sea-level oscillations as a particularly characteristic phenomenon of that geologic period. Schlanger found 6 major transgressions during the Late Cretaceous, with a periodicity of 5.5 Ma. Their frequency and rate suggest that those changes cannot be accounted for by seafloor processes, nor by other geological processes which have been invoked to explain the sea-level changes. Even though we are ignorant of the cause of those transgressions, their economic implications are clear: The genesis of the source-beds for much of the oil produced on both sides of the Atlantic, for example, has been related to oxygen-deficiency in seawater when the Cretaceous cratons were flooded.

The following article by the editor summarizes the evidence for an unusual event across the Cretaceous/Tertiary boundary. The biotic and geochemical records give clear indications of an environmental catastrophe, lasting for several millenia at least. Trace-element studies suggest that the impact of a large meteorite triggered the dramatic global changes in the oceans.

The article by H. Oberhänsli and K. J. Hsü on the Paleocene-Eocene oceans reviews the sedimentary and paleontological data relevant to paleoceanography. The ocean climate was relatively cool in the beginning, but reached an optimum in Early Eocene. Then the temperatures started to decline and decreased stepwise, with a dramatic drop at the end of the Eocene. Climatic and tectonic changes during the Eocene may have been the factors leading to a major reorganization of the ocean-current patterns. The driving force of the bottom-circulation had been dominantly halokinetic before it was changed to to thermohaline during the Late Eocene and Oligocene. This intrusion of cold bottom waters resulted in the establishment of a psychrosphere benthic fauna to replace the thermospheric fauna of the earlier Paleogene.

B.H.Corliss and L. D. Kergwin have taken a closer look at the Eocene/Oligocene boundary. They note an acceleration of environmental changes across the transition, but they could find no evidence of mass extinction. There may or may not have been a large meteorite-impact near the end of the Eocene, but no major environmental catastrophe has been triggered even if a bolide did hit.

The Miocene Ocean has been investigated by a major project on Cenozoic Paleoceanography (CENOP) and the results published in a monographic treatise. J. P. Kennett prepared an extended abstract of that work for our volume. The latitudinal and vertical thermal gradients in ocean waters were steepened after the development of Circumantarctic current toward the end of the Oligocene. A crucial stage in the development of global paleoceanography was reached during the Middle Miocene when much of the Antarctic ice sheet formed. Biogeographic patterns of planktic microfossils underwent distinct change during the Miocene. Microfaunas, different in the eastern and western Pacific earlier, became similar across the entire tropical Pacific in Late Miocene, when gyral circulation was strengthened. Large changes in benthic communities were also registered, with the rapid replacement of numerous Oligocene and older species by forms that dominate the Neogene and modern deep-sea environments.

In their article on the terminal Miocene event, M. B. Cita and J. A. McKenzie discuss the paleo-oceanographic implications of the desiccation of the Mediterranean Sea. They report indications of extended regression on continental margins, and isotopic evidence for more than one episode of lowered sea-level. There was also a carbon-isotope signal suggestive of a reorganization of global circulation in Late Miocene. The initiation of the Mediterranean desiccation may in fact have been triggered by a lowering of the sea-level, consequent upon an expansion of the Antarctic Ice Cap. The resubmergence of the Mediterranean after the Strait of Gibraltar was opened again at the beginning of the Pliocene may have been a critical factor in causing the rapid retreat of the Antarctic Ice.

The final article on the Plio-Quaternary Oceans is contributed by N. J. Shackleton. He summarizes the oxygen-isotope evidence for sea-level changes during the Plio-Quarternary, and the carbon-isotope evidence relating the variations in atmospheric carbon dioxide to the plankton production in the oceans. The article illustrates the novel uses of the stable-isotope data in gaining insights into the global climate system.

In conclusion, I would like to express my gratitude to all members of ICL WG 7 for their efforts, particularly the authors of the dozen articles in this volume, which summarize succinctly our present knowledge of the late Mesozoic and Cenozoic oceans and have superbly illustrated the applications of the multi-disciplinary methodologies, ranging from computer modelling, to isotope and trace-element geochemistry, to sedimentology and micropaleontology, in the study of ancient oceans.

Kenneth J. Hsü
Zurich,Switzerland

PHYSICAL PALEOCEANOGRAPHY: A STATUS REPORT

Eric J. Barron[1]

National Center for Atmospheric Research, Boulder, Colorado 80307

Abstract. Observational studies have provided an inventory of major trends and events in the history of the oceans. The present status of physical paleoceanography is one of tremendous potential to address the problems of past ocean circulations and the role of the oceans in climate with physical models based on fundamental laws thought to govern the atmosphere and ocean. As yet the application of physical models to the study of ancient oceans has been limited to inferences from simple models and from climate models which lack explicit ocean formulations. Although limited in scope these models provide some important insights into the nature of the surface ocean circulation, the thermohaline circulation and the role of ocean heat transport during earth history.

Introduction

Physical paleoceanography has two goals. The first goal is to recreate past oceanic conditions, particularly the large-scale mean three-dimensional structure of ocean currents and related temperature and salinity fields. To date this objective has been approached almost entirely from the interpretation of observations. These observations incorporate the entire spectrum of research in paleoceanography including geochemistry, paleobiology, sedimentology and stratigraphy. There are numerous qualitative reconstructions of ancient circulation patterns based on data such as biogeography (e.g. Berggren and Hollister, 1974) but as yet there are no comprehensive mathematical models of past circulations.

The second goal in physical paleoceanography is to understand the role of the oceans as a governing factor in climate and to determine the sensitivity of oceanic conditions to external forcing such as changes in geography and atmospheric conditions. The record of isotopic paleotemperatures, variations in sediment character and sedimentation rates, variations in the spatial and temporal distribution of floras and faunas, variations in latitudinal and vertical gradients in the ocean and unusual conditions such as anoxic "events" have focused our attention on many different problems. These problems include the role of heat transport by the oceans in climate change and the differences in circulation intensities, bottom water formation and the uniformity of water masses between the modern and ancient oceans. Our understanding of the sensitivity of oceanic conditions to external forcing has been limited largely to correlations of "causes" and effects (e.g. an oceanographic change associated with the removal of a geographic barrier) and inferences based on theoretical arguments.

The objectives in physical paleoceanography require consideration of the entire global atmosphere, hydrosphere, cryosphere, lithosphere system. Physical oceanography and the atmospheric sciences have evolved rapidly during the time period of extensive data collection in paleoceanography and there is a distinct lag in the integration of modern concepts into geological thought. With the increased empirical knowledge of ancient oceans derived from the ship-based sampling programs of the last two decades and the increased ability to reconstruct the configuration of ancient oceans, quantitative studies have become more feasible and more important. The opportunity now exists to examine past oceanic conditions as a function of governing equations (e.g. equations of motion, continuity, conservation laws) and of forcing at the ocean surface (heat and water balances, winds etc.). Theoretical studies in paleoceanography are in their infancy. Physical models are limited to one pioneering laboratory study (Luyendyk et al., 1972), the contribution by Seidov in this volume, and inferences from simple one-dimensional models or from climate models which lack explicit ocean formulations. Although limited in scope these models are beginning to provide some important insights in paleoceanography.

A "paleoceanographic inventory" of major trends and events in the history of ocean circulation and climate exists as a result of extensive data collection (Berger, 1979). This tremendous body of

[1]Now at Rosenstiel School of Marine & Atmospheric Science, University of Miami, Miami, Florida 33149

information will not be addressed in this report although clearly a comprehensive review of physical paleoceanography should touch on the entire spectrum of observational studies. Instead the focus here will be a status report on theoretical studies of ancient oceans. There are three natural topics on which physical models have some bearing for ancient oceans: 1) the surface circulation characteristics of ancient oceans, 2) the thermohaline circulation and 3) poleward heat transport by the ocean. The application of quantitative models to paleoceanography provides some important constraints and insights but is best described as an area of, as yet, unrealized potential.

The Surface Ocean Circulation

Since the adoption of plate tectonics as a working hypothesis of crustal evolution, paleoceanographers have recognized that ocean basin size, shape and paleolatitude will alter the nature of the surface circulation. There are some excellent correlations between geography, ocean circulation and climate. The opening of the Drake Passage, formation of the Antarctic circumpolar current and extensive glaciation of Antarctica (e.g. Kennett, 1977) is perhaps one of the most frequently cited examples of such a correlation. Ocean gateways and barriers have received particular attention for their role in modifying the surface circulation and influencing climate (Berggren, 1982). Ocean current patterns (e.g. Gordon, 1973; Berggren and Hollister, 1974) have been reconstructed based on paleogeography, biogeography and assumptions on the nature of the global wind patterns. The reconstruction of currents from biogeography, particularly from the distribution of planktonic fossils is often practiced, but rarely is the distribution of samples sufficient for comprehensive reconstructions.

In theory the surface circulation can be reconstructed based on mean atmospheric winds. Two well-known relationships must be taken into consideration. First, as noted by Ekman (1905), the transport of the water is at right angles to the driving force of the winds as a consequence of the earth's rotation and the frictional interaction between the atmosphere and ocean and water layers beneath the surface. Within a hypothetical basin under the influence of the planetary wind field, anticyclonic vorticity is generated in the oceanic region beneath the low latitude easterlies and cyclonic vorticity beneath the westerlies and polar easterlies. This yields gyre-like circulation which would be symmetrical except for a second consideration, the tendency to conserve angular momentum as the water mass changes latitude. Fluid columns moving equatorward on the eastern sides of the mid-latitude ocean basin acquire cyclonic vorticity, opposite that derived from wind-driven motions. Poleward moving columns acquire anticyclonic vorticity adding to the wind-driven effect. This fact explains the strong asymmetry of water motion and the intensified western boundary currents such as the Gulf Stream (Stommel, 1955).

The vorticity theory of the wind-driven ocean surface circulation has been extended to laboratory models by von Arx (1957). The key element of the laboratory models and some analytical and numerical models of the ocean circulation is that a reasonable simulation of surface currents can be generated by consideration of the earth's rotation and the zonal wind stress without regard to heat and moisture balances or bathymetry. The model of von Arx (1957) was the basis for a pioneering laboratory experiment of Luyendyk et al. (1972) based on an early reconstruction of mid-Cretaceous geography. This study considered such interesting aspects of the circulation as geographic controls on cross-polar flow and the circumglobal Tethys current. Unfortunately theoretical studies of past ocean circulation patterns have not been extended much beyond the laboratory model of Luyendyk et al. (1972). Seidov (this volume) applies zonal winds in modeling Cenozoic circulation patterns. The atmospheric forcing applied to the upper ocean surface has been derived from assumptions of how the mean winds and moisture balances may have changed during the Cenozoic cooling trend.

The relationship between global wind patterns and the surface circulation of the ocean has generally been adapted to paleoceanography in a qualitative manner. Either of two assumptions typically underlies the prediction of past surface ocean circulations: 1) global winds are assumed to follow closely the present pattern of low latitude easterlies, mid-latitude westerlies and polar easterlies given in relationship to surface pressure in Figure 1, or 2), the wind patterns are assumed to shift poleward during periods of reduced equator-to-pole surface temperature gradient. However, the potential for different global wind patterns during earth history has not been adequately assessed.

A qualitative physical argument can be presented which suggests that wind patterns must have been different in the past. Energy and angular momentum balances maintain the atmospheric circulation. Consider briefly the angular momentum balance. Different frictional torques exerted over land and sea and associated with topography suggest that different continental and topographic distributions must alter the angular momentum exchange between the surface and the atmosphere, and hence the atmospheric motions which maintain the angular momentum balance. The potential importance of geography as an influence on the global circulation is illustrated in a series of sensitivity experiments using a General Circulation Model of the atmosphere with a simple energy balance ocean and mean annual solar insolation (Barron, 1985). Model experiments were completed for the present day, and for the mid-Cretaceous with assumptions of flat or mountainous and high or low sea level continents. Figure 2 shows the large differences in surface pressure

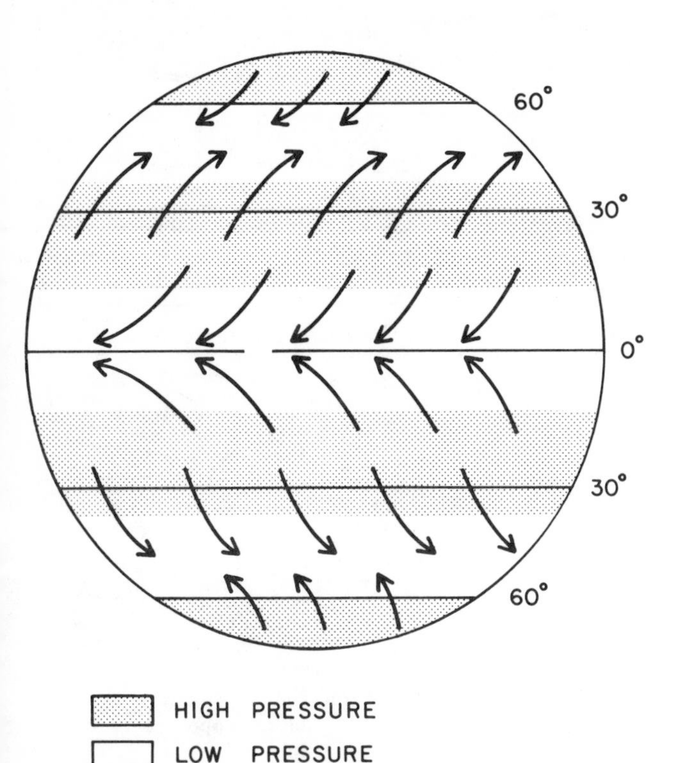

Figure 1. An idealized surface atmospheric circulation giving the major surface pressure zones and easterly and westerly wind belts.

patterns due to geography assumptions. These differences are extreme enough to change the sign of the mean winds at high latitudes. Conversely the low lati-tude circulation appears to be fairly stable. Topography was the most important variable, at least from the viewpoints of as yet limited experiments.

Such experiments illustrate the potential for a variable surface ocean circulation due to atmospheric wind patterns. Given the probability that ocean basin evolution will also alter the nature of the circulation, the research status of the prediction of surface current patterns is one of understanding qualitatively the importance of geography and the role of global winds and perhaps of delineating particular scenarios of ocean circulation through time. However, there exists little physical insight into the sensitivity of the ocean system to specific variables.

The surface circulation of ancient oceans has been addressed partially in terms of current directions alone, but not as vectors or stream functions. An important qualitative assumption about the intensity of the circulation during warm geologic time periods is that the intensity of the atmospheric circulation, and hence the wind-driven surface ocean circulation, will decrease for a decrease in equator-to-pole surface temperature gradient. This assumption is really an overextension of the thermal wind relationship. The thermal wind relationship relates the horizontal temperature gradient to the vertical shear of the geostrophic wind and indicates that the intensity of the winds aloft will decrease if the vertically integrated meridional temperature gradient decreases. However, a change in the surface temperature gradient may not be associated with an equivalent change in the vertically integrated meridional temperature gradient.

The geologic version of the thermal wind relationship is not general because any change in tropical sea-surface temperature is amplified in the upper troposphere. The heat of condensation of water vapor can play an important role in maintaining the vertically integrated meridional temperature gradient during warm climates if tropical temperatures increase even slightly. The key element is saturation vapor pressure, defined as the pressure exerted by water vapor when the air is saturated with respect to a plane surface of pure water. The saturation vapor pressure is a strongly nonlinear function of temperature. At cooler (polar) temperatures a 1°C increase in temperature changes the saturation vapor pressure only slightly, while at tropical temperatures the same increase produces a dramatic increase in saturation vapor pressure. For warmer climates in which tropical sea-surface temperatures are the same or cooler than at present, a decrease in the intensity of the atmospheric winds and the ocean surface circulation is likely. If tropical temperatures can warm slightly, then the intensity of atmospheric winds may be maintained or even increase (Barron and Washington, 1983a).

The intensity of the ocean circulation has not yet been adequately addressed from the viewpoint of the oceanic record. For example, Rea and Janecek (1981) show the interesting result of an eolian mass accumulation rate which was at peak levels during the Cretaceous, supposedly a time of weak circulation. This is unlikely to be explained by changes in the area of the desert source but fits well with the stronger zonal winds in Cretaceous simulations (Barron and Washington, 1983a). Conversely, Thiede (1981) suggests that the absence of sediment reworking in Cretaceous oceanic sediments implies relatively sluggish deep-sea currents. Although not clearly related to the surface circulation the implications extend to the world ocean in general. There is much work to be done to distinguish between circulation patterns which are "different" as opposed to sluggish vs. active designations which are conceptually simple but physically more problematic.

The contribution from physical models on the subject of the ocean circulation intensity is as yet limited to a more precise definition of the implications of reduced equator-to-pole surface temperatures gradients. This focuses our attention on the record of tropical sea surface temperatures during earth history.

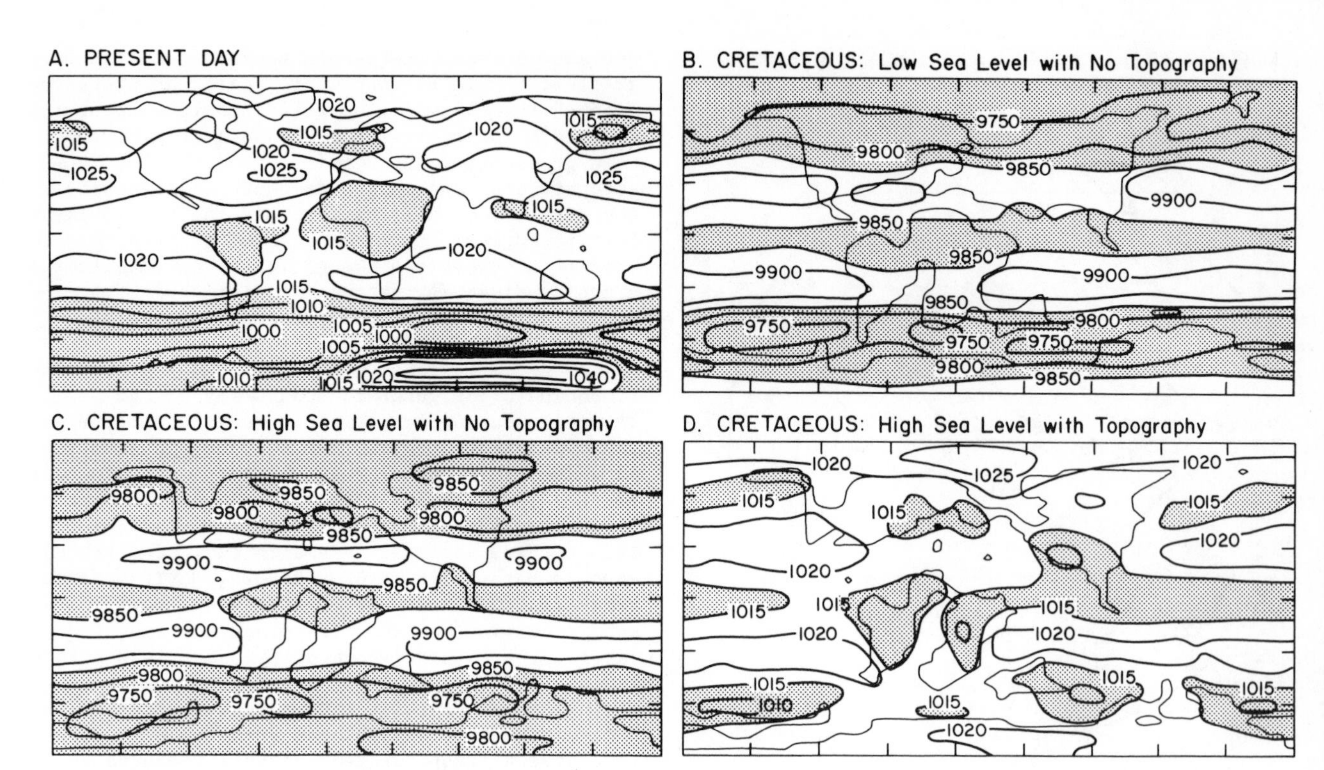

Figure 2. A comparison of the predicted surface pressure distribution for a present-day control and Cretaceous sensitivity experiments with low sea level and no topography, high sea level and no topography and high sea level with topography using a general circulation model of the atmosphere (Barron, 1984). Continental outlines are given. Relative low pressure regions are shaded.

The Thermohaline Circulation

The thermohaline circulation, the component of the ocean circulation driven by density differences, is important in paleoceanography for several reasons. First, the present-day ocean heat transport by the surface circulation in the mid-latitudes opposes the thermohaline circulation. A reversal of the thermohaline circulation could result in a somewhat greater poleward oceanic heat transport.

Second, the mechanisms required to maintain warm polar temperatures are problematic. The warmest estimates for the Cretaceous (≈ 17°C) are based on bottom-water isotopic paleotemperature estimates from benthic Foraminifera (e.g. Savin, 1977) and the assumption that bottom water is formed by sinking of cool dense waters at high latitudes as it is today. A reversal of the thermohaline circulation, as first suggested by Chamberlin (1906), and formation of deep water by sinking of warm, saline subtropical water offers a different explanation of warm benthic paleotemperatures.

Third, present-day deep water formation is associated with high latitude cooling and salt exclusion during sea ice formation. During warm climates high latitude deep water formation may have been inhibited. If the wind driven surface circulation was also reduced, a sluggish deep ocean circulation is indicated. The concept of a stagnant ocean has also been presented as a key to the interpretation of oceanic sedimentation during warm time periods (e.g. Degens and Stoffers, 1976; Schlanger and Jenkyns, 1976).

Finally, thermohaline processes form distinctive water masses and may be the dominant forcing for vertical exchange in the ocean. Thus our conceptions of the nature of the thermohaline circulation have implications for the uniformity of ocean properties.

Until recently the importance of thermohaline circulation in modeling ocean circulation has been overshadowed by research on the nature of the wind-driven surface ocean circulation. Present research efforts focus on the processes of deep water formation and mixing. Efforts to predict present or past conditions are rather limited. The research effort of Peterson (1979) is of particular interest because of the implications for paleoceanography. Peterson (1979) developed a one-dimensional steady thermohaline convection model driven by turbulent buoyant plumes. Steady flow (a buoyant plume) results from the continuous release of a small source of fluid which has a density different from that of the environment. Density differences result from interaction of the ocean surface with the atmosphere and can be accentuated in regions which are partially isolated from the main oceans (e.g. the Weddell Sea).

In Peterson's (1979) model the flow is driven by the action of gravity on the density contrast between the plume and the environment. The density contrast is

$$g' = \frac{g}{\rho o} (\rho_e - \rho) \quad (1)$$

where g is the acceleration by gravity, ρo is a reference density, ρ_e is the environmental density and ρ is the plume density. The plume interacts with the interior ocean by turbulent entrainment. In the case of multiple plumes it is the plume with the greatest buoyancy flux which penetrates to the bottom. The buoyancy flux, F, is the product of the volume flux, Q, and density contrast.

$$F = g' Q \quad (2)$$

The waters formed in the Weddell Sea presently have the greatest buoyancy flux. However, a comparison of the Weddell Sea and the Mediterranean Sea as described by Kraus et al. (1978) shows that the waters of the Weddell Sea are not the densest, but they become the bottom waters by virtue of a strong volume flux. Brass et al. (1982) stress that the areas, locations and configurations of marginal seas have changed considerably through time. An appropriate picture of the thermohaline circulation for any time period may be that of numerous competing plumes from different sources and with different strengths, terminating at different depths depending on the source strength. The source with the greatest buoyancy flux at a given time could conceivably be a low latitude saline source.

Recent isotopic data from Inoceramus, an epibenthic bivalve, support the concept of a variable thermohaline circulation for the Late Cretaceous. Saltzman and Barron (1982) interpret isotopic values from different basins and depths for the Late Cretaceous as indicative of deep waters formed both by cooling in polar regions and by evaporation in the subtropics. The warm Cretaceous oceans thus may have been characterized by quite diverse water masses.

The major implication of these results is that a vigorous thermohaline circulation and water mass variability may have characterized much of the oceans' history. Only the location and strengths of different sources may have varied. Of course changes in the number, location and strengths of deep water sources have numerous implications for the oceanic record. Cretaceous anoxic events may reflect the decreased oxygen solubility in warm saline waters (e.g. Arthur and Natland, 1979; Brass et al., 1982) rather than stable stratification or a sluggish circulation. Changes in global sea level which affect marginal basin geometry and degree of isolation from the main ocean and climatic warming or cooling can modify the sources of deep water. This could produce step-like changes in the deep water isotopic and sedimentologic record, especially if the bottom water source switched from high to low latitudes (e.g. Berger, 1979; Brass et al., 1982). Continental runoff patterns will influence stratification in marginal seas and could have a global oceanographic effect. Barron et al. (1985) even suggest that orbital cycles in the Cretaceous, which modulate land-sea thermal contrast and continental margin precipitation, could produce cyclic deep-sea sedimentation. The potential effects of changes in the thermohaline circulation on poleward ocean heat transport are much more difficult to evaluate.

The results from the one-dimensional thermohaline model clearly have important implications for the deep water circulation, however they tell us only a little about the evolution of the thermohaline circulation. Three-dimensional general circulation models of the oceans, either with an interactive atmosphere or with fixed atmospheric winds and evaporation-precipitation balance derived from atmospheric simulations of past time periods, have considerable potential for predicting bottom water sources and source strengths. Although a number of models (e.g. Holland, 1977) incorporate the thermohaline circulation, convection typically is not handled explicitly in the large-scale models. Vertical mixing is usually parameterized as a function of static stability (e.g. Semtner, 1974; Bryan and Lewis, 1979). For an unstable change in density with depth, both temperature and salinity are completely mixed over the unstable parts of the water column. There remains much to be learned about the processes of bottom water formation and mixing and this aspect of ocean modeling is likely to become more sophisticated.

Ocean Heat Transport

The transfer of heat by the world ocean is an important component of the global heat balance and potentially a major mechanism of climatic change. Numerous authors have hypothesized that land at the poles, which inhibits poleward oceanic heat transport, may be major factor in causing glacia-maintenance of warm polar temperatures from the Cretaceous to the Eocene has been attributed to greater warmth transported by the oceans (e.g. Frakes, 1979), especially in light of the assumption that poleward atmospheric heat transport would decline during periods of reduced equator-to-pole surface temperature gradient (e.g. Brooks, 1928). Apparently much of the burden of maintaining the climatic extremes during earth history has been placed on the ability of the ocean to transport heat poleward.

Climate model experiments suggest more complex relationships between oceanic heat transport and the maintenance of climatic extremes. There are distinctive differences even between warm geologic time periods. Surface temperatures for both the Cretaceous (Barron, 1983a) and the Eocene

(Shackleton and Boersma, 1981) have been reconstructed which indicate that the latitudinal gradient was less than half the value of today. However there are two important distinctions. Mean global Eocene warmth estimated by Shackleton and Boersma (1981) is only about 2°C higher than today with tropical surface temperatures estimated to be as much as 5°C cooler than today. This point has been disputed by Matthews and Poore (1980) as an artifact of assumed no-ice volume.

In contrast, the Cretaceous warming is estimated to be 6 to 12°C higher than today with tropical temperatures similar to or somewhat warmer than modern values. The Cretaceous warming requires a large climatic forcing while the Eocene warmth, if properly interpreted by Shackleton and Boersma (1981), could conceivably result from a redistribution of energy. For both periods the basic assumption has been that the oceans transported more heat poleward because the atmosphere must have been much less effective due to the diminished equator-to-pole surface temperature gradient. As will be shown, Cretaceous and Eocene paleotemperature records have quite different implications. This point is also clear from the discussion of the surface ocean circulation.

Poleward ocean heat transport during periods of reduced equator-to-pole surface temperature gradients has not been investigated, as yet, with explicit ocean formulations. Rather the insights are based on atmospheric climate models with implied or specified oceanic heat transport. Although somewhat crude, inferences concerning the ocean based on the atmosphere are not farfetched. The best modern estimates of oceanic heat transport are not derived from direct observations, but rather they are calculated as a residual by measuring the meridional atmospheric heat transport and the net radiation at the top of the atmosphere (Oort and Vonder Haar, 1976). The point is that there is a great deal more confidence in our knowledge of the atmosphere than there is for the oceans. The inferences from climate models provide a number of interesting insights for Eocene and Cretaceous oceans.

Barron (1983b) specified 10°C as a minimum oceanic surface temperature in an atmospheric general circulation model with an energy balance ocean and Cretaceous geography. In an energy balance ocean the surface temperatures are computed based on the balance of incoming and outgoing energy fluxes. A minimum temperature specification which is greater than the temperature which would be computed from the model energy balance implies an ocean heat transport convergence (the term convergence refers to heat transport into and out of a region such as a latitude zone, rather than across a latitude line). The point of the calculation is to estimate the implied ocean heat transport required to maintain warm polar temperatures. The atmospheric heat transport, implied ocean heat transport, and incoming and outgoing energy fluxes must balance. Therefore, the implied ocean heat transport can be computed as a residual of the atmospheric meridional heat transport and the net radiation at the top of the atmosphere.

The question is whether the poleward heat transport by the ocean required to maintain 10°C polar ocean temperatures is reasonable given known mechanisms and what is known of the geologic record. For a January simulation the 10°C minimum would require that $6x10^{15}$ Watts be transported by the ocean across 60°N latitude. For perspective, an assumption that water at 15°C is transported to the poles and then allowed to cool to 10°C would require a volume flux of 300 Sverdrups ($10^6 m^3 s^{-1}$) to achieve a heat transport of $6x10^{15}$ Watts. The present volume transport of the Gulf Stream through the Florida Straits is an order of magnitude smaller than this value. This estimate ignores any contribution from summer heat storage in an ice-free Arctic, but this factor can only account for a small fraction of the required energy.

Upwelling of warm waters over broad polar areas could be a plausible means of reducing the required current intensities. However, the maximum current velocities wil still be dependent on the size of the gateways entering the Arctic Basin. Smaller gateways imply much stronger flow. Neither the Eocene nor the Cretaceous are characterized by sufficiently wide and deep Arctic connections so as to reduce current strengths above 60°N to realistic values.

Interestingly the simulation with specified 10°C ocean surface temperature minima and a much reduced equator-to-pole surface temperature gradient did not result in a large decrease in the poleward heat transport by the atmosphere nor did it reduce the intensity of the atmospheric winds. The key element in this result is that tropical temperatures were slightly warmer in the simulation (≈ 1-2°C). To a large degree, increased latent heat transport tends to compensate for decreased sensible heat transport in the atmosphere. Held et al. (1981) noted this compensation between latent and sensible poleward heat transport for both increases in carbon dioxide and the solar constant for present-day model simulations. If this compensation did not exist in the Cretaceous simulations described above the required ocean heat transport would be even greater.

There are two implications of the results of the model experiments described above. First, the oceanic heat transport required to maintain 10°C polar temperatures is excessive. The burden of maintaining climatic extremes of warm and glacial climates cannot be placed solely on the ability of the ocean to transport heat. This is not to say that changes in ocean heat transport are not an important element of climate change, only that there are limits to the capabilities of the ocean. Second, the heat transport by the atmosphere is not strictly dependent on the temperature gradient at the surface. A number of factors are important, one of which is the sign of

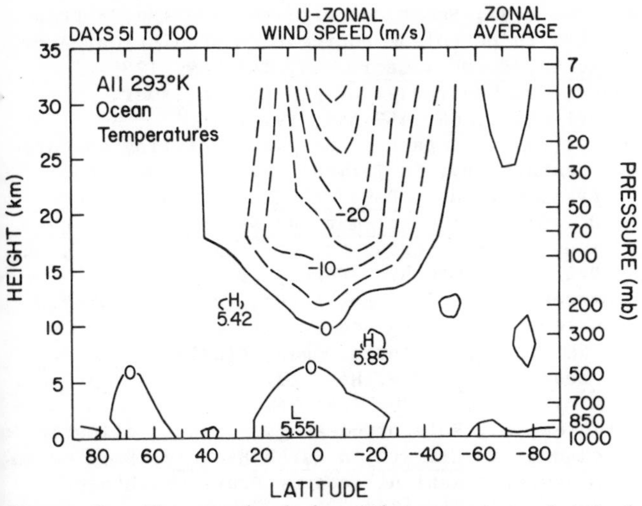

Figure 3. The zonal winds with respect to latitude for a Cretaceous sensitivity experiment with instantaneous ocean mixing (uniform 20°C) using a general circulation model of the atmosphere. Note the lack of the familiar mid-latitude jets which today are on the order of 30 to 40 m/sec.

the temperature change in the tropics. Increased tropical temperatures of even a few degrees in the Cretaceous could maintain the atmospheric poleward heat transport despite the reduced equator-to-pole surface temperature gradient.

For the Eocene temperatures described by Shackleton and Boersma (1981), the role of the atmosphere must decrease substantially. Decreased tropical temperatures with a decreased equator-to-pole temperature gradient are problematic because substantially more efficient poleward ocean heat transport is required. This is especially enigmatic because these conditions also imply a weaker wind-driven ocean circulation, which is the major component of poleward heat transport by the oceans today.

Let us assume for the sake of argument that the ocean is capable through unknown mechanisms, of substantially more efficient poleward heat transport. Schneider et al. (1985) have considered the extreme case of instantaneous ocean mixing and uniform 20°C ocean temperatures. One of the interesting results of this simulation is that continental interiors at higher latitudes remain substantially below freezing during much of the year despite adjacent warm oceans. Combined with the 10°C limit experiment of Barron and Washington (1983b), the simulations indicate that colder continental interior winter temperatures should accompany greater implied ocean heat transport. There is a limit to the ability of the atmosphere to advect warm air from the oceans into a continental interior, and this ability is dramatically decreased in the model simulation with uniform ocean temperatures. Figure 3 illustrates the large decrease in atmospheric motion for this case. The familiar present-day mid-latitude jets which approach 30 to 40 meters per second are not evident in the uniform ocean temperature experiments. So even if the ocean is capable of substantially more efficient poleward heat transport, in the absence of other external mechanisms to maintain warmth this condition implies cold continental winters and a greater amplitude of the seasonal cycle in continental interiors. This does not appear to be the case for the Eocene.

Summary

A "paleoceanographic inventory" of observations, major trends, and events in the history of ocean circulation and climate exists as a result of extensive data collection. With this tremendous volume of data, theoretical studies are more feasible and hence become an important element of paleoceanographic studies. The theoretical studies are a new and growing aspect of paleoceanography. As yet, explicit ocean formulations have been applied only in a limited sense to the investigation of ancient oceans. Rather the results are based on inferences from models of modern processes or from climate models with highly simplified oceans. Although limited in scope these models are beginning to provide some important insights. Some of the conclusions derived from physical models are summarized below.

1) The reconstruction of the ocean surface circulation has been limited to interpretation of data and qualitative recognition of the role of basin configurations. Typically the atmospheric winds which drive the surface circulation are considered to be unchanging. Atmospheric General Circulation Models suggest that geography, particularly topography, may alter the atmospheric circulation sufficiently to modify the ocean surface circulation. We must therefore consider the atmosphere and the geography of ocean basins to be important variables.

2) The intensity of the atmospheric and ocean surface circulation cannot be derived simply from knowledge of the equator-to-pole surface temperature gradients. The actual surface temperatures, particularly the sign of the temperature change in the tropics, are important. The circulation intensity may be maintained despite the reduced equator-to-pole temperature gradient if tropical temperatures increase.

3) Simple thermohaline circulation models indicate that the buoyancy flux (volume flux times the density contrast) govern the formation of deep water. The concept of competing plumes with different locations and source strengths rather than an "on" or "off" high latitude bottom water source may be the most appropriate way of viewing ancient oceans.

4) Results from atmospheric models with implied ocean heat transport indicate that the climatic extremes during earth history should not be attributed solely to changes in ocean heat flux. Warm poles of even 10°C would require excessive poleward ocean heat flux. Additionally the interpre-

tations of the Eocene and Cretaceous warmth, with quite different changes in tropical temperatures, have quite different implications for the role of the ocean although both are characterized by reduced temperature gradients. Without additional climate forcing factors, a much greater poleward ocean heat flux implies colder continental interiors at high latitudes with a greater amplitude of the seasonal cycle in continental interiors.

Physical models may be modifying our understanding of the evolution of the oceans, but the most appropriate description of the status of theoretical studies in paleoceanography is still one of great potential. There is considerable potential to investigate ancient oceans with explicit ocean models and to examine the role of winds, the moisture balance, heat balance and geography. Combined with the observational studies, models can at least provide paleoceanography with new insights and a more physically rigorous perspective. Models are likely to further the goals of paleoceanography, first to recreate the large-scale, mean three-dimensional structure of ocean currents and related temperature and salinity fields, and second to understand the role of the oceans as a governing factor in climate and to determine the sensitivity of the oceans to external forcing.

Acknowledgments. The National Center for Atmospheric Research is sponsored by the National Science Foundation. The author gratefully acknowledges a review by L.A. Frakes.

References

Arthur, M.A., and J.H. Natland, Carbonaceous sediments in the North and South Atlantic: The role of salinity in stable stratification of early Cretaceous basins, in Deep Drilling Results in the Atlantic Ocean: Continental Margins and Paleoenvironments, M. Talwani, W. Hay and W. Ryan, eds., Maurice Ewing Series, Vol. 3, Amer. Geophys. Union, p. 375-401, 1979.

Barron, E.J., A warm, equable Cretaceous: The nature of the problem, Earth Sci. Revs., 19, 305-338, 1983a.

Barron, E.J., The oceans and atmosphere during warm, geologic periods, Proceedings Joint Oceanographic Assembly, August 1983, Halifax, Nova Scotia, in press, 1983b.

Barron, E.J., and W.M. Washington, The atmospheric circulation during warm, geologic periods: Is the equator-to-pole surface temperature gradient the controlling factor, Geology, 10, 633-636, 1982a.

Barron, E.J., and W.M. Washington, Cretaceous climate: A comparison of atmospheric simulations with the geologic record, Palaeogeog. Palaeoclim. Palaeoecol., 40, 103-133, 1982b.

Barron, E.J., M.A. Arthur and E.G. Kauffman, Cretaceous rhythmic bedding sequences: A plausible link between orbital variations and climate, Earth Planet. Sci. Letters, in press, 1985.

Barron, E.J., Numerical climate modeling: A frontier in petroleum source rock prediction, Amer. Assoc. Petrol. Geol., 69, 448-459, 1985.

Beaty, C., The causes of glaciation, Amer. Scientist, 66, 452-459, 1978.

Berger, W.H., Impact of deep-sea drilling on paleoceanography, in Talwani, M., Hay, W., and Ryan, W., eds., Deep Drilling in the Atlantic Ocean: Continental margins and paleoenvironment, Maurice Ewing Series, Vol. 3: Washington, D.C. Amer. Geophys. Union, 297-314, 1979.

Berggren, W., and C. Hollister, Paleogeography, paleobiogeography and the history of circulation in the Atlantic Ocean, Studies in Paleo-Oceanography, W.W. Hay, ed., Soc. Econ. Paleon. Min. Sp. Pub., 20, 126-186, 1974.

Berggren, W., Role of ocean gateways in climatic change, in Climate in Earth History, Studies in Geophysics Nationl Academy Press, Washington, D.C., 118-125, 1982.

Brass, G., E. Saltzman, J. Sloan, J. Southam, W. Hay, W. Holzer and W. Peterson, Ocean Circulation, Plate Tectonics and Climate, in Climate in Earth History, Studies in Geophysics, National Academy Press, Washington, D.C., pp. 83-89, 1981.

Brooks, C.E.P., Climate Through the Ages, A Study of Climatic Factors and Their Variations, McGraw-Hill, New York, N.Y., 395 pp., 1928.

Bryan, L., and L.J. Lewis, A water mass model of the world ocean, J. Geophys. Res., 84, 2503-2517, 1979.

Chamberlin, T.C., On a possible reversal of deep-sea circulation and its influence on geologic climates, J. Geol., 14, 363-373, 1906.

Crowell, J., and L.A. Frakes, Phanerozoic glaciation and the causes of the ice ages, Amer. J. Sci., 268, 193-224.

Degens, E.T., and P. Stoffers, Stratified waters as a key to the past, Nature, 263, 22-27, 1976.

Ekman, F.W., On the influence of the earth's rotation on ocean currents, Arkiv. f. Matem., Astr. o. Fysik (Stockham), 2(11) 53 pp, 1905.

Frakes, L.A., Climates Throughout Geologic Time, Elsevier, Amsterdam, pp. 1-310, 1979.

Frakes, L.A., and E. Kemp, Influence of continental positions on early Tertiary climates, Nature, 240, 97-100, 1972.

Gordon, W.A., Marine life and ocean surface currents in the Cretaceous, J. Geol., 81, 269-284, 1973.

Held, I.M., D.I. Linder and M.J. Suarez, Albedo feedback, the meridional structure of the effective heat diffusivity, and climate sensitivity: Results from dynamic and diffusive models, J. Atmos. Sci., 38, 1911-1927, 1981.

Holland, W.R., Oceanic General Circulation Models, The Sea, 6, 3-46, 1977.

Kennett, J.P., Cenozoic evolution of Antarctic glaciation, the circum-Antarctic Ocean, and their impact on global paleoceanography, J. Geophys. Res., 82, 3843-3860, 1977.

Kraus, E.B., W. Peterson and C. Rooth, The thermal evolution of the ocean, Evolution des

Atmospheres Planetaires et Climatologie, Centre National d'Etudes Spatiales, France, pp. 201-211, 1979.

Luyendyk, B., D. Forsyth and J. Phillips, An experimental approach to the paleocirculation of oceanic surface waters, Geol. Soc. Amer. Bull., 83, 2649-2664, 1972.

Matthews, R.K., and R.Z. Poore, Tertiary $\delta^{18}O$ record and glacio-eustatic sea level fluctuations, Geology, 8, 501-504, 1980.

Oort, A., and T. Vonder Haar, On the observed annual cycle in the ocean-atmosphere heat balance over the Northern Hemisphere, J. Phys. Oceanogr., 6, 781-800, 1976.

Peterson, W., A steady thermo-haline convection model, Unpubl., Dissertation, University of Miami, Miami, Fl., 1979.

Rea, D.K., and T.R. Janeck, Late Cretaceous history of eolian deposition in the mid-Pacific mountains, central North Pacific Ocean, Palaeogeo.Palaeoclim.Palaeoecol., 36,55-67,1981.

Saltzman, E., and E.J. Barron, Deep circulation in the Late Cretaceous: Oxygen isotope paleotemperatures from Inoceramus remains in DSDP cores, Palaeogeo. palaeoclim. Palaeoecol., 40, 167-182, 1982.

Savin, S., The history of the earth's surface temperature during the past 100 million years, Ann. Rev. Earth Planet. Sci., 5, 319-355, 1977.

Schlanger, S.O., and H.C. Jenkyns, Cretaceous anoxic events: Causes and consequences, Geologie en Mijnbouw, 55, 179-184, 1976.

Schneider, S.H., S.L. Thompson, and E.J. Barron, Mid-Cretaceous, continental surface temperatures: are high CO_2 concentrations needed to simulate above freezing winter conditions, in The Carbon Cycle and Atmospheric CO_2: Natural Variations Archean to Present, E.T. Sundquist and W.S. Broecker, eds., Geophys. Monogr. Series 32, Amer. Geophys. Union, in press, 1985.

Semtner, A.J., An oceanic general circulation model with bottom topography, Tech. Rept. 9, Dept. Meteorology, University of California, Los Angeles, 99 pp., 1974.

Shackleton, N., and A. Boersma, The climate of the Eocene Ocean, J. Geol. Soc. London, 138, 153-157, 1981.

Stommel, H., The Gulf Stream, University of California Press, Berkeley and Los Angeles, 248 pp., 1955.

Tarling, D.H., The geological-geophysical framework of ice ages, Climatic Change, J. Gribben, ed., Cambridge Univ. Press, Cambridge, pp. 3-24, 1978.

Thiede, J., Reworking in Upper Mesozoic and Cenozoic central Pacific deep-sea sediments, Nature, 289, 667-670, 1981.

von Arx, W., An experimental approach to problems in physical oceanography, Progress in Physics and Chemistry of the Earth, Pergamon Press, New York, 2:1-29, 1957.

NUMERICAL MODELLING OF THE OCEAN CIRCULATION AND PALEOCIRCULATION

D.G. Seidov

Shirshov's Institute of Oceanology, USSR Academy of Sciences, Krasikova, 23, 117 218 Moscow, USSR

Abstract. A numerical model of the ocean circulation has been developed and applied to the problem of paleocirculation reconstruction. The physical basis of the model is presented as well as a sketch of the numerical approach as a method are discussed. The model has been tested in numerical experiments aimed to reproduce modern circulation and hydrology patterns. Then, the model has been used to reconstruct schemes of currents and hydrology of the Late Mesozoic and Cenozoic eras. The atmosphere has been assumed to be zonal, and continents drifted after the breakup of Pangaea as it is proposed by geodynamics. The results of the numerical experiments are discussed and major feedbacks are identified which regulate the climatic changes of the ocean.

Introduction

There has always been great interest among paleoecologists, paleoclimatologists and geologists in geologic history of the earth's climate and the paleoenvironment of the World ocean. However, paleogeography has operated mainly with currents reconstructed using indirect data, i.e. paleoecological data (e.g. Gordon, 1973; Westermann, 1980; Berggren and Hollister, 1974) or using analogies between modern and ancient circulations of the atmosphere (Parrish and Curtis, 1982). Although such reconstructions are helpful to some extent in paleogeographical studies, they too often are arbitrary or inconsistent with hydrophysics. The data are unevenly distributed in time and space and generally insufficient for reconstructions of the global or large-scale current systems. The reconstructions themselves are not based on the laws of physical oceanography such as conservation of mass (i.e. the mass of water which flows into certain volume of fluid should be equal to the mass which flows out), hydrostatic stability, conservation of angular momentum, etc. The only study of the ocean paleocirculation by physical methods has been carried out using a laboratory model (Luyendyk et al., 1972). We should also mention here a laboratory reconstructions performed by Lazarev (1950), although these were based on a fixed continental position approach. There is a lot of evidence now that continents move over the earth's surface. It is absolutely clear that for different position of continents there would be accompanying different resulting patterns of the ocean circulation. We are not going to discuss any aspect of plate tectonics or geodynamics technique. We will simply deal with a hypothesis that the continents were situated in very different places than they are at present time. We will also use geodynamical reconstruction of previous continental configurations on the globe. Inferences from this hypothesis (using paleobiogeographical data and geological records) have already been exploited by some authors (Gordon, 1973; Kennet, 1977 and others; see for example Lisitsin, 1980). The role of continents may be easily seen from their present constraints on the ocean circulation. The earth's atmosphere generally is able to circulate around the globe since the air has no meridional borders on its circumzonal way. The ocean currents evidently are in very different situation. There is only one circumglobal ocean current at present time, namely the West Wind Drift current. Meridional ocean boundaries produce most general features of the circulation such as subtropical anticylcones or subpolar cyclonic gyres. This major characteristic quality of the circulation patterns gave a basis for well known idealized theoretical studies (e.g. Stommel, 1948; Munk, 1950; Veronis, 1966 and others). In these studies rectangular basins with flat bottoms and with homogeneous fluid approximations were used as conceptual models. The only real geophysical features were rotation of the earth, its spherical geometry (parameterized by meridional change of the Coriolis parameter), and the wind stress which itself was highly idealized and even had been introduced into the model as a zonal field (e.g. Stommel, 1984). Significant success of such enterprises may be interpreted only as a proof of the priority of the meridional boundaries plus differential rotation (or spherical geometry) in formation of the ocean circulation patterns. It should be stressed, however, that specific current systems are formed with different specific causes, and especially under effects of the specific geometry of different oceans.

At the same time one has available the possibility of reconstructing the ocean circulation of past geologic periods using numerical simulation techniques. This possibility emerges from certain encouraging results achieved in numerical studies of present ocean circulation. At least major features of global ocean circulation are reflected by these models in qualitative agreements with observations, or with existing images of present day circulation and hydrology patterns. If one raises not too rigorous demands and does not wait for detailed mapping of the currents, or exact reconstructions of the hydrological fields using numerical models, then useful results might be obtained right away. In any case such reconstructions seem to be more sophisticated and reliable than those inferred from educated guesses or than reconstructions made using indirect methods combined with geographical intuition. Numerical reconstructions are at last hydrodynamically consistent.

The atmospheric parameters (such as equatorial to polar temperature differences) may be estimated with the aid of paleoclimatic and/or theoretical studies (Monin, 1982; Monin and Shishkov, 1978). Using them together with paleogeodynamical reconstructions of the continental positions (Gorodnitizki and Zonenshine, 1979) provides the possibility of reconstructing of the paleocirculation with some degree of traditional physical oceanographic sophistication. We should also point it out that for every specific time in geologic history specific equilibrium condition existed, i.e. that the currents and hydrological fields were totally adjusted to basin's configuration and the atmospheric forcing. In other words, we must carry out a great deal of numerical experimentation, each time with a spin up to equilibrium state that suits specific and specified external conditions. In the framework of such a task an extended series of numerical experiments must be performed. Therefore a numerical model should be a rather simple one in order to keep within reasonable computational limits. Another important point is that if one uses only estimations of the atmospheric parameters, and does not utilize paleoecological or geological data directly in computations, there is, in fact, an independent approach to the paleocirculation problem. In addition to that, the smallest possible set of paleoclimatic input has been used in this model study. Clearly, at this stage, taking into account these restrictions, the results should be considered mainly as an illustration of the numerical modelling approach. In any case, the study should be considered as very preliminary.

Model

The mathematical model of the ocean circulation was constructed with the aim to stimulate the schemes of the global ocean current system and general hydrological structure of the ocean. These fields are formed on spherical earth's surface due to heat, mass and momentum exchange between the ocean and the atmosphere. The geometry of ocean basins and bottom topography are considered, and sea-ice formation and melting are parameterized (Seidov, 1984; Seidov and Enikeev, 1983, 1984 a,b). The numerical modelling techniques have been presented, with more details, in many publications and monographs (e.g. Sarkisyan, 1977; Marchuk, 1974; Bryan, 1969; Mesinger and Arakawa, 1976; Haney, 1974; Huang, 1978 and other).

Those who are interested mainly in paleoceanographical aspects and in illustrations of applicability of numerical approach may skip this section. The initial equations of the thermohydrodynamics of the ocean written in the left spherical coordinate system after certain simplification (Kamenkovich, 1973; Sarkisyan, 1977) are:

$$\frac{\partial u}{\partial t} + \frac{u}{a\cos\phi}\frac{\partial u}{\partial \lambda} + \frac{v}{a}\frac{\partial u}{\partial \phi} + w\frac{\partial u}{\partial \phi} - \frac{uv}{a}\tan\phi - 2\Omega\sin\phi\, v = -\frac{1}{\rho_0 a\cos\phi}\frac{\partial P}{\partial \lambda}\quad A_M \Delta u + K_M\frac{\partial^2 u}{\partial z^2}\;; \tag{1}$$

$$\frac{\partial v}{\partial t} + \frac{u}{a\cos\phi}\frac{\partial v}{\partial \lambda} + \frac{v}{a}\frac{\partial v}{\partial \phi} + w\frac{\partial v}{\partial z} + \frac{u^2}{a}\tan\phi + 2\Omega\sin\phi\, u = -\frac{1}{\rho_0 a}\frac{\partial P}{\partial \phi}\quad A_M \Delta v + K_M\frac{\partial^2 v}{\partial z^2}\;; \tag{2}$$

$$\frac{\partial P}{\partial z} = \rho g\;; \tag{3}$$

$$\frac{1}{a\cos\phi}\frac{\partial u}{\partial \lambda} + \frac{1}{a\cos\phi}\frac{\partial}{\partial \phi}(v\cos\phi) + \frac{\partial w}{\partial z} = 0\;; \tag{4}$$

$$\frac{\partial T}{\partial t} + \frac{u}{a\cos\phi}\frac{\partial T}{\partial \lambda} + \frac{v}{a}\frac{\partial T}{\partial z} + w\frac{\partial T}{\partial z} = A_T\Delta T + K_T\frac{\partial^2 T}{\partial z^2}\;; \tag{5}$$

$$\frac{\partial S}{\partial t} + \frac{u}{a\cos\phi}\frac{\partial S}{\partial \lambda} + \frac{v}{a}\frac{\partial S}{\partial \phi}\quad w\frac{\partial S}{\partial z} = A_s\Delta S + K_s\frac{\partial^2 s}{\partial z^2}\;; \tag{6}$$

$$\rho = \rho\,(T, S, P) \tag{7}$$

The symbols are: λ is longitude; ϕ is latitude $(-\frac{\pi}{2} \le \phi \le \frac{\pi}{2})$; a is the earth's radius; u, v, w, are flow velocity components along λ, ϕ and z respectively; Ω is the earth's rotation velocity; T, S, P and ρ are temperature, salinity, pressure, and density of the sea water; A_M, A_T, A_S are coefficients of the horizontal turbulent exchange of momentum, temperature, and salinity; K_M, K_T, K_S are analogues coefficients for vertical exchange; Δ is horizontal Laplace operator is spherical coordinates.

Turbulent exchange coefficients are used to parameterize small-scale phenomena in the ocean processes. From a numerical modelling point of view all processes with a scale smaller than the size of the numerical grid should be treated as small-scale processes. Obviously, every numerical experiment with grid size greater than 2° in the horizontal direction totally ignores the true nature of the synoptic scale processes, i.e. oceanic eddies, jet currents, etc. These processes are important general ocean circulation components. They are paramaterized as "turbulence" using the mentioned coefficients (A, K). Therefore, non-linear terms should be neglected everywhere except for the advective part of the equations of heat and mass (salt) transport. Since short-term variability is not reproduced in coarse-grid resolution experiments, the geostrophic relations for the velocity field may be used with great reliance. In other words the dynamical part of the paleocirculation model may be constructed as a linear.

At the lateral boundaries and at the bottom we specify T and S or make the heat and salt fluxes vanish, i.e.:

$$\frac{\partial T}{\partial n} = \frac{\partial S}{\partial n} = 0 \tag{8}$$

where n is a normal to the confining surface. All side boundaries are rigid walls with no water fluxes across them. Water slips freely along these boundaries except for passages which provide circumzonal flow (in this case periodical boundary conditions are used instead of no-flux, free-slip conditions). If the sea surface is covered by ice then at this surface the phase transition temperature and corresponding salinity are specified, and heat, and mass fluxes between air and water across the sea surface vanish. When the surface is free from ice the heat and mass (salt) fluxes are calculated according conventionally adopted technique (Haney, 1971; Huang, 1978). These conditions may be combined as follows:

$$(1 - \delta_1) T_0 + \delta_1 \frac{\partial T}{\partial z}\Big|_{z=0} = \delta_1 \gamma (T_0 - T_A) + (1 - \delta_1) T_I \tag{9}$$

$$(1 - \delta_1) T_0 + \delta_1 \frac{\partial T}{\partial z}\Big|_{z=0} = \delta_1 (P - E) S_0 + (1 - \delta_1) S_I \tag{10}$$

$$\frac{\partial u_0}{\partial z} = - \frac{\delta_1 \tau_\lambda}{\rho_0 K_M}; \qquad \frac{\partial v_0}{\partial z} = - \frac{\delta_1 \tau_\phi}{\rho_0 K_M} \tag{11}$$

where T_0, S_0, u_0 and V_0 are temperature, salinity, and horizontal velocity components, respectively, at the surface (T_0 and S_0 are constant in the surface homogeneous layer); $\delta_1 = 1$ if there is no ice at the surface; and $\delta_1 = 0$ if the ocean surface is covered by ice (i.e. if the surface temperature dropped just below the freezing point); T_I, S_I are −1.88 °C and 34.5°/₀₀ (Seidov and Enikeev, 1984 b); τ_λ, τ_ϕ are the wind stress components; (P - E) is the difference of precipitation and evaporation rates. Here γ is an empirical coefficient which parameterizes heat exchange of all kinds (i.e. radiative, sensible and latent heat fluxes) as has been done by Haney (1974).

At the bottom (for $z = H(\lambda, \phi)$) the kinematic condition is introduced in a form:

$$w_B = \frac{u_B}{a \cos\phi} \frac{\partial H}{\partial \lambda} + \frac{v_B}{a} \frac{\partial H}{\partial \phi} \tag{12}$$

where u_B, v_B are velocity components near the bottom at the upper boundary of the bottom Ekman's layer (Kamenkovich, 1973; Sarkisyan, 1977).

At the surface we place the so called "rigid lid" condition (Bryan, 1969), i.e.

$$w = 0 \tag{13}$$

After neglecting non-linear terms and local accelerations in (1) and (2), and after cross-differencing and subtracting (1) from (2), integrated from the surface to the bottom using (12) and (13) one gets a vorticity equation

$$A_M \Delta\omega - \varepsilon\omega = - \frac{\delta_1}{\rho_0} \operatorname{rot}_z \tau + \frac{2\Omega}{a} \frac{\partial \Psi}{\partial \lambda} + \frac{2\Omega \sin\phi}{H} J(H, \Psi - \Psi_D) + \frac{\varepsilon g}{2\Omega \sin\phi\, \rho_0} \int_0^H z \Delta\rho^1 dz \tag{14}$$

Where Ψ is a total stream function such as:

$$\hat{u} = \frac{1}{H} \int_0^H u dz = - \frac{1}{aH} \frac{\partial \Psi}{\partial \phi}; \qquad \hat{v} = \frac{1}{H} \int_0^H v dz = \frac{1}{aH \cos\phi} \frac{\partial \Psi}{\partial \lambda} \tag{15}$$

The equation (14) should be solved together with Poisson's equation for

$$\Delta \Psi = \omega \tag{16}$$

In (14) we have $\varepsilon = \frac{f}{2 \alpha H}$; $\alpha = \left(\frac{|f|}{2 K_M}\right)$ (Sarkisyan, 1977). Symbol J denotes Jacobian; $f = 2\Omega \sin\phi$ is the Coriolis parameter. The function Ψ_D determines geostrophic vertically averaged flow calculated using the dynamical method with the reference level (level of no motion) placed at the depth H_0. In other words Ψ_D is the function calculated using the formula (Sarkisyan, 1977):

$$\Psi_D = - \frac{g}{\rho_0 f_1} \int_0^H z \rho^1 dz \tag{17}$$

where f_1 is the Coriolis parameter averaged over each hemisphere. In the described version of the model we made some other simplifications which we believe do not affect the quality of the reconstructions of paleocurrents. First we assumed that below H_0 water is homogeneous with abyssal density ρ_0 = const, and above this layer $\rho = \rho_0 + \rho^1$, where $\rho^1 = \rho(\lambda, \phi, z)$. Always $H_0 \leq H(\lambda, \phi)$, i.e. bottom irregularities lie below the depth of the baroclinic layer. This assumption allows a simplification of (14) and produces some other technical conveniences, although the joint effect of baroclinicity and bottom topography is not directly considered and role of bottom topography is distorted to some extent. At this time we tested an advanced version of the model where bottom topography penetrates the baroclinic layer, and the role of the topography is contained in the model in the proper way. The global modern circulation, presented on Merkator projection was evaluated in the framework of this advanced version (see next sections). Nevertheless, since the major pool of the results has been obtained using simplified version we restrict ourselves to this version.

Current velocities, according to Bryan (1969), are divided into two components which are vertically averaged (or barotropic) and deviations from average (or shear velocities)

$$u = \hat{u} + u^1; \qquad v = \hat{v} + v^1 \tag{18}$$

Shear velocities are calculated using geostrophic relations (with Ekman's currents taken into consideration). The relations for u^1 and v^1 can be obtained if we integrate

TABLE 1. Parameters of the Zonal Climat

Latitude	Jurassic 160 Ma			Cretaceous 100 Ma			Eocene 50 Ma		
	T_A	τ	P-E	T_A	τ	P-E	T_A	τ	P-E
13°	20	-1.5	-5	22 (22)	-2.0 (-1.3)	-7 (-33)	24	-2.5	-[illegible]
20°	18	-1.5	-30	21 (21)	-2.0 (-1.3)	-45 (-67)	23	-2.5	-6[illegible]
27°	16	-1.0	-70	19 (18)	-1.3 (0)	-90 (-67)	22	-1.6	-11[illegible]
34°	14	1.0	-30	17 (16)	1.3 (1.3)	-45 (-100)	20	1.6	-6[illegible]
41°	12	1.0	0	14 (13)	1.3 (2.0)	0 (-33)	16	1.6	[illegible]
48°	10	2.0	25	11 (10)	2.7 (1.3)	33 (0)	11	1.6	4[illegible]
55°	8	1.2	25	7 (8)	1.6 (0)	33 (33)	7	2.0	4[illegible]
62°	6	0	25	4 (6)	0 (-0.7)	33 (67)	2	0	4[illegible]
69°	4	-0.2	25	1 (2)	-0.3 (-0.9)	33 (33)	-1	-0.4	4[illegible]
76°	2	-0.7	25	-2 (0)	-0.9 (-0.9)	33 (33)	-6	-1.2	4[illegible]
83°	0	0.2	25	-7 (-3)	-0.3 (-0.9)	33 (7)	-15	0.4	4[illegible]

T_A – air temperature near the sea surface (°C)
τ – wind stress amplitude (d in cm^{-2})
P-E – difference between precipitation and evaporation rates (cm $year^{-1}$)

Northern hemisphere values in brackets.

from the surface to the bottom the equation of motion (1) and (2) (with neglected local accelerations, non-linear terms, and horizontal friction) with the upper boundary condition (we do not consider the bottom Ekman layer anywhere except for in w in equation (14)) and subtract the product from equations (1) and (2). Omitting the details (see Sarkisyan, 1977; or Bryan, 1969) the formulas are:

$$u^1 = -\frac{1}{fa\rho_0}\frac{\partial P^1}{\partial \phi} - \frac{\delta_1 \tau_\lambda}{f\rho_0 H} + \delta_1\delta_2 u_E \tag{19}$$

$$v^1 = \frac{1}{fa\rho_0 \cos\phi}\frac{\partial P^1}{\partial \lambda} + \frac{\delta_1 \tau_\phi}{f\rho_0 H} + \delta_1\delta_2 v_E \tag{20}$$

where

$$\delta_2 = \begin{cases} 1 & \text{for } z = 0 \\ 0 & \text{for } z > 0 \end{cases} \tag{21}$$

Shear components are the sum of gradient velocities, Ekman's gradient velocities, and Ekman's drift velocities. Drift velocities are calculated (only at the surface) from:

$$u_E = \frac{\alpha}{f}(\tau_\lambda + \tau_\phi) \; ; \qquad v_E = \frac{\alpha}{f}(\tau_\lambda - \tau_\phi) \tag{22}$$

The pressure deviations P^1 may be obtained by subtracting the vertically averaged pressure $\hat{P}$ from P using the hydrostatic equation (3)

$$P^1 = P - \hat{P} = -\frac{g}{H}\int_0^H (H - z)\,\rho^1 dz + g\int_0^z P^1 dz \tag{23}$$

Vertical velocity components are evaluated from

$$f\,\frac{\partial w}{\partial z} = \frac{2\Omega \cos\phi}{a}\,v^1 - \frac{\delta_1}{\rho_0}\,\text{rot}_z\left(\frac{\tau}{H}\right) + \frac{\delta_1\delta_2}{H}\,\text{rot}_z\,\tau \quad (24$$

This equation should be integrated from z to H with boundary condition (12).

When vertical hydrostatic instability arises, i.e. more dense water forms above less dense water, a hydrostatic adjustment mechanism is started. Due to convective mixing T, S and ρ become the same or equal to the averaged values of these variables over the layer where instability took place. This convective mixing may be interpreted as a parametrization of bottom water formation due to cooling or increased salinity of the uppe layer waters (Bryan, 1969).

Since some continents and large islands are surrounde by water, the region becomes a multiconnected one. At the shores of such land masses the boundary condition is

$$\Psi \,|_{\Sigma_i} = C_i \qquad i = 1, \ldots, N \quad (25$$

where L_i is the index of the land mass and Σ_i denotes the boundary of this mass; C_i are the constants. The advance and exact technique for calculation of the C_i has been developed (Kamenkovich, 1973; Bryan, 1969), but here a more simplified method is used. The basis of the simplified approach lies in the empirical fact that Ψ_D approximates the solution Ψ (Sarkisyan, 1977). As soon as we have (25) it may be written

$$\frac{1}{L_i}\oint_{\Sigma_i} (L_i - \Psi_D)\,|_{\Sigma_i}\, ds = \mu(\Sigma_i) \quad (26$$

for Late Mesozoic and Early Cretaceous Eras

	Oligocene			Early Miocene			Late Miocene			Modern Period		
	30 Ma			20 Ma			10 Ma					
Latitude	T_A	τ	P-E	T_A	τ	P-E	T_A	τ	P-E	T_A	τ	P-E
13°	24	-2.6	-8	25	-2.8	-8	26	-2.9	-8	26 (26)	-3.0 (-2.0)	-10
20°	23	-2.6	-60	24	-2.8	-65	25	-2.9	-65	26 (26)	-3.0 (-2.0)	-75
27°	22	-1.7	-120	23	-1.8	-130	24	-1.9	-130	25 (23)	-2.0 (0)	-150
34°	20	1.7	-60	21	1.8	-65	22	1.9	-65	23 (20)	2.0 (2.0)	-75
41°	16	1.7	0	17	1.8	0	18	1.9	0	18 (15)	2.0 (3.0)	0
48°	11	3.5	44	11	3.6	46	12	3.8	48	12 (10)	4.0 (2.0)	50
55°	7	2.1	44	7	2.2	46	6	2.4	48	6 (7)	2.5 (0)	50
62°	1	0	44	0	0.0	46	0	0.0	48	0 (5)	0 (-1.0)	50
69°	-1	-0.4	44	-2	-0.5	46	-3	-0.5	48	-4 (0)	-0.5 (-1.5)	50
76°	-7	-1.3	44	-8	-1.4	46	-9	-1.4	48	-10 (-5)	-1.5 (-1.5)	50
83°	-15	0.4	44	-16	0.5	46	-18	0.5	48	-20 (-10)	0.5 (-1.5)	50

where L_i is the length of Σ_i. So, with the accuracy of $\mu(\Sigma)$ we have

$$C_i = \frac{1}{L_i} \oint_{\Sigma_i} \Psi_D ds \tag{27}$$

From numerical modelling experience the smallness of μ may be expected. Nevertheless, for more sophisticated studies the advanced method (Kamenkovich, 1973; Bryan, 1969) is recommended. Since Ψ solutions contain an arbitrary constant which is eliminated by chosing certain C_i as specified values, the reference C should be chosen at any closed boundary line. For this reason (27) should be rewritten as

$$C_i = \frac{1}{L_i} \oint_{\Sigma_i} \Psi_D \Big|_{\Sigma_i} ds - \frac{1}{L_1} \oint_{\Sigma_i} \Psi_D \Big|_{\Sigma_i} ds \tag{28}$$

$i = 2, \ldots, N$

where L_i is the length of the reference perimeter (Σ_i) (i.e. perimeter of the Antarctic for modern ocean circulation (Seidov and Enikeev, 1984 b)). The model has been also presented in recent publication (Seidov, 1984). It should be stressed from the very beginning that this model was not constructed as a competitive instrument for comparison with other models. A number of such models are highly advanced ones (Sarkisyan, 1978; Bryan et al, 1975, Marchuk, 1974; Bryan, 1969; Haney, 1974, Huang, 1978 and so on). This model is an instrument for qualitative evaluation of schemes of ocean circulation and paleocirculation, i.e. general features of water mass motion and comparative studies of modern and paleocirculation and hydrology.

Numerical scheme

Numerical schemes for study of dynamics of the ocean and the atmosphere are described in detail in many monographs and papers (e.g. Sarkisyan, 1977; Marchuk, 1974; Mesinger and Arakawa, 1976; Bryan, 1969; Haney, 1974; Crowley, 1968 and so on). Let us mention only that equations (14) and (16) are solved using iterative procedures. Equations of the heat and salt balance (5) and (6) are integrated with time using a splitting procedure. All other relations are diagnostic and produce no difficulties for calculations. The splitting method (Marchuk, 1974) is applied in a manner when the evolution process is subdivided in two different steps, i.e. diffusive and advective steps treated separately. More detailed description of this procedure may be found in other papers (Leith, 1965; Crowley, 1968 a; Seidov, 1980). Here we will observe briefly an illustration of the technique using Cartesian coordinate system and two-dimensional heat transport with diffusion as an example

$$\frac{\partial T}{\partial t} + u \frac{\partial T}{\partial x} + v \frac{\partial T}{\partial y} = A \Delta T \tag{29}$$

Equation (29) is split and solved in two steps

$$\text{I} \quad \frac{\partial T}{\partial t} \quad u \frac{\partial T}{\partial x} + v \frac{\partial T}{\partial y} = 0 \tag{30a}$$

$$\text{II} \quad \frac{\partial T}{\partial t} = A \Delta T \tag{30b}$$

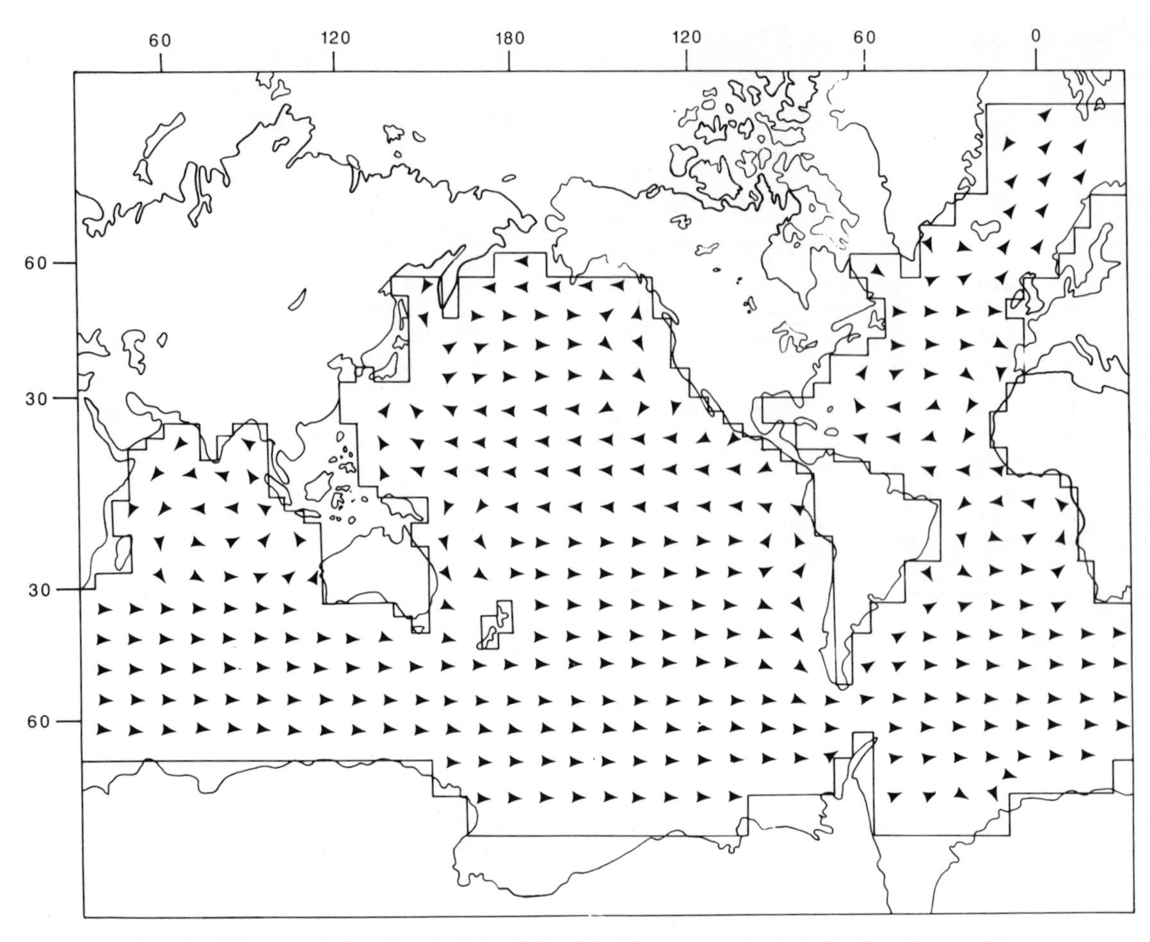

Fig. 1. Scheme of the circulation of the World ocean calculated using the model.

The first step can be represented as:

$$\frac{dT}{dt} = 0 \qquad (31)$$

which states that temperature of a given particle does not change during its motion along its trajectory. If at the moment t^{n+1} the particle comes to the grid point with (ij) indexes, at δt time interval ago this particle had the same temperature that it brought to the point (ij). In other words we only ahve to evaluate temperature at the point which is situated at $-V_{ij}^{n}\ \delta t$ distance from the (ij) - point at t^n (upper index denotes the time step number). For this we write the Taylor's series

$$T^*_{ij} = T^n_{ij} - U_{ij}\,\delta t \left.\frac{\partial T}{\partial x}\right|^n_{ij} - v_{ij}\delta t \left.\frac{\partial T}{\partial y}\right|^n_{ij} + \frac{u^2_{ij}\,\delta t^2}{2}\left.\frac{\partial^2 T}{\partial x^2}\right|^n_{ij} + \frac{v^2\delta t^2}{2}\left.\frac{\partial^2 T}{\partial y^2}\right|^n_{ij} + \frac{uv\delta t^2}{2}\left.\frac{\partial^2 T}{\partial x \partial y}\right|^n_{ij} \qquad (32)$$

where T^*_{ij} is temperature in (ij) point after advective step, i.e. at t^{n+1} moment. After replacement of the partial derivatives in (32) by their finite-difference analogs one obtains the algorithm for the first advective step. Seidov (1978) has shown the equivalance of the Leith-Richmayer scheme (Crowley, 1968) with splitting the advective step into consecutive steps completed in the direction of the coordinate axes (Leith, 1965), and the scheme described above. Actually, this splitting technique has also been used in the model described above but the interpretation is clearer in a Lagrangian approach rather than Eulerian. The solution of the advective step T* is an initial condition for the diffusive step.

So, one has the schemes for (5) and (6) which may second order approximation in space (as soon as central differencing for space partial derivatives are employed) but only first order in time. It has been proved that this scheme's viscosity is smaller than the real one (Seidov, 1978).

Numerical Experiments

The majority of the experiments discussed below have been carried out on a rough numerical grid with 31 x 11 x 4 points for each hemisphere. There is no hope in obtaining real transport and velocities of currents usin such space resolution. So, significance should be attributed only to a comparative analysis of modern and paleocirculations. Modern circulation was collated using the global model as well as separated hemischerical models; paleocirculations were evaluated only for the

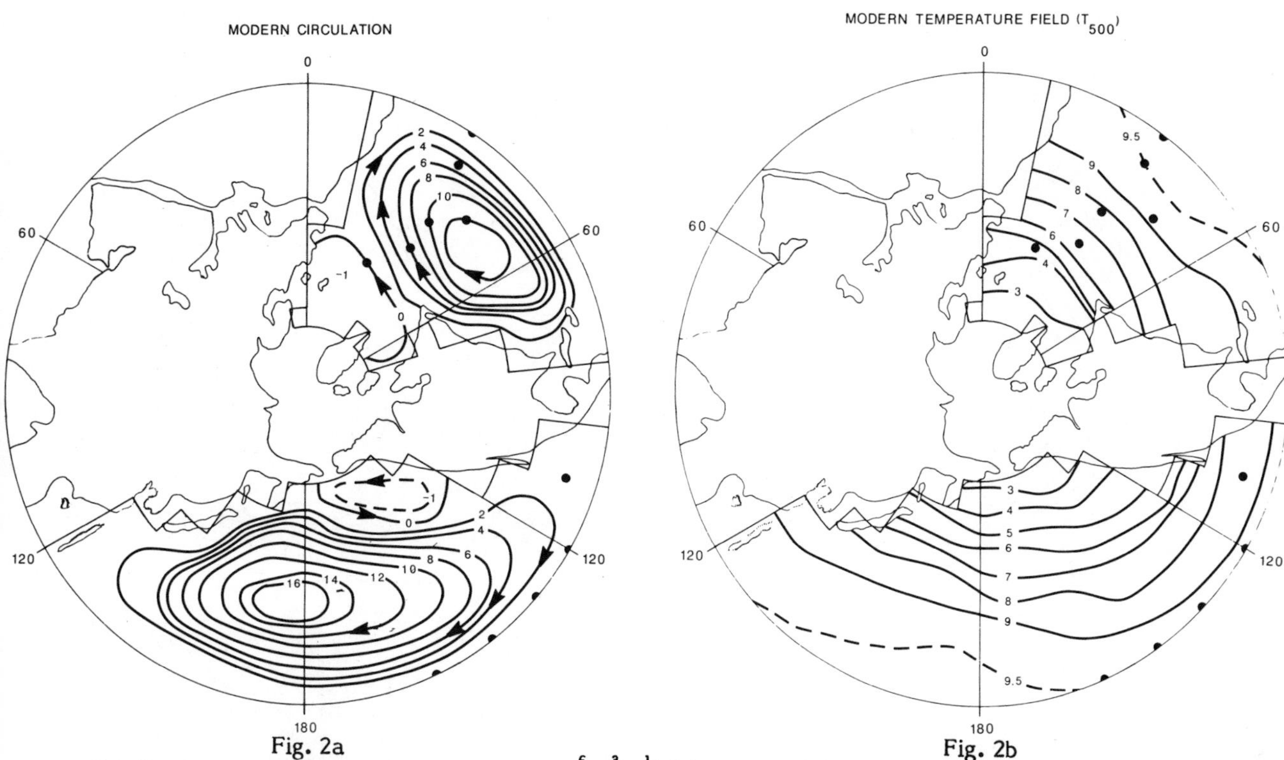

Fig. 2. Circulation (total flux function in $10^6 m^3 s^{-1}$) (a), and T_{500} (b) in the Northern hemisphere for modern period (circles show the oceanic ridges).

atter. History of the southern hemisphere ocean circulaion is considered in more detail than the northern one.

The air temperature near the sea surface T_A, wind tress τ, and difference between precipitation and vaporation rates (P - E) are introduced into the model as pecified atmospheric parameters. Many researchers ssume that if a big land mass moves to the pole, this nass is cooled and becomes glaciated (Monin, 1982; Monin nd Shiskov, 1979; Gordon, 1973; see also Schopf, 1980, or Crowley, 1983). The increase in temperature difference etween the equator and the poles should be accompanied y increased wind and moisture exchange processes. It as also been assumed that cooling of the subpolar space akes place mainly due to a decrease in ocean poleward eat transport. Though this might be true concerning the cean regions it is not as obvious for continental glaciaions. Recent studies using atmospheric models have hown (Barron and Washington, 1982; Barron, 1983) that he continental glaciation does not correspond with seaurface temperature so straightforwardly. At least, pecifying high temperature at the subpolar ocean surface as not seriously affected continental glaciation of Antarctic.

Although it has been shown (Barron and Washington, 982) that atmospheric paleocirculation might have ignificant azonal components, the main feature of the cean circulation, as it has been already mentioned, is a roximity of the ocean's reaction to zonal and real azonal) atmospheric forcing. The success of laboratory nodelling of the paleocirculation with a zonal wind stress attern proves the correctness of this assumption (Luyendyk et al., 1972). As an ocean-atmosphere model study of paleoclimate has not been done yet, we performed our experiments based on the hypothesis of polar continents resulting in regional cooling. In addition to this assumption, we, for the time being, proceed with zonal atmospheric parameters in the model. It should be mentioned that zonal atmospheric fields are used for both modern and paleocirculation modelling. Therefore, paleocirculation deviates from the real one due to atmospheric zonality in the same way as the modern circulation does, i.e. the distortions caused by atmospheric zonality may be at least roughly estimated. Table 1 represents atmospheric parameters of the study. They have been introduced using estimations made by Monin (1982). New investigations (e.g. Barron and Washington, 1982; Parrish and Curtis, 1982; see also Crowley, 1983) should be taken into account in future studies.

At 13°N and 13°S latitudes rigid walls have been placed, and, therefore, we considered the two hemispheres separately. It should be kept in mind that the equatorial area is not considered here since we put walls at the trade wind bands, so it is possible to reconstruct only part of the Tethys ocean, and circumequatorial currents of the time of the Pangaea's breakup are not presented. The parameters of the model are following:
$A_M = 9 \times 10^9$ cm^2 s^{-1}; $A_T = A_S = 10^8$ cm^2 s^{-1}; $K_M = 10$ cm^2 s^{-1}, and $K_T = K_S = 5$ cm^2 s^{-1}. Model equations were integrated for 10 years of model time. In the upper layers (model has 4 levels or 5 layers) the model approaches equilibrium, yet in abyssal waters a trend still

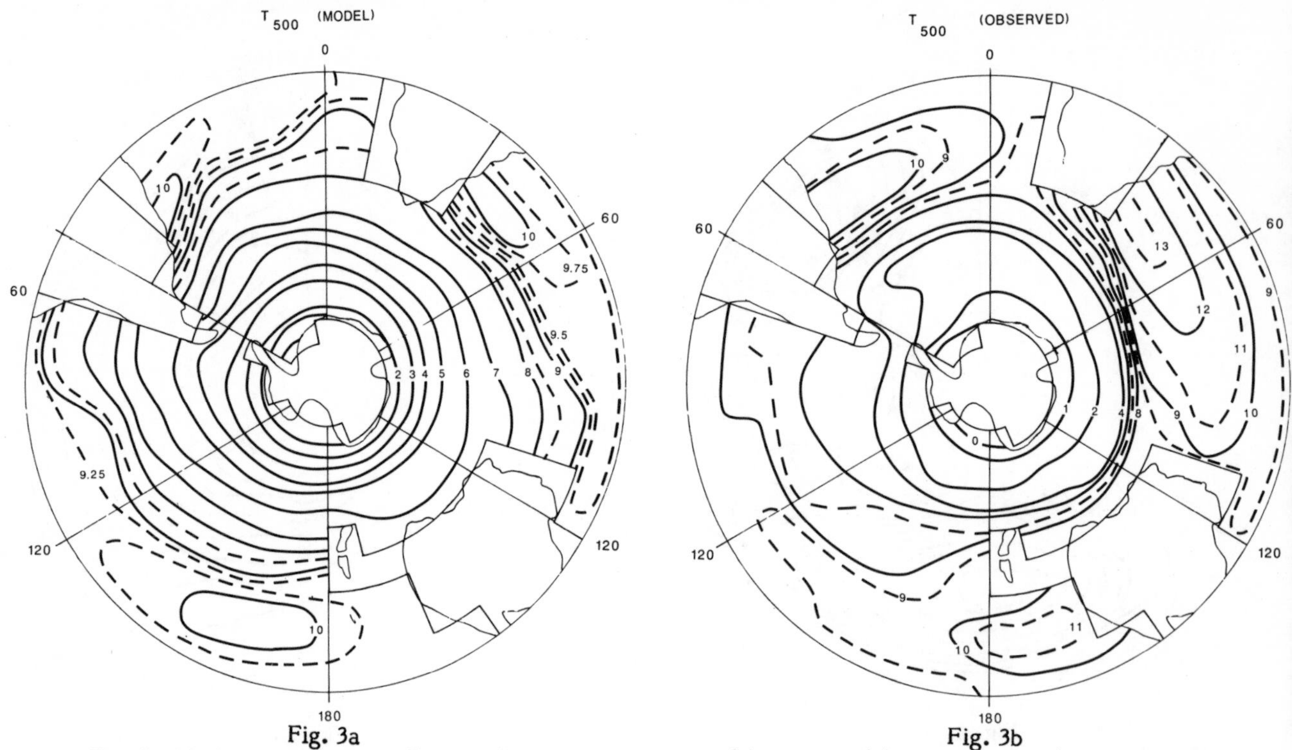

Fig. 3. Modern temperature T_{500} of Southern hemisphere (a) - model; (b) -observations (from Atlas of Oceans, 1974, 1977).

existed. It has been shown that such a trend generally is not significant, and vanishes only after hundreds of years of model time (Bryan et al, 1975).

As for bottom topography, it should be noted that paleobathymetry is still highly simplified, and we only introduce into the model the mid-ocean slope for both modern and paleocirculations. The bottom is flat and 4.5 km deep everywhere except for ridges and slopes (minimal depth of these regions is 3.5 km; baroclinic layer stretches to 2.8 km).

Results are presented in the next section. Usually paleogeographers prefer to deal with surface or bottom currents and surface temperature. Physical oceanography, on the other hand, more often consider total water fluxes and hydrology of the main thermocline. Throughout this paper currents will be presented either as schemes without specifying the values of the vectors or as a total flux function maps. Thermocline temperature will be given at the 500 m level. Total stream function is given in special units - sverdrups (1 sv = 10^6 m^3 s^{-1}). Temperature at 500 m (T_{500}) seems to be more interesting as the surface T is under a more direct control of zonal atmosphere temperature. The heat content of the thermocline seems to be climatically the most important variable.

Positions of the continents were taken from the reconstruction studies performed by Gorodnitski and Zonenshine (1979) for the Jurassic and the Cretaceous. The Cenozoic continental positions have been postulated on the assumption that the positions of the continents have been continuously and monotonically changing from the Cretaceous until modern times. True geodynamical reconstructions should be used in future studies. The detailed mapping of the paleobasins is not available, however, for this preliminary study. There may be a number of inconsistencies from a geodynamical point of view. For example, in the model the Drake passage was opened at the late Cretaceous. Paleogeographers can fin other inaccuracies. Nevertheless, it is hoped that the geographic simplifications do not seriously modify the results.

Results

Modern period

First, there are maps of present day currents and hydrology. Figure 1 shows the scheme of calculated currents of the World ocean (this is the only Mercator projection, i.e. the only experiment with calculations for the World ocean as a whole). It can easily be seen that a general features of the modern ocean circulation have been relatively well reproduced by the model (especially since the atmosphere is zonal; see Table 1). The main ocean gyres and well-known current systems are recogniz able, for example the Gulf Stream and the Kurosio, Labrador and California currents in the northern hemisphere, Brazil, Agulhas, Australian and Perry currents in the southern one. The Antarctic circumpolar current or the West Wind Drift also is correctly simulated. Total stream function Ψ and temperature T_{500} in northern hemisphere are shown on Figure 2. Figure 3

ABLE 2. Transports (in 10^6 m^3s^{-1}) of Some Currents rom the Model, From Bryan et al. (1975) Results and From Observations

urrents	1	2	3
ulfstream	12	14	100
razil	10	12	-
ast-Australian	15	24	20-43
urosio	14	20	80-90
est Wind Drift	40	22	0-230

1 - from the model presented in this paper
2 - from joint ocean – atmosphere model
3 - estimates
, 3 - are given as in (Bryan et al., 1975)

resents T_{500} obtained using the model (a), and T_{500} aken from the Atlas of the Ocean (1974, 1977), i.e. the bserved temperature (b). This comparison is the basis or concluding that the model is reliable enough for tilizing the results of paleoreconstructions. Total fluxes ppear to be two or three times smaller than the bserved" values. This deficiency is inherent in almost ll models and calculations with coarse numerical grids. able 2 contains the total fluxes obtained in this model, ryan et al (1975) results, and the "observed" fluxes.

Paleocirculations

The northern hemisphere has been mainly continental since the Jurassic. It is not as impressive for demonstrating the changes of the circulation due to continental drift. That is why only the Cretaceous of the northern hemisphere is under discussion here.

Figure 4 shows the scheme of the currents (a) and T_{500} (b) for the Cretaceous. For comparison, the currents obtained in the laboratory model (Luyendyk et al., 1972) are shown in Figure 5. Without detailed discussion of the results, two interesting peculiarities can be inferred from these maps. The Paleo-Pacific appears to have been noticeably cooler than at the present time. At the same time, the Mesozoic North Atlantic was a rather warm ocean compared to the relatively cold modern waters of this area. Warm waters of the Paleo-Atlantic were formed due to a specific pattern of the current system, i.e. because the ocean was too narrow for subpolar cyclonic gyre to be formed, and warm subtropical waters freely penetrated subpolar region. This has happened mainly because the subtropical gyre has not been blocked from penetrating into high latitudes. Later on we will see that this is the most important element of the ocean climate balance. The model shows Pacific waters to be some 3 - 4°C cooler than Atlantic waters in thermocline for the same latitudes in subpolar areas. Although we can easily explain this from the point of view of physical oceanography, this result should be carefully compared with existing paleoclimatological and paleoecological data.

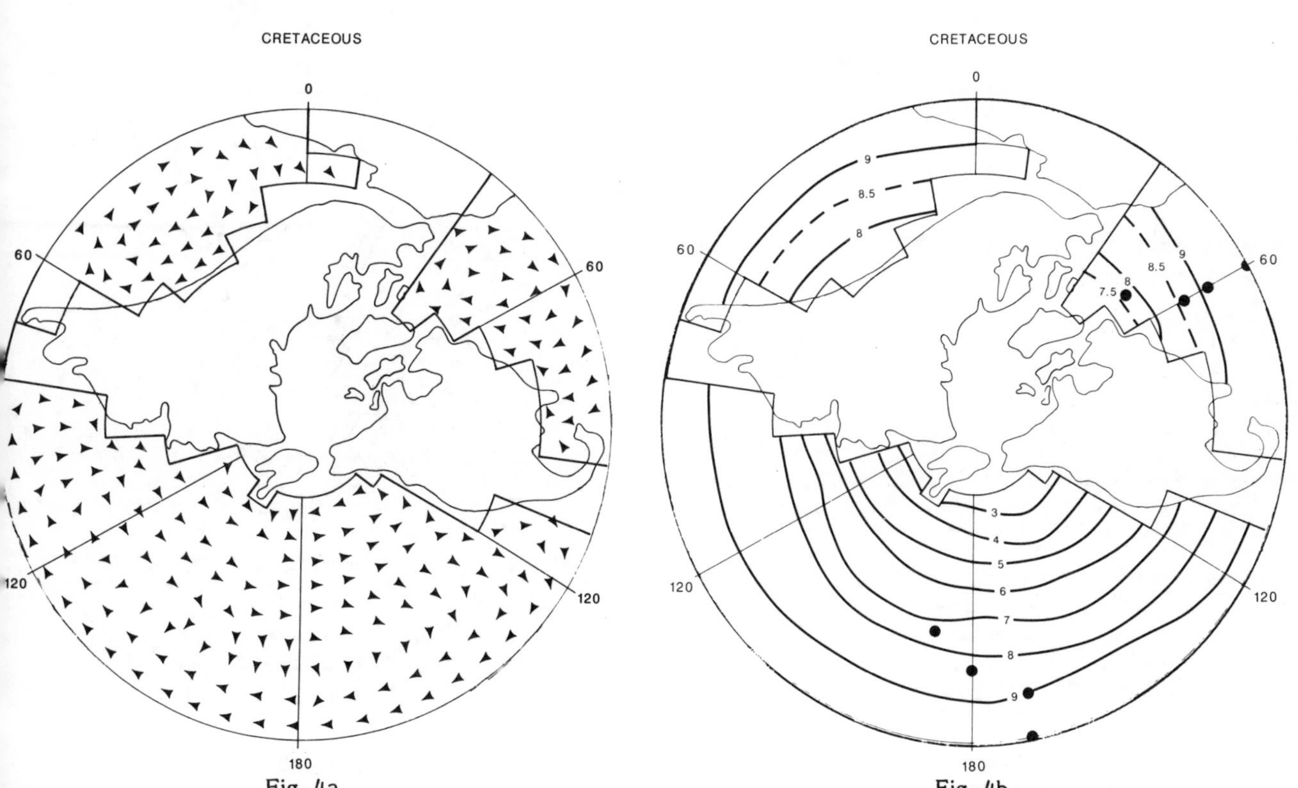

Fig. 4a Fig. 4b

Fig. 4. Scheme of currents (a), and T_{500} (b) of the Northern hemisphere oceans in the Cretaceous.

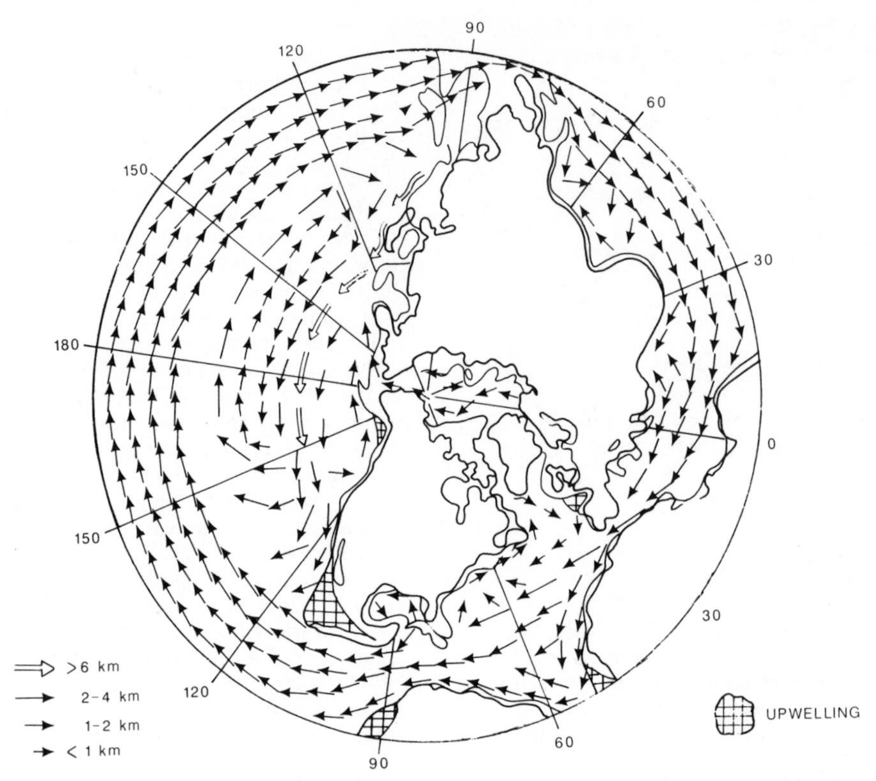

Fig. 5. Currents obtained in laboratory model for the Cretaceous of the Northern hemisphere (adopted from Luyendyk et al., 1972).

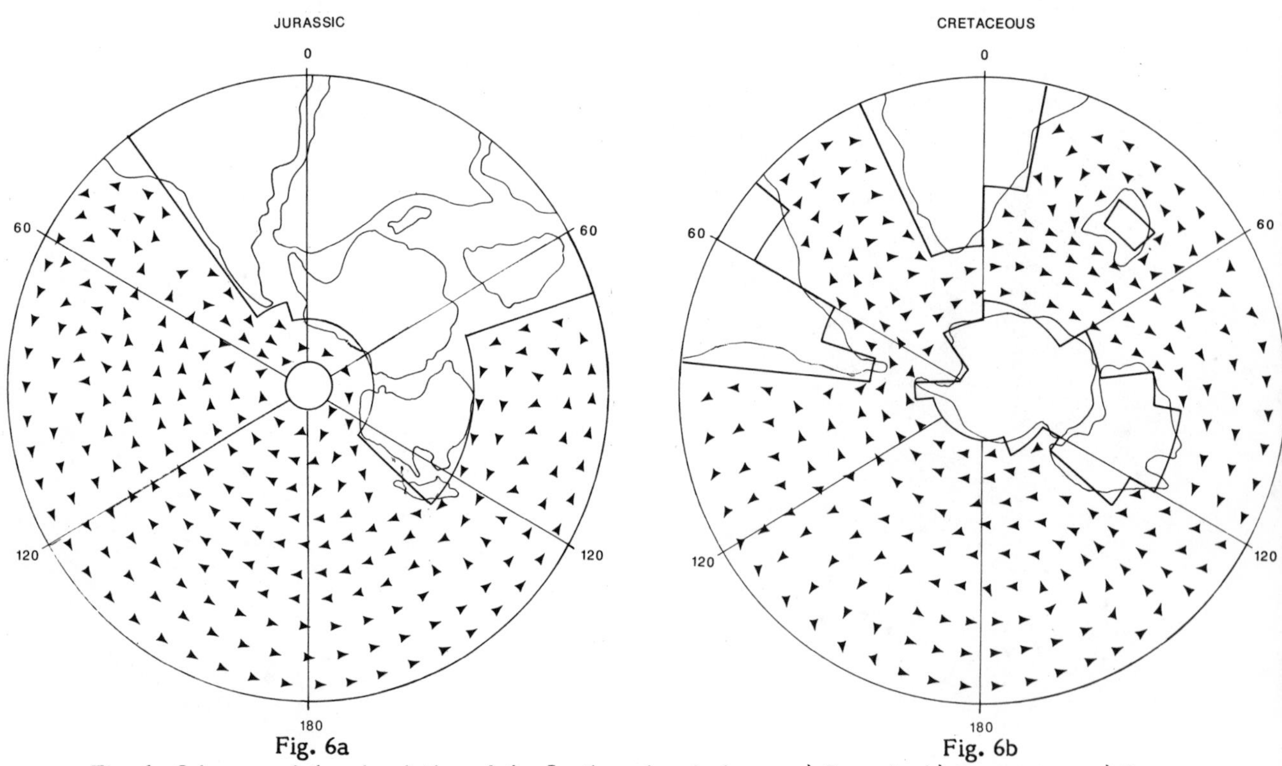

Fig. 6a

Fig. 6b

Fig. 6. Schemes of the circulation of the Southern hemisphere: a) Jurassic; b) Cretaceous; c) Eocene; d) Oligocene; e) Early Miocene; f) Late Miocene.

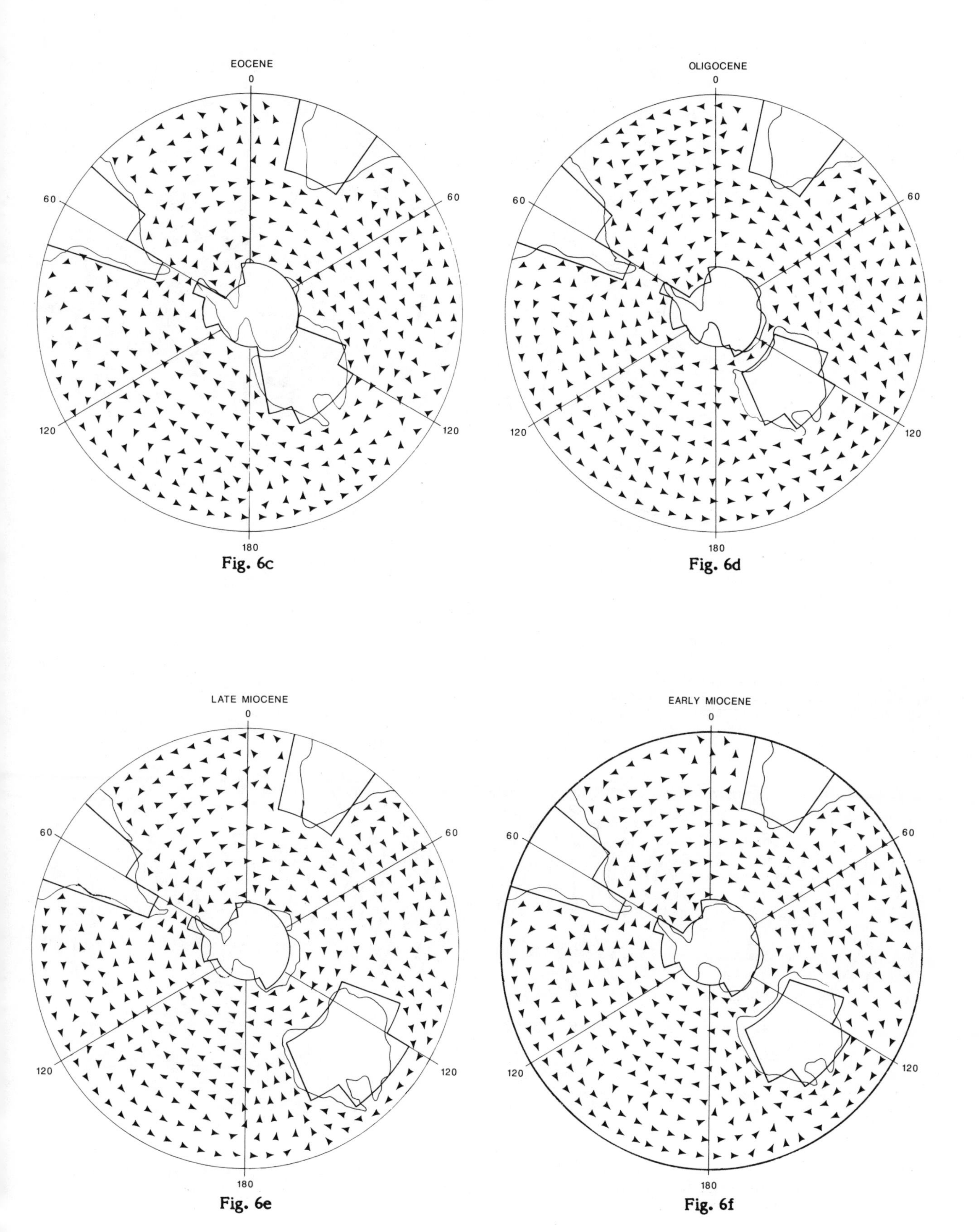

Fig. 6c

Fig. 6d

Fig. 6e

Fig. 6f

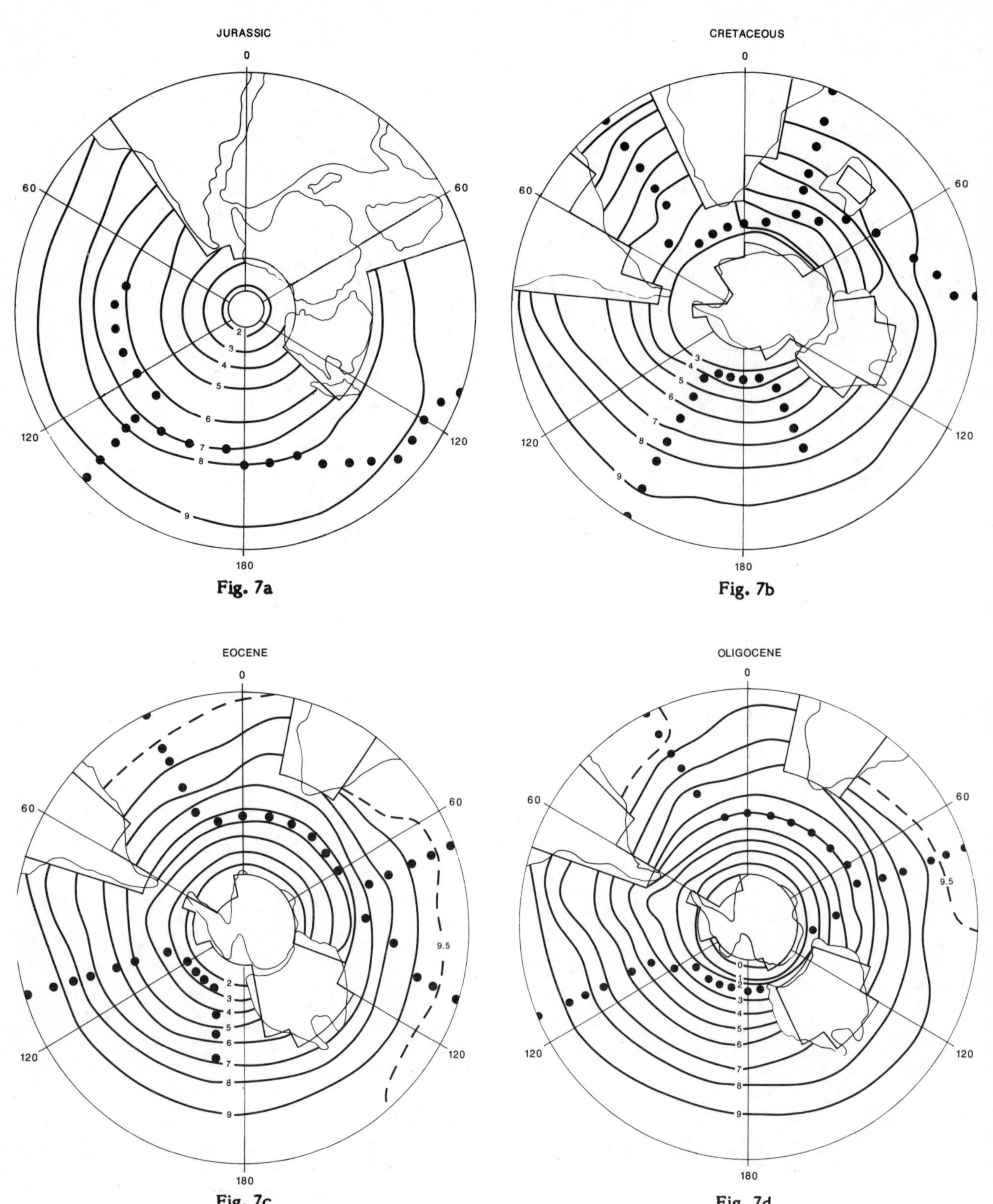

Fig. 7. History of temperature field of the Southern hemisphere: a) Jurassic; b) Cretaceous; c) Eocene; d) Oligocene; e) Early Miocene; f) Late Miocene.

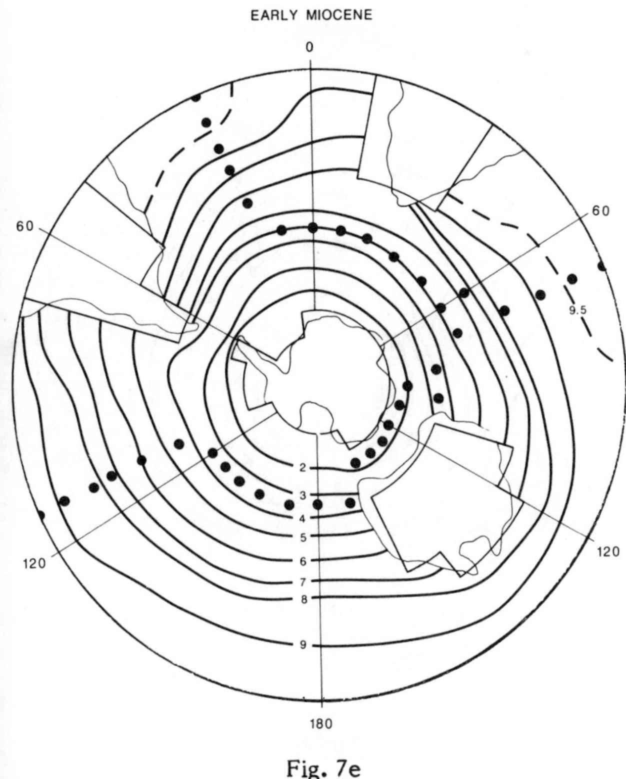

Fig. 7e

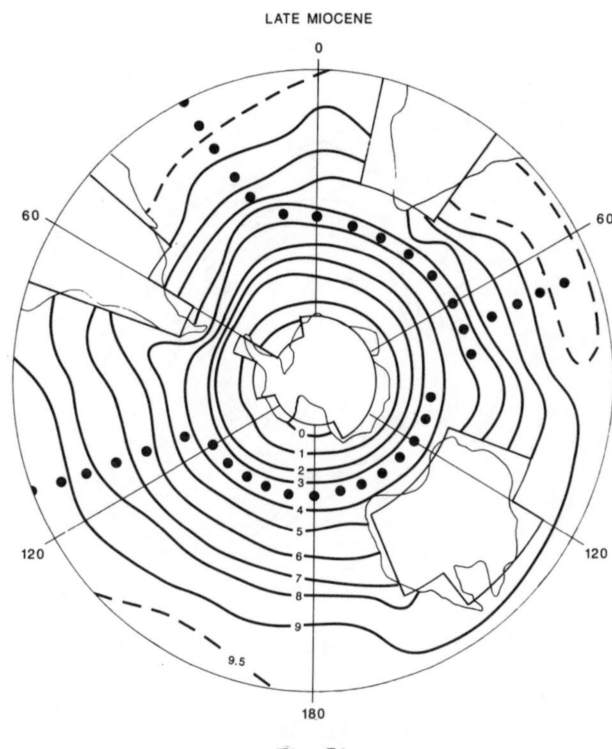

Fig. 7f

Next, let us look at the history of the southern hemisphere's circulation from Jurassic to the modern period. Figure 6 shows the series of schemes and Figure 7 represents the temperature field through this time interval. A generally warmer climate was found for the Mesozoic when the polar areas were free from land. During the Jurassic there was one giant continent - Gondwanaland. The breakup of this supercontinent began in the late Jurassic - early Cretaceous. Before that time a slow and gigantic circulation gyre existed in the only ocean of that period - the southern Paleo-Pacific. The model gives very warm waters in high latitudes, with almost zonal isotherms. Weak vertical circulation led to inefficient ventilation of the deep waters. It should be noticed here, that the model circulation is in good agreement with currents outlined using Jurassic biogeographical provinces (Westermann, 1980).

After the breakup of Gondwana (100 - 80 my b.p.), circulation changed radically. Assuming the Drake Passage was open at that time (although it is not so; see above) one gets a global flow which had to be formed in any case. The major feature of this circumglobal current from the Cretaceous until the Oligicene epoch is the absence of a gateway betwen Australia and Antarctica. Current flowed over Australia, i.e. went up to equatorial regions before coming back to the West Antarctic coast. That is why there was not over-all glaciation on the shelf of Antartica until the passage between Australia and Antarctic opened and created a place for a true circumpolar current which is known now as a West Wind Drift. The South Atlantic remained cold because it was too narrow for strong subtropical gyre to be formed, and cold waters of the circumzonal current intruded into this ancient ocean. Tranpsort through Drake Passage was rather small-almost half today's transport (please note the model transport for today, however). Here one can see the dramatic role of gateways and isthmuses for paleoclimate of the ocean. If we assumed Drake Passage be closed in the Cretaceous and some later periods we would obtain a warm South Atlantic (as in the case of the North Atlantic) since the anticyclonic gyre had to be formed with tropical water transport to Western Antarctic. It is important to stress here that other regions would not be seriously affected since the other parts of the southern oceans would be in any case be dominated by almost circumglobal current. This result proves the importance of the role of gateways and isthmuses (e.g. Berggren, 1982). During the Eocene epoch pronounced subtropical gyres were already formed, and they trapped a considerable amount of heat in that area. We consider this to be a consequence of the intensified circumzonal flow which partly blocked the low latitudes from the high ones. At that time Antarctic bottom water began to form in the Weddel Sea; where a stationary cyclone had formed. Accordingly to Stommel (1958, 1962) this process provides the abyssal water formation in this same region (as well as in the area of the Ross Sea, where this process had not started till the Oligocene epoch due to warm water supply by the current which flowed to the West Antarctic). The most dramatic change of Southern Ocean circulation took

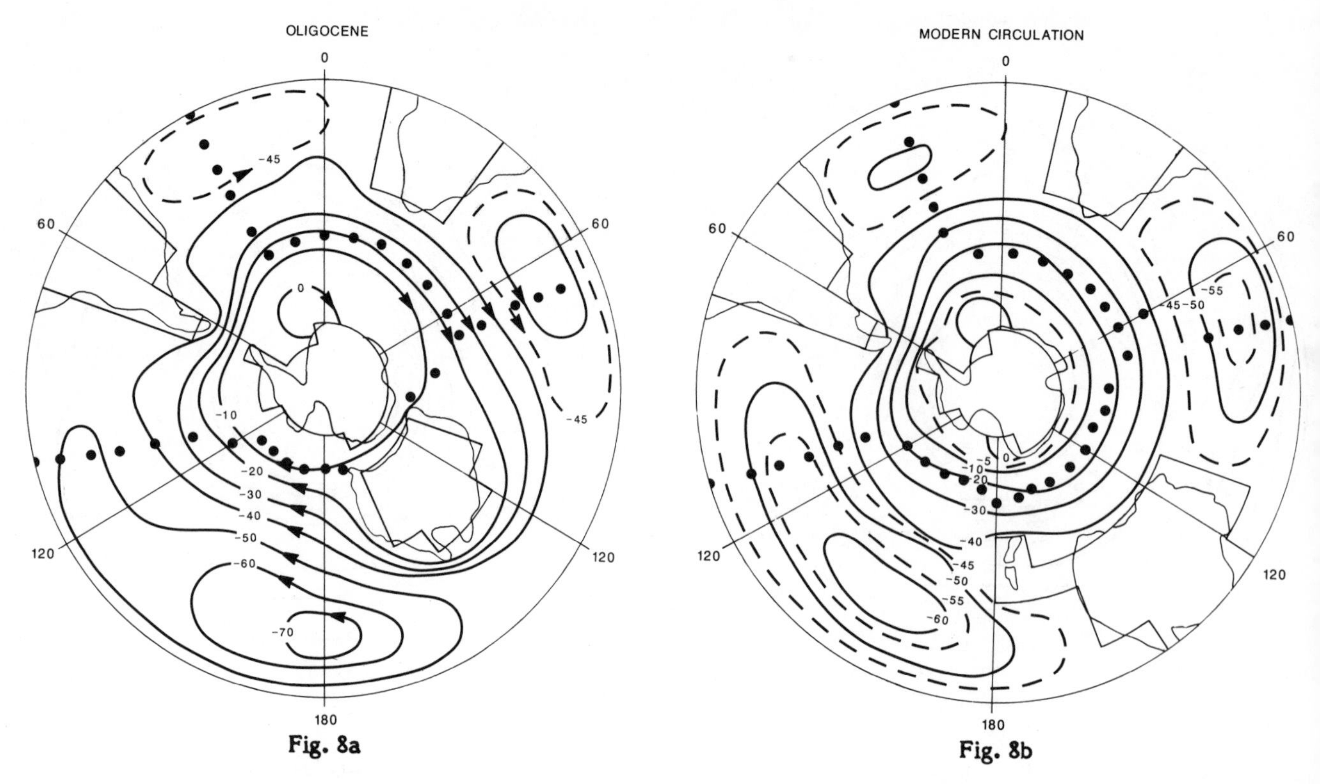

Fig. 8. Total stream function and of the oceans of the Oligocene epoch (a), and of the modern Southern hemisphere oceans (b).

place in the Oligocene (30 - 35 my b.p.). Australia separated from Antarctica, and the true circumpolar current partly developed. At this time sea ice appeared over the whole Antarctic area. These results enable us to believe that the Oligicene Epoch was the turning point in the history of the southern hemisphere's climate as it was the beginning of the total glaciation of this contient. This conclusion corresponds with other (indirect) data (Kennett, 1977).

After the Oligocene minimum the model gives a short-term Miocene warming at high latitudes which compares well with hypothesis proposed here about the crucial role of blocking zonal flows and subtropical gyres. In these gyres, blocked away from high latitudes by zonal current (in the southern ocean by circumzonal West Wind Drift) the heat is stored in abundance. This feedback is a positive one - stronger zonal current leads to more effective heat storage in gyres, which in turn leads to greater temperature and salinity differences between lows and highs, it again intensifies zonal currents, and so on. As the circumzonal current was not strong enough at the Early Miocene epoch, the heat stored in subtropical gyres was more easily transported to high latitudes. But, after further opening of the strait between Australia and Antarctica the West Wind Drift became sufficiently strong for the heat blockade of the subpolar regions. Also, the part of the circumglobal current which overflowed Australia from the north became weak, and therefore did not bring enough heat to the Western Antarctic area.

The Early Miocene model warming should be viewed as an essential proof of the usefulness of the numerical modelling method, and as showing that the nontrivial results might be obtained in the framework of this approach. The warming originated due to internal dynamics of the system rather than as a result of an external forcing. The air temperature profile, wind and mass exchange evolved monotonously through the Cenozoic with continuous cooling in high latitudes. It might be expected that since Antarctic atmosphere became cooler the water temperature should follow this trend. But the circulation and hydrology exhibited a "non-linear" benaviour with warming Antarctic water - not cooling, which is consistent with certain data interpretations (e.g. see Crowley, 1983; Schopf, 1980). As the model is extremely simple from the view of the numerical geophysical hydrodynamics (although it is rather complicated for geographical standings), the only conclusion to be employed is that the major regulating mechanism has been correctly identified, i.e. the main feature is the dynamical balance between the gyres and zonal blocking currents. The negative feedback which counterbalances the above mentioned positive one is a stabilizing exchange with the atmosphere and stronger zonation of the current system. As the passage between Australia and Antarctic had become wider and circumpolar current had intensified, the Late Miocene cooling of the high latitude took place again. In order to be a little bit more specific about the quantitive aspects of the changes there are the maps of total stream function

TABLE 3. Transport (in $10^6 m^3 s^{-1}$) of Some Paleocurrents From the Model* (see Table 2)

Paleocurrents	Cretaceous	Eocene	Early Oligocene	Miocene
Gulfstream	2	-	-	-
Brazil	did not exist	10	10	10
East-Australian	20	30	20	20
Curosio	12	-	-	-
West Wind Drift (in Drake Passage)**	20	30	40	40
Agulhas	5	10	10	5

* Paleocurrents have names of their analogs.
** It has been assumed that the Drake Passage was already opened in the Cretaceous (see text).

Figure 8) presented for the Oligocene (a) and for the modern period (b) of southern hemisphere ocean circulation. Table 3 presents water fluxes (obtained in the model) in the major current systems.

Conclusions

The ocean paleocirculation study has been carried out using a numerical modelling approach. The main goal was the demonstration of the applicability of such an approach and the role of the continental drift as a global geological climatic factor.

The main inference from the study, which we believe however has its own oceanographical value, is the understanding of general dynamical balance betwen the gyres and the zonal blocking currents. This balance is based on the feedback between the ocean - atmosphere exchange and gyres-jets interaction. The role of the gateways and isthmuses has proved again to be a major factor, but this time using a numerical model of the ocean circulation. The model gave results which generally are in qualitative agreement with laboratory modelling and with data based interpretations. The study has a lot of deficiencies at this stage. It should be mentioned, that in addition to several paleogeographical errors (Drake Passage's opening at the wrong time, as well as others), the ocean stayed rather cold in the deeper levels. This is in contradiction with a lot of estimations from geological records for the Late Mesozoic era (see, for example, the discussion in (Crowley, 1982; Schopf, 1980; Kennett, 1977; Monin and Shishkov, 1979 and so on). But, it was difficult to bring into conformity the low level of evaporation in gyres with convective adjustment in the cooler regions. The convection in cooler regions easily overpowered the convection due to salinity increase in subtropics. If we had increased evaporation (as in modern Mediterranean (Hay, 1983)) and, also, had increased subpolar air temperature, we would end up with rather warm abyssal waters. Nevertheless, this aspect should be clarified in future studies. Other shortcoming in the presented experiments is an ignorance of the sea level changes, i.e. epicontinental seas or the shoreline changes were not presented. Although these deficiencies obviously affect the paleogeographical part of the investigation, the latter should be observed as a preliminary test of the approach itself. On the other hand, the main feedback evaluation could be of some use for paleogeographers as a guideline inferred from geophysical hydrodynamics. At least, one might gain additional competence when studying ocean environment changes on the geological scale.

Acknowledgements. The numerical experiments have been carried out together with V. Enikeev who contributed basic participation in computational part of the whole study. The help of Professor K.J. Hsü is sincerely appreciated. Professor Hsü edited the paper and made significant stylistic and language corrections of the text. Anna Belyh provided a great deal of help in preparing this paper. Many colleagues showed interest and discussed the results with the author and V. Enikeev. The author is greatly indebted to all of them.

References

Atlas of the Oceans, 1974, 1977 (S. Gorshkov et al.); Moscow. GUNIO.

Barron, E.J. (1983). A warm, equable Cretaceous: the nature of the problem. Earth Sci. Revs., 19, 305-338.

Barron, E.J., Washington, W.M. (1982). Cretaceous climate: a comparison of atmospheric simulations with the geologic record. Palaeogeogr., Palaeoclimatol., Palaeocol., 40, N 1-3, 103-133.

Berggren, W.A., Holister, C.D., (1974). Paleogeography, paleobiogeography and history of circulation in the Altantic Ocean. In: Studies in paleoceanography (W.W. Hay, Ed.), 126-186.

Berggren, W. (1982). Role of ocean gateways in climatic change. In: Climate in Earth History. NAS, Wash., D.C., 118-125.

Bryan, K. (1969). A numerical method for study of the circulation of the World ocean. J. Comput. Phys., 4, 347-376.

Bryan, K., Cox, M.D. (1967). A numerical investigation of the ocean general circulation. Tellus, 19, 54-80.

Bryan, K., Manabe, S., Pacanowski, R.C. (1975). A global ocean-atmosphere climate model. Part II. The ocean circulation. J.Phys. Oceanogr., 5, 30-46.

Crowley, T.J. (1983). The geologic record of climate change. Rev. Geophys., 21, 828-877.

Crowley, W.P. (1968a). A global numerical ocean model. Part I. J.Comput. Phys., 3, 111-147.

Crowley, W.P. (1968b). Numerical advection experiments. Mon. Weather Rev., 96, 1-11.

Gordon, W.A. (1973). Marine life and ocean surface currents in the Cretaceous. J. Geol., 81, 269-284.

Gorodnitsky, A.M., Zonenshain, L.P. (1979). Paleogeodynamical reconstructions for the Phanerozoic. Geophysics of the ocean, v.2, Moscow: Nauka, 338-369 (in Russian).

Haney, R.L. (1971). Surface thermal boundary condition for ocean circulation models. J.Phys. Oceanogr., 1, 241-248.

Haney, R.L. (1974). A numerical study of the response of an idealized ocean to large-scale surface heat and momentum flux. J.Phys. Oceanogr., 4, 145-167.

Heath, G.R. (1979). Simulations of a glacial paleoclimate by three different atmospheric general circulation models. Palaeogeogr., Palaeoclimatol., Palaeoecol., 26, 291-303.

Huang, J.C.K. (1978). Numerical simulation studies of oceanic anomalies in the North Pacific basin I. The ocean model and the long-term mean state. J.Phys. Oceanogr., 8, 755-778.

Hay, W.W. (1983). The global significance of regional Mediterranean neogen paleoenvironmental studies. In: Reconstruction of marine paleoenvironments (J.E. Meulenkamp, Ed.),30, 9-23.

Kennett, J.P. (1977). Cenozoic evolution of Antarctic glaciation, the circum-Antarctic ocean currents and their influence on global paleoceanography. J.Geophys. Res., 82, 3843-3860.

Kamenkovich, V.M. (1973). Fundamentals of the ocean dynamics. Leningrad, Gidrometeoizdat (in Russian).

Lazarev, P.P. (1950). Papers on the laboratory modelling of the ocean currents. Collections of papers, v.3, Geophysics. Nauka, Moscow (in Russian).

Leith, C.E. (1965). Numerical simulation of the Earth's atmosphere. In: Methods in computational physics, v.4, Acad. Press, N.Y. , 1-28.

Lisitsin, A.P. (1980). Paleoceanology. In: Oceanology. Geologic history of the ocean. Moscow, Nauka, 386-406 (in Russian).

Lisitsin, A.P. et al. (1980). The history of the Mesozoic-Cenozoic sedimentation in the World ocean. In: Oceanology. Geologic history of the ocean. Moscow, Nauka, 407-427.

Luyendyk, B.P., Forsyth, D., Phillips, J.D. (1972). Experimental approach to the paleocirculation of the ocean surface waters. Bull. Geol. Soc. Amer., 83, 2449-2464

Marchyk, G.I. (1974). Methods of computational mathematics. Novosibirsk, Nauka, (in Russian).

Mesinger, F., Arakawa, A. (1976). Numerical models used in atmospheric models. GARP Publ. Ser., v.17.

Monin, A.S. (1982). Introduction to the theory of climate. Leningrad, Gidrometeoizdat (in Russian).

Monin, A.S., Shishkov, Yu.A. (1979). The history of climate. Leningrad, Gidrometeoizdat (in Russian).

Munk, W.H. (1950). On the wind-driven ocean circulation. J.Meteorol., 7, 79-93.

Parrish, J., Curtis, R. (1982). Atmospheric circulation, upwelling, and organic rich rocks in the Mesozoic and Cenozoic eras. Palaeogeogr., Palaeoclimatol., Palaeocol., 40, N 1-3, 31-66.

Sarkisyan, A.S. (1977). The diagnostic calculations of a large-scale oceanic circulation. In: The Sea, v.6 (E.D. Goldberg, Ed.), 363-458.

Schopf, J.M.(1980). Paleoceanography. Harvard Univ. Press

Seidov, D.G. (1978). Numerical scheme for study of synoptic scale oceanic eddies. Izv of the USSR Acad. Sci., 14, 757-767.

Seidov, D.G. (1980). Synoptic eddies in the ocean. Numerical experiment. Izv. of the USSR Acad. Sci., Physics of atmosphere and ocean, 16, N1, 73-87 (in Russian).

Seidov, D.G. (1984). Model of global circulation of ocean. Izv. of the USSR Acad. Sci., Physics of atmosphere and ocean, 20, N4, 287-296 (in Russian).

Seidov, D.G., Yenikeev, V.H. (1983). Numerical modelling of the Mesozoic and Cenozoic ocean paleocirculation. Ocean modelling, N53, 5-8.

Seidov, D.G., Yenikeev, V.H. (1984). Numerical modelling of paleocirculation of Late Mesozoic and Cenozoic oceans. Oceanology, 24, N4, 656-663 (in Russian).

Seidov, D.G., Enikeev, V.H. (1984). The southern ocean circulation model. Meteorology and Hydrology, N6, 51-60 (in Russian).

Stommel, H. (1948). The westward intensification of the wind-driven ocean currents. Trans. Amer. Geophys. Union, 29, 202-206.

Stommel, H. (1958). The abyssal circulation. Deep-Sea Res., 5, 80-82.

Stommel, H. (1962). On the smallness of sinking region in the ocean. Proc. Nat. Acad. Sci., 48, 766-772.

Veronis, G. (1966). Wind-driven ocean circulation. Part 2, Numerical solutions of the non-linear problem. Deep-Sea Res., 13, 31-55.

Westermann, G.E.G. (1980). Ammonite biochronology and biogeography of the circum-Pacific Middle Jurassic. In: Systematics Association Spec. Vol., N18 (Eds. M.R. House and J.R. Senior), Acad. Press, 459-498.

MODELLING OF THE QUATERNARY PALEOCIRCULATION IN THE NORTH ATLANTIC

V.H. Enikeev

Shirshov's Institute of Oceanology, Academy of Sciences, USSR

Quaternary climatic changes are of great interest for paleogeographers and paleoecologists, because the last glacial period occurred so recently on a geological time scale and substantial paleoclimatic data are available (Barash, 1974, 1981). One of the important reconstructions of the Quaternary climate was made in the framework of the CLIMAP project (1981).

Pleistocene temperatures were investigated with a variety of methods of paleotemperature analysis. It is necessary to mention paleotemperature reconstruction which has been performed by M.S. Barash (1974, 1981). Nevertheless, for the purpose of reconstructed World ocean surface temperature, existing paleotemperature data are not enough. We have several schemes of paleocurrents reconstructed using indirect data (Barash, 1974; Gordon, 1973), but it is necessary to have an independent method for verification of different schemes.

At the present time we have some schemes for intercomparison. Numerical models may be used for simulation of the ocean circulation and distribution of the hydrological characteristics. Analogous investigations for the atmosphere have been already carried out (Heath, 1979). Reconstructions of the Quaternary global circulation were performed in numerical experiments using three different models of the atmospheric circulation (Heath, 1979).

D.G. Seidov (1984) suggested a three-dimensional hydrodynamical model of the global ocean circulation. This model has been applied to the calculation of the modern circulation in the Southern ocean (Seidov and Enikeev, 1984) and for the reconstruction of the hydrological characteristics of the Mesozoic and Cenozoic oceans (Seidov and Enikeev, 1984).

In this paper we present a numerical investigation of the modern and Quaternary circulation of the North Atlantic in the framework of this model. We concentrated on the ocean circulation for the Pleistocene glacial period. The numerical solution for the equations of the model gives the equilibrium state of temperature, salinity and current fields in response to the effects of the heat, mass and momentum fluxes through the sea surface. The balance equation of the integral vorticity, heat and salt and Poissons equation for the total stream function are solved for the ocean with an arbitrary geometry of the shore line and bottom topography. At the surface we have specified a zonal field of the wind stress, the air temperature and the difference between evaporation and precipitation rates. All these parameters of the model vary only with latitude. It is possible to specify the temperature of the surface water instead of the heat flux, proportional to the different surface water temperature and the air temperature. The numerical model and numerical scheme are the same as in Seidov, 1984.

The temperature of the air has been taken from two sources: the Atlas of the ocean (1974, 1977) and the results of numerical simulations with an atmospheric model (Verbitsky and Chalikov, 1982). If we have radiative balance at the boundary of the atmosphere and the distribution of the continents on the latitude circle then this simple atmospheric model allows one to calculate a zonal distribution of the air temperature.

All numerical atmospheric experiments have been carried out with certain annual average zonal distribution of the short wave radiation flux on the upper boundary of

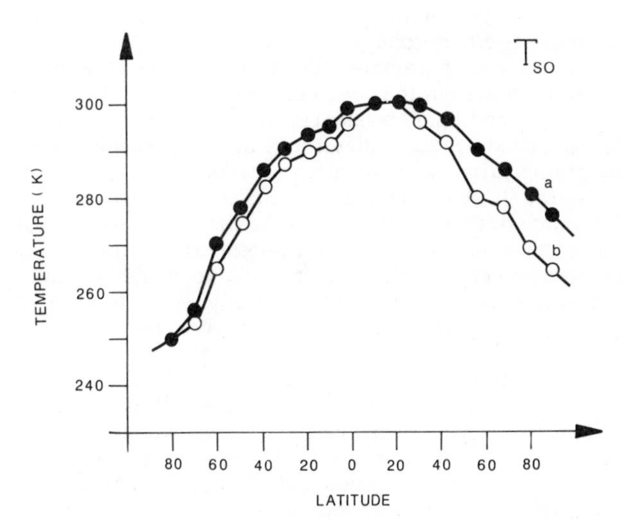

Fig. 1. Profile of the mean ocean surface temperature (zonal) at T_{S0}; a) modern period; b) Würm glacial period.

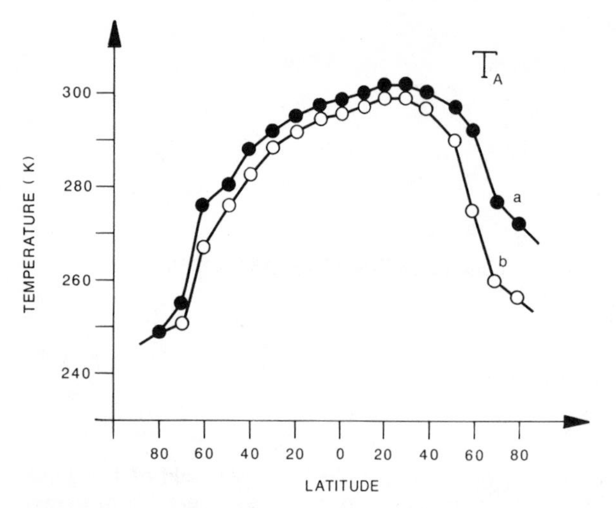

Fig. 2. Profile of the annual average atmospheric temperature, T_A, at the upper part of the boundary layer: a) modern period; b) Würm glacial period (results of the numerical simulation).

the atmosphere, specified surface temperature of the ocean and given distribution of the continental and ocean ice. The boundary of the continental ice sheet in the northern hemisphere is lowered to 40°N.

On Figure 1 one can see a profile of annually averaged ocean surface temperature at the present time and for the last glacial period. The calculated model temperature of the air T_A at the upper part of the boundary layer is presented on Figure 2. One can see the similarity of curves at low latitudes. The difference between the modern atmosphere temperature and that 18,000 years ago at 60°N amounted to 12°K. Isotherms drop off in the southern and northern hemispheres with different slopes. This is connected with different amounts of continental ice in the two hemispheres.

Atmospheric parameters derived from the atmospheric model were used in the ocean block of the model. Salinity on all rigid and liquid boundaries and at the ocean surface was specified. A zonal difference between evaporation and precipitation was taken from Atlas of the oceans (1974, 1977). A temperature flux is proportional to the difference between temperature of the upper layer of the ocean and temperature of the atmosphere. If the surface temperature drops below the freezing temperature the flux of the heat and wind stress become equal to zero. Temperature and salinity of the surface water below ice become -1.88°C and 34.5°/oo accordingly. The wind stress for the modern period is calculated by the pressure field taken from the Atlas of the oceans (1974, 1977). Bottom topography of the North Atlantic was taken from the bathymetric map of the World ocean. The numerical grid has a 4° step in latitude and 6° step in longitude. Calculations are made at four levels: 0 m, 500 m, 1500 m and 2500 m.

Figure 3 shows the total flux of water ψ in special units: sverdrups (1 sv= $10^{12}cm^3s^{-1}$). The modern velocity field is presented in Figure 4; modern temperature is shown in Figure 5. The total flux in the Gulf Stream is about 30 sv. The maximum current velocity reaches 30 cm s^{-1}. The axis of the North Atlantic currents is in the 46°-50° latitudinal belt.

Figure 6 shows a total flux function for the last glacial period. The Quaternary circulation differs from the modern. This paleocirculation is more zonal than the modern system. The boundary of the sea ice is 12° further south. The continental ice sheets of North America and Western Europe are united at 60°N latitude. A visible difference in the temperature fields of the modern and the Würm ocean may be seen in Figure 8. They are about 5°-6°C. The circulation seems to be similar at all depths, but the Quaternary velocity (Figure 7) at depths of 500 m and 1500 m are different from modern current field. Total flux is changed only in the Gulf Stream area, and is about 80% of the modern value. This result confirmed the hypothesis about the greater significance of the current configuration than the current intensity for the Earth's climate.

Numerical experiments on the reconstruction of the Quaternary paleocirculation of the North Atlantic, presented in this paper, are preliminary. It is necessary to compare the results of calculations with other paleogeographical data and interpretations.

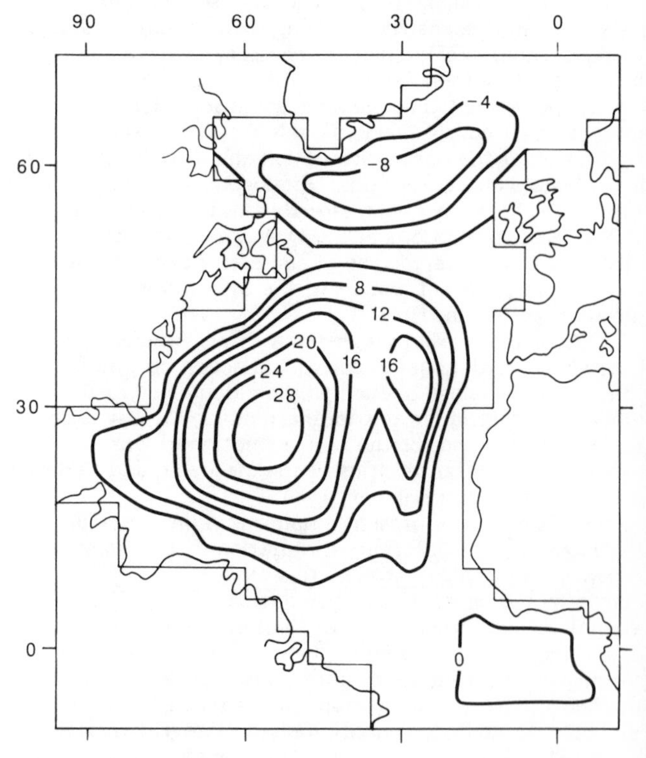

Fig. 3. Total flux function ψ in the modern North Atlantic.

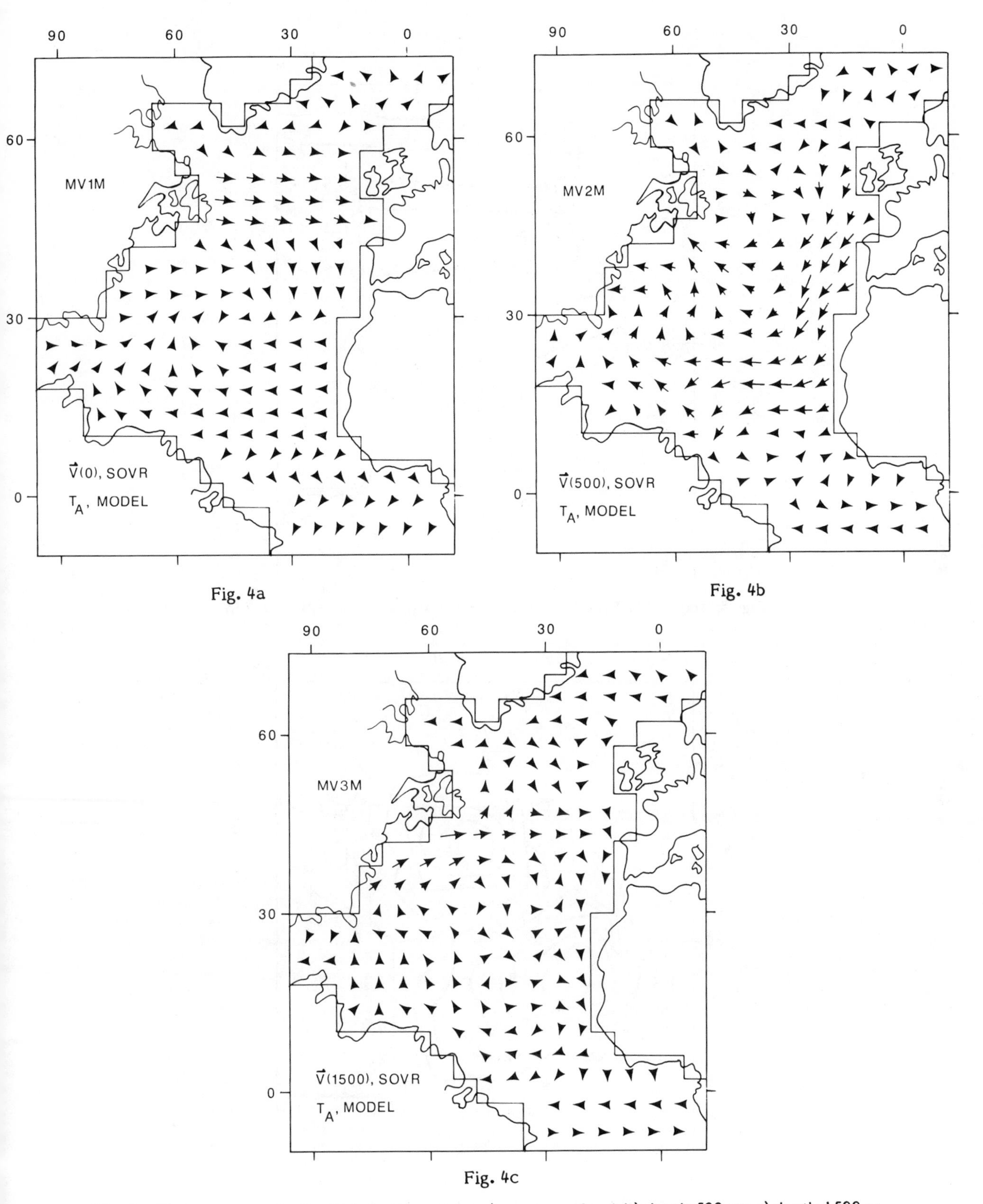

Fig. 4. Modern current velocities in the ocean: a) ocean surface; b) depth 500 m; c) depth 1500 m.

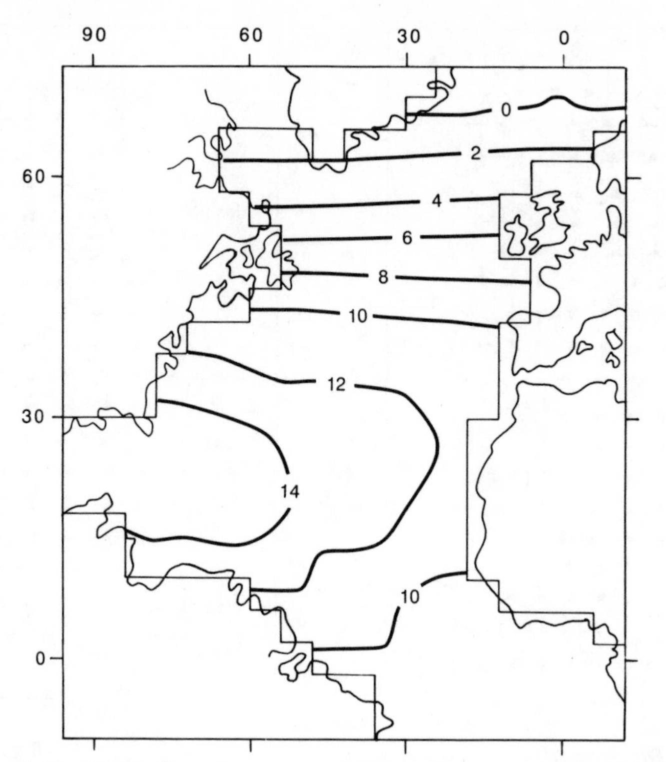

Fig. 5. The field of modern ocean temperature at a depth of 500 m

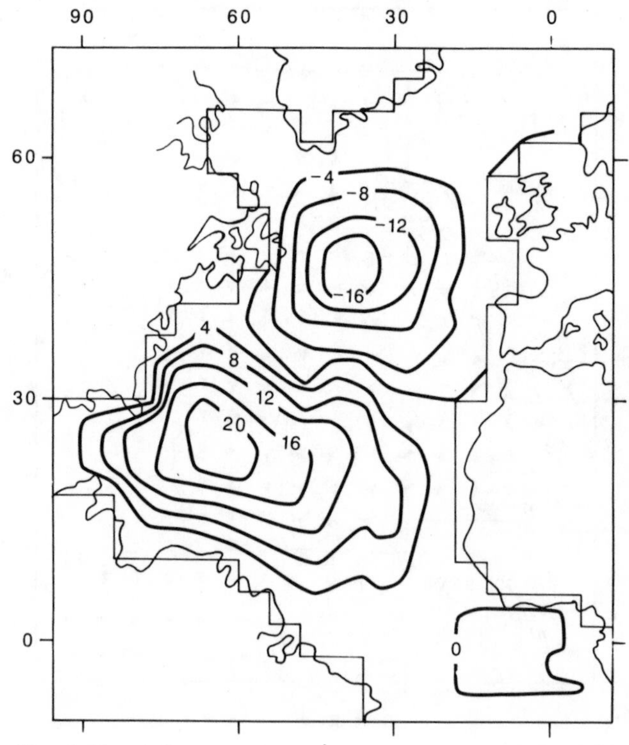

Fig. 6. Total flux function ψ for the last glacial period.

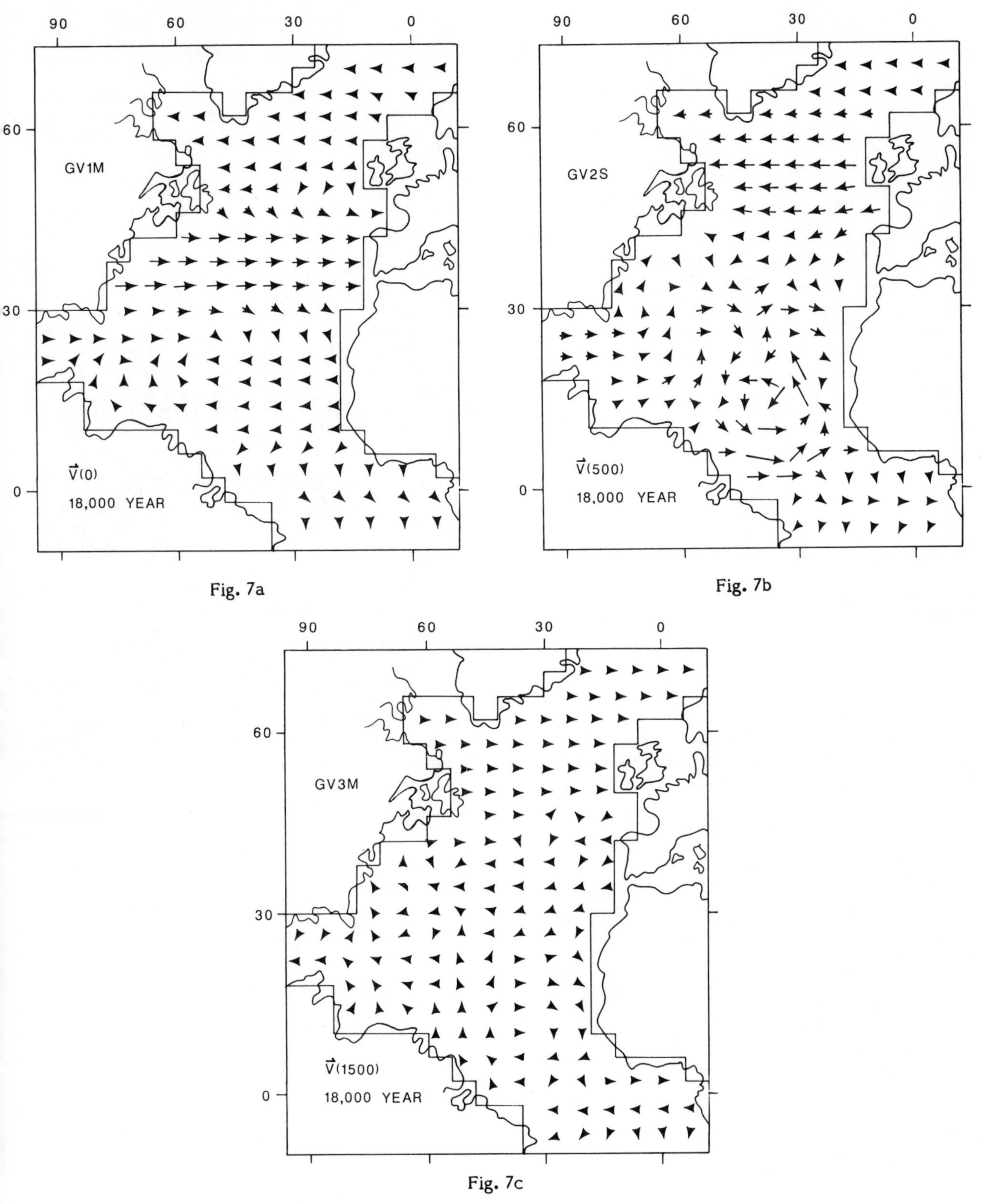

Fig. 7. Würm current velocities in the North Atlantic: a) ocean surface; b) depth 500 m; c) depth 1500 m.

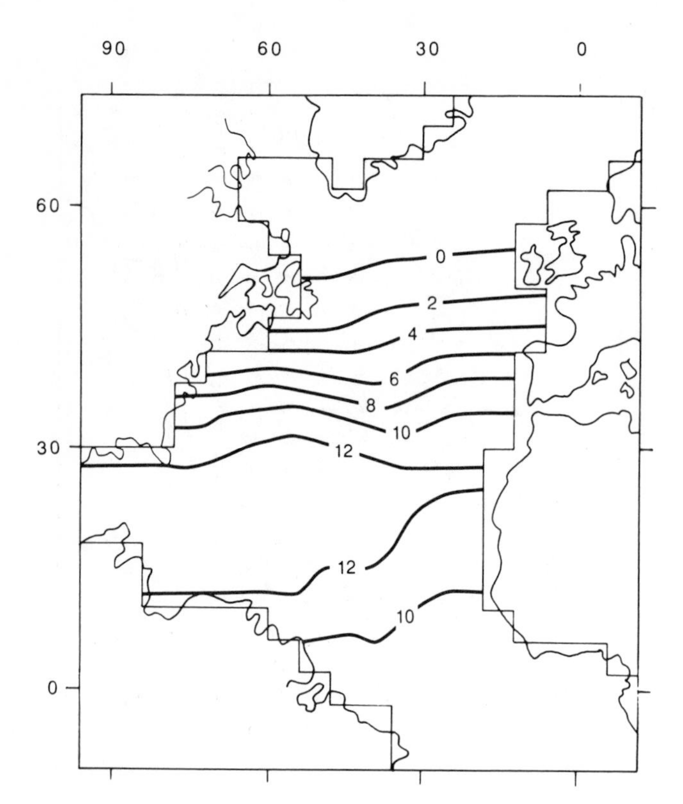

Fig. 8. The field of Würm ocean temperature at a depth of 500 m.

Acknowledgements. In conclusion, I want to express gratitude to Dr. D.G. Seidov for the support and help in this work. I appreciate V.N. Stepanov's assistance in the computational phase of the study.

References

Atlas of the Oceans, 1974, 1977 (S. Gorshkov et al.); Moscow. GUNIO.

Barash, M.S., 1974, Migration of climatic belts of the Altantic ocean in the upper quarternary period, Dokl. of the USSR Acad. of Science, 216, N5, 1158-1160 (in Russian).

Barash, M.S., 1981, Climatic zonality of the Altantic ocean in the Quaternary period (on planktonic foraminifera), in Climatic zonality and sedimentation, Moscow, Nauka, p. 126-139 (in Russian).

CLIMAP project members, 1981, Seasonal reconstructions of Earth's surface at the last glacial maximum, Geol. Soc. Amer. Map and Chart Ser., MC-36.

Gordon, W.A., 1973, Marine life and ocean surface currents in the Cretaceous; J. Geol., 81, 269-284.

Heath, G.R., 1979, Simulation of glacial paleoclimate by three different atmospheric circulation models, Palaeogegr., Palaeoclim., Palaeoecol., 26, 291-303.

Monin, A.S., and Yu.A. Shichkov, 1979, The history of climate, Leningrad, Gidrometeoizdat, (in Russian).

Seidov, D.G., 1984, Model of global circulation of oceans, Izv. of the USSR Acad. of Sci., Physics of atmosphere and ocean, 20, N4, 287-296 (in Russian).

Seidov, D.G., and V.H. Enikeev, 1984, Numerical modelling of paleocirculation of Late Mesozoic and Cenozoic oceans, Oceanology, 24, N4, 656-663 (in Russian).

Seidov, D.G., and V.H. Enikeev, 1984, The southern ocean circulation model, Meteorology and hydrology, N6, 51-60 (in Russian).

Sergin, V.Ja., and S.Ja. Sergin, 1978, A systematic analysis of the problem of the big climate and Earth glaciation variation, Leningrad, Gidrometeoizdat, 279 (in Russian).

Verbtisky, M.Ya., and D.V. Chalikov, 1982, A one-dimensional Atmospheric model as a bloc a climatic system ocean-atmosphere-ice; Izv. of the USSR Acad. of Sc., Physics of an atmosphere and ocean, 18, N10, 1011-1018 (in Russian).

MESOZOIC-CENOZOIC CLIMATIC HISTORY AND CAUSES OF THE GLACIATION

L.A. Frakes

Department of Geology and Geophysics, University of Adelaide, Adelaide, S.A.

Abstract. Over the interval since the late Permian, the earth has seen a recovery from Paleozoic glaciation followed by an anomalously warm Mesozoic and a punctuated decline in mean global temperature in the late Cretaceous-Cenozoic. The main changes in thermal condition occurred in the middle Triassic (strong warming = W); late Jurassic (W); post-Santonian late Cretaceous (strong cooling = C); late Paleocene-early Eocene (W); middle Eocene (C); end Eocene (C); middle Miocene (C); end Miocene (C); middle Pliocene (C) and end Pliocene (C). Intervening times were characterized by low amplitude fluctuations and/or gradual warming or cooling trends. Abrupt climate changes have been explained by a variety of mechanisms, including changes in paleogeography, oceanic circulation, structure of the water mass, volcanic dust input, and recently, volcanism-related CO_2 levels.

From matching of a Phanerozoic curve of atmospheric CO_2 concentration, based on estimates of deposition of carbonate on the continents, with a sub-parallel curve for abundance of continental volcanic rocks, Budyko and Ronov concluded that global climate changes are related to the variable output of CO_2 from volcanic activity. This conclusion is supported by most Phanerozoic climate trends. However, their CO_2 derived temperature curve suggests warming in the late Cretaceous and in the Miocene; these two trends are against the geological evidence. To explain the Miocene, a substantial though gradual warming in the early Miocene might be overly weighted in construction of the CO_2 curve. The known cooling in the late Cretaceous is not so easily explained, nor is the fact that CO_2 - derived temperatures in the Tertiary are consistently higher than the present mean global temperature.

It is suggested that the warming effect of CO_2 produced by volcanism normally outweighs the cooling tendency arising from the dust veil index from volcanism. However, CO_2 - induced warming was not effective in the post-Santonian Cretaceous, probably because of the long-term decrease in seafloor spreading and the consequent falls in sea level. The result was instead increased global albedo and hence cooling. The general decline in oceanic volcanism and climatic deterioration in the Cenozoic appears to have been related to the slowed spreading through the mechanism of lowered CO_2 input to the atmosphere- hydrosphere. However, abrupt changes in paleogeography and density structure of the oceans via tectonic activity probably brought about most marked climatic coolings seen in the geologic record.

Introduction

Since the initiation of the Deep Sea Drilling Project, the study of global climates has increased enormously. We now have a fairly reliable and complete record of climate trends over the interval since the middle Jurassic. Much of this new information is derived from seafloor materials and hence relates to oceanic climates, but the stimulus has extended beyond the shoreline and hence, there has been a quantum jump in our knowledge of how recent earth climates have varied on both land and sea. This is not to say that our understanding of the workings of ancient climates has progresed to such an extent.

For two reasons, the causal mechanisms of climate evolution in the Mesozoic and Cenozoic should be relatively easily deciphered. First, the glaciation which developed was simple in that it was polar. Second, for these comparatively recent times the information available is at the maximum: more of the sedimentary products are preserved. It is possible because of this abundance of information to explore the physiology of the glaciation in all its complexity. We are still at an early stage of this investigation into the feedback-dominated climate system and as a consequence, the more obvious variables are under scrutiny -- CO_2 effects, land-sea configurations, water-mass structure, sea level change, etc. It may be that the likelihood of causality of any of these for glaciation will not be fully comprehended until much more data are at hand. And, since we cannot yet place full trust in any single explanation for climatic deterioration in the Mesozoic - Cenozoic, we are certainly justified in asking the hard questions about mechanisms already proposed.

The Nature of the Record

The data on Mesozoic-Cenozoic climates are scant for the interval before the middle Jurassic, but some indication of Triassic and early Jurassic conditions is gained from study of rocks of these ages located on the continental margins and on the continents themselves. Here, non-marine rocks play a more important part than in younger sequences because diagenetic alteration has been pervasive in marine carbonates and oxygen isotope studies for paleotemperatures accordingly are futile. Early determinations on belemnites and other molluscs probably are useful in determining trends in changing environments, but any calculated paleotemperature should be considered as a maximum for a given sample. More recent studies on belemnites give more trustworthy results. A further problem arises from the fact that early Mesozoic marine strata are common only at depth on continental margins, most having been deposited during low stands of sea level or having later sunk as a result of margin foundering. On land, evidence on climates of this time derives largely from landplant paleobiogeography, diversity patterns, and distributions of indicator rock-types. Despite these shortcomings, the record yields a fairly comprehensive picture of climates in the early Mesozoic.

By mid-Jurassic time, eustatic sea level rise covered a substantial part of the continental margins and marine rocks became more widespread. Also, in slightly younger rocks, carbonates are less affected by diagenesis and isotope studies begin to find applicability (although there are still only a few modern studies of Cretaceous isotopes). All such work suffers from uncertainties about the relative contributions of paleotemperature and ice-volume effects to isotopic composition of the ocean. Pelagic sediments of Jurassic and early Cretaceous age are progressively more abundant, allowing investigations on many of their aspects related to climate. For the Cenozoic, of course, the full range of paleoclimatic tools can be brought to bear on the widespread and variable sedimentary cover, and our understanding of processes and events is increased by the fact that thermal regimes and land-sea configurations are more similar to those of the present.

For all periods in earth history, interpretations of the global climate state are made difficult by two factors of enormous importance. First there is the problem of strict synchroneity of the available data: even closely spaced data points may not be synchronous if climatic, depositional and erosional processes were fluctuating widely. Yet, the understanding, and indeed, the proper definition of past climates requires an abundance of synchronous data spread over a large area. Second, we are sufficiently ignorant of the workings of the present climate that in many cases our extension of present meteorological, climatological and oceanographic relationships to what may have been very different circumstances in the past, is simply not justified. Geologists, while aware of the first problem, are forced to generalize on the basis of assumed synchroneity in order to explain their data. Fortunately, almost on a daily basis, climate modelling is now telling us more about modern climate, so that sound application to past climates is becoming less difficult.

Mesozoic Climates

It is almost an axiom that with regard to the Mesozoic, the climates were warm and dry. An abundance of evidence testifies to warm conditions and the general absence of indicators of cold climates provides further support. In detail, however, it is more accurate to describe most Mesozoic climates as warmer than present, at least in the middle latitudes, while in latitudes higher than about 50° and extending at least as far as 70°, rainfall appears to have been very abundant. Taken together with evaporite abundance indicating extraordinary rates of low latitude evaporation, these circumstances imply strong poleward transport of heat in the atmosphere via the mechanism of latent heat of evaporation. This in turn suggests that Mesozoic circulation patterns differed markedly from those of the present, particularly in the location and extent of the Hadley cell/subtropical low-pressure system. The mechanisms of this altered global state are now being investigated by means of mathematical modelling in several institutions, as is the relative importance of oceanic heat transport, another critical parameter in Mesozoic climates.

Absence of Mesozoic glaciation anywhere on the globe is inferred from what appears to be a lack of glacially deposited rocks. The presence of ice is indicated by occurrences of boulder mudstones in high-latitude Cretaceous sequences, but these have been interpreted as resulting from non-glacial ice (shore-ice, river ice, etc.). However, eustatic changes of sea level took place throughout the Mesozoic and it is matter of interpretation whether a component due to the construction of continental ice is represented along with the tectonic contribution. This will not be fully resolved in the absence of good evidence for the activity of glaciers.

In broad outlines the global climate state during the Mesozoic passed through, first, a long and somewhat irregular warming which began (at

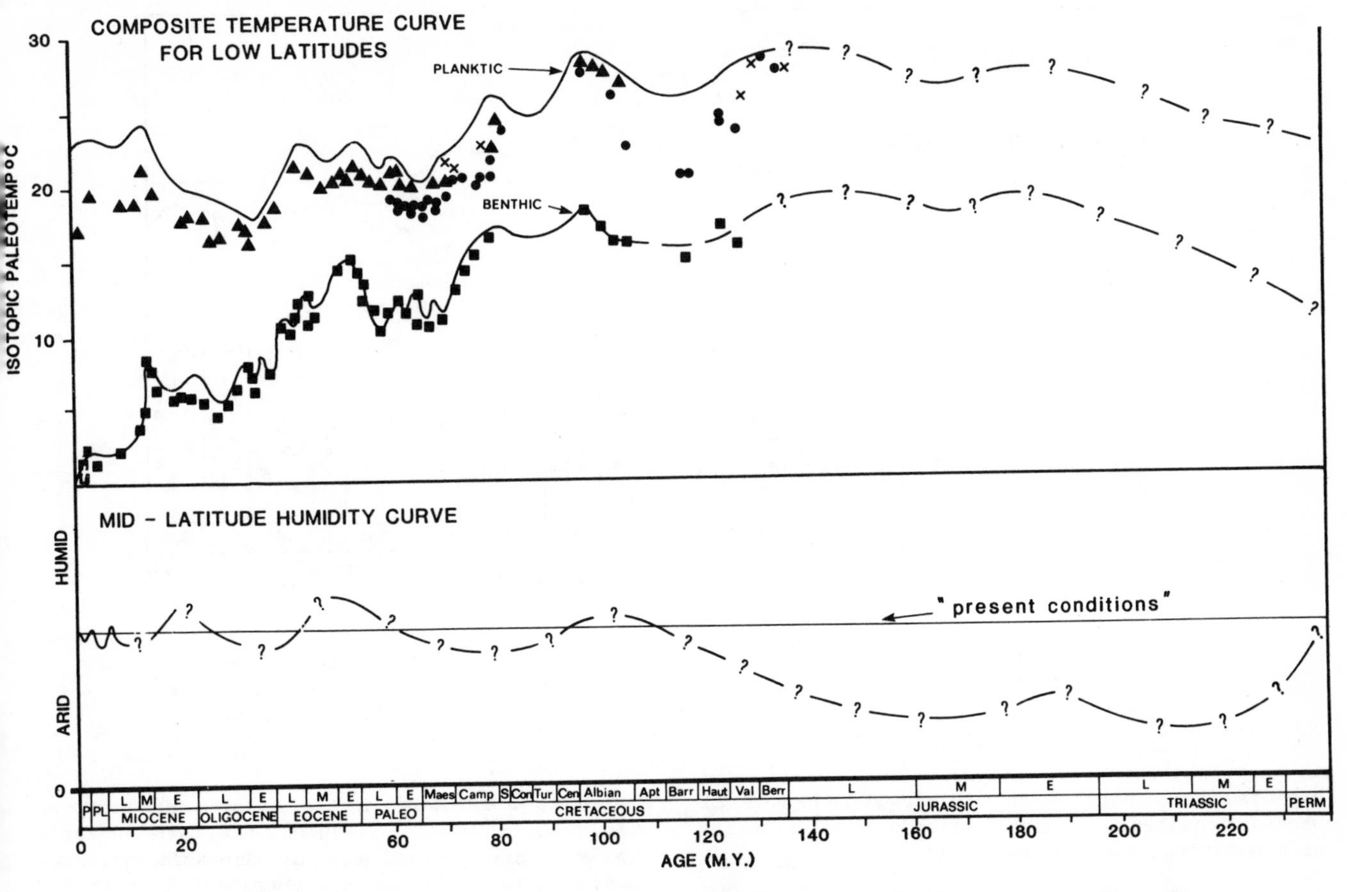

Fig. 1. Low-latitude temperature and mid-latitude humidity curves for the Mesozoic-Cenozoic. Oceanic temperature curves from oxygen isotope determinations on planktic (surface) and benthic (bottom) foraminifera (from summary by Douglas and Woodruff, 1981). The humidity curve and Pre-Cretaceous parts of the temperature curves are qualitative estimates from global conditions.

the end of the Gondwana glaciation) in the middle Permian and culminated in the late Cretaceous; and second, a succeeding long interval of punctuated cooling continuing to the present (Fig. 1). The global climate can thus be said to have evolved to its present state over a complete cycle of some 250 million years.

Triassic

Triassic climates are dominated by the effects of low sea level and the formation of the greatest volume of marine evaporites recorded for any geologic period. The record of the early Triassic is too scant to allow more than broad generalizations. It is likely that the post-glacial warming of the late Permian continued until the middle Triassic, when there was a marked warming and the initiation of large-scale formation of evaporites. At this time the Dominion Coal Measures accumulated at about 65° paleolatitude in Antarctica, signifying high latitude humidity and, presumably, winter conditions suitable for the growth of a variety of plants. The late Triassic was characterized by dry-climates locally reaching to about 50° latitude, but with a region of monsoon rainfall extending to 50° latitude in the eastern Tethys (Robinson, 1973).

Interpretations of Triassic climate (Fig. 1) and its internal variations are handicapped by a lack of oxygen isotope data, and estimation of conditions is further limited by the fact that sea level curves are largely generalized (Fig. 2). Ager (1981) summarizes sea level changes as follows: regression at the beginning of the Triassic and transgression beginning in the middle Triassic (Ladinian) and culminating in the late Triassic (Norian). It is noteworthy that these gross changes in the level of the sea roughly correspond to the prevailing temperature conditions outlined above. However, it is not possible to relate temperature changes to ice-volume effects; the thermal state could be

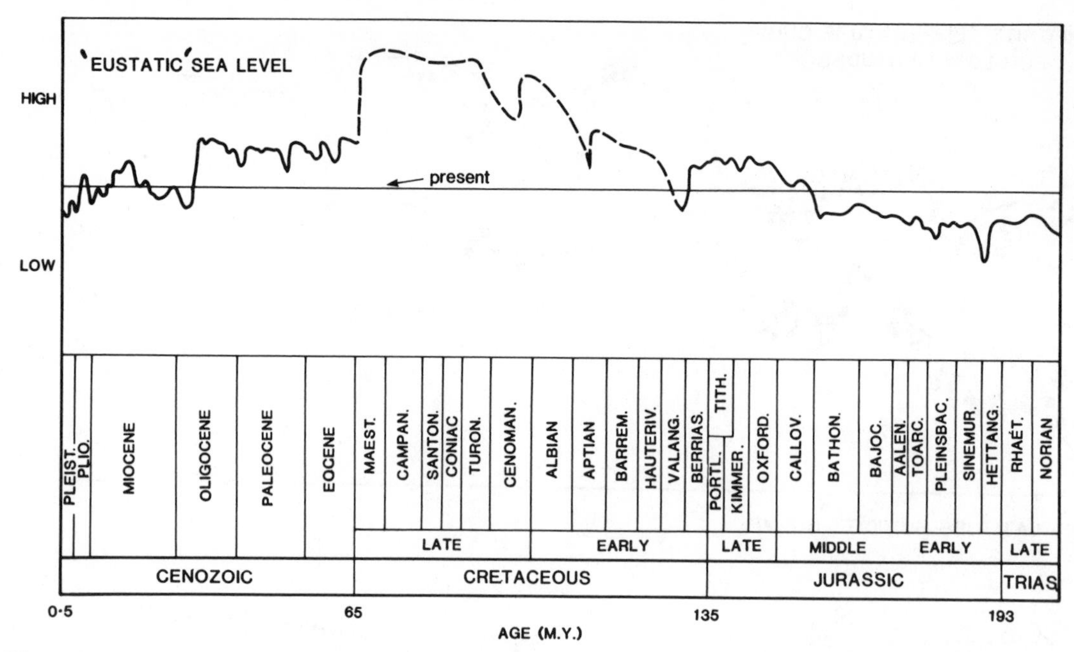

Fig. 2. Estimated "eustatic" sea level for the Mesozoic-Cenozoic. The curve represents departures from present mean sea level as determined by coastal offlap sequences (Vail et al., 1977; Vail and Hardenbol, 1979; Vail et al., in press).

expected to vary with sea-level change of whatever causes, as global albedo would operate as a positive feedback mechanism.

Jurassic

An abundance of oxygen isotope determinations is available for the Jurassic Period but unfortunately these are not widely cited because at the time the studies were carried out, the extent of diagenetic alteration in Jurassic molluscs was not fully appreciated. There is perhaps justification in using some of these data to delineate trends in paleotemperature, particularly where isotope trends conform approximately with trends from other sources. On global data summarized by Stevens (1971), Jurassic paleotemperatures range from about 10° to 27°C; however, due to possible alteration effects, these are best considered as representing the maximum temperatures under which mollusc shell material may have formed. The data also show great variability of temperature relative to geographic location and stratigraphic position, but other climatic indicators suggest that climatic fluctuations were subdued throughout the period (Hallam, 1982).

An estimate of these gentle changes is provided by shifts in floral boundaries in the U.S.S.R. (Vakhrameyeev, 1964, 1982). According to the most recently published interpretations of these data, climate warmed slightly from the late Triassic into the late early Jurassic (Toarcian), cooled to a minimum in the middle Jurassic, and warmed again to a plateau in the Late Jurassic. The most striking aspect of Jurassic climates, however, is that warm, possibly subtropical, conditions reached to roughly 60°N. latitude, and global temperatures in the late Jurassic may have averaged as much as 7° warmer than at present (Frakes, 1979). Another important feature was the continued existence of widespread dry climates as indicated by abundant evaporite basins of middle and late Jurassic age.

Vail et al. (in press) have presented a Jurassic sea-level curve which conforms fairly well to the temperature scheme above (gentle rise of sea level through the period but with a culmination in the latest Jurassic, and with short-term fluctuations throughout).

Cretaceous

In the Cretaceous Period the long warming trend began in the late Permian culminated and what may be called the initial coolings of the Cenozoic glaciation began. A variety of evidence defines these trends, not the least important of which is the data from oxygen isotopes. For reasons cited earlier, the isotope curves based on belemnites in particular are considered less reliable than the more recently derived deep-sea curves based on benthic and planktonic foraminifera (Douglas and Woodruff, 1981). However, most data points (Fig. 1) are from forams from the northwestern Pacific, and hence, are poorly representative of global paleotemperatures. These limitations restrict somewhat our interpretations of global climates.

Paleobiogeographic and biostratigraphic distributions show that there was no large temperature change across the Jurassic-Cretaceous boundary. Therefore, the isotopic paleotemperature estimates for the earliest Cretaceous can also be used to provide an estimate for the late Jurassic. Low latitude surface-water temperatures were approximately the same or possibly slightly higher than at present (∿28° vs. 21° - 28°C; depending on longitudinal position in the Equatorial Current). Sub-polar surface temperatures (∿16°C), taken as equivalent to low latitude bottom temperatures, similarly fall near the upper limit of temperature for surface waters in the modern sub-polar Pacific (16° to -1°C). I have previously estimated that late Jurassic sub-polar waters may have attained 14°C in at least one site (Frakes, 1979); this derives from isotope measurements on belmnites and hence represents a maximum temperature of formation. If we assume that equatorial bottom waters formed only in the coldest part of sub-polar regions during the Cretaceous, then the temperature difference with respect to the present was on the order of 15°C and a significant cause for retention of heat in the climatic system is to be sought. This assumption may not be justified in view of the likelihood that Cretaceous bottom waters did not form in association with polar ice shelves as today.

Before the begining of the long temperature decline into the Cenozoic glaciation in the Campanian, the world ocean cooled by 2-3°C in the early third of the Cretaceous (Berriasian to Barremian), judging from results on benthic foraminifera. Warming then took place in the Aptian toward a major Cretaceous peak in the Albian. The Cenomanian to Coniacian record lacks isotopic data but the cool interval indicated in Fig. 1 is supported by development of boreal and austral nannofossil assemblages beginning in the Cenomanian (Roth and Bowdler, 1981). Certainly the end of the Santonian saw a precipitous drop in ocean temperatures. The Maestrichtian was characterized by further cooling but included a slight warming phase in high latitudes (benthic curve).

Estimates of the temperature gradient between the subtropical and sub-polar zones are an important measure of the global thermal state. In Fig. 1, it is seen that a gradient of about 10°C was maintained until at least the Albian. By Santonian time this apparently had decreased to about 8°C. On these data, it is not unreasonable to postulate that the warmest interval in the Cretaceous was from the Albian to the Santonian inclusive, an interval of about 20 million years. Interestingly, the interval embraces the 3 Cretaceous "anoxic events", times of accumulation of oceanic sediments rich in organic carbon (Schlanger and Jenkyns, 1976).

The cause for anomalously warm climates in the Cretaceous has been the subject of much discussion, though few hypotheses have eventuated (Schwarzbach, 1974; Donn and Shaw, 1977; Frakes, 1979; Manabe and Hahn, 1977). Decreased albedo resulting from very high stands of sea level in the early Cretaceous has been shown to allow an increase of 2.3-3.9% in absorbed solar radiaton, but this would account for a global increase of only about 6°C, or 40% of the global temperature differential relative to the present (Barron and others, 1981). It is possible that Mesozoic subpolar high temperatures were maintained largely by unusually high heat transport in the oceans (Frakes 1979); this idea is now utilized in some model studies (Barron and Washington, 1982). The lack of independent evidence for increased solar emission and/or atmosphere CO_2 accumulation until now has discouraged speculation on either of these as causes for warm Mesozoic climates.

Whereas the sea-level curve of Vail et al. (1977) generally conforms with global temperature trends throughout the Phanerozoic, Cretaceous sea levels are at variance with the paleotemperature records. Late Cretaceous cooling coincides with the highest sea-levels recorded (see Hancock and Kaufman, 1979). This may or may not change with the release of additional proprietary information on sea levels.

Cenozoic Climates

The major Cenozoic coolings in oceanic climates took place in the middle and latest Eocene, the mid-Miocene, and the Plio-Pleistocene. The Cenozoic record also reveals the occurrence of important warming phases in the late Paleocene and the Oligocene. However, the nature of the changes was not consistent in that they affected surface and bottom water temperatures in different ways, as is apparent in isotope curves. Surface-to-bottom (=subtropical to sub-polar) temperature gradients provide an index of these changes. Because of its global significance, such an index is more useful than either $\delta^{18}O$ or calculated paleotemperatures. Together with lithologic and paleobiogeographic information, the gradient index is used to document Cenozoic climates in the following sections.

Paleocene

For the early Cenozoic, it appears that the vertical temperature gradient in the tropical Pacific (Fig. 3) ranged from 6 to 12°C; the oceans in general were some 5°C colder than earlier. This time of moderately cool conditions (Fig. 1) began in the Maastrichtian and, with fluctuations, extended into the late Paleocene

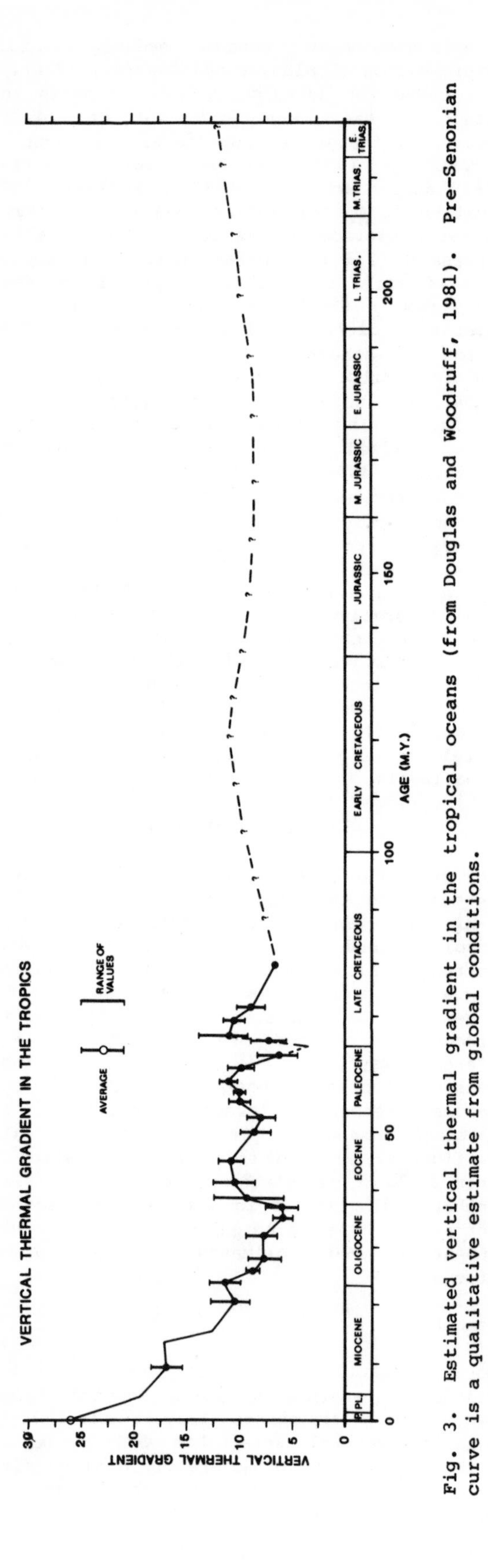

Fig. 3. Estimated vertical thermal gradient in the tropical oceans (from Douglas and Woodruff, 1981). Pre-Senonian curve is a qualitative estimate from global conditions.

(Haq, 1982), when strong warming took place in bottom/polar water-masses (Fig. 3). The data for surface waters suggest only slight warming in the subtropics. A cool interval (between ∿70 and 60 million years ago) is suggested by the curve for benthic foraminifera.

It is known from other evidence that Paleocene climates also are characterized by increasing levels of atmospheric humidity (Frakes, 1979). Whether this cool snap, with the possibility of attendant high precipitation rates in polar regions, could have led to the initiation of Antarctic glaciation is uncertain. It is important to recognise the Maastrichtian to early Paleocene as the time of the first protracted cooling from elevated temperature regimes which prevailed earlier.

Moore et al (1978) recognized a major concentration of oceanic hiatuses at the Cretaceous-Tertiary boundary. They attributed this to rapid opening of the Atlantic Ocean and Norwegian Sea and the gradual closing of the Tethys. Related tectonic-eustatic effects may have contributed to the oscillatory rises of sea level in the middle and late Paleocene (Vail and Hardenbol, 1979). The pronounced sea level fall at the end of the Danian appears, however, to have been in concert with a marked cooling observed by Haq et al. (1977). The late Paleocene warming can be considered as a somewhat delayed response to rising sea levels through albedo feedback.

Eocene

The Eocene isotope record (Figs. 1, 3) indicates that temperatures in the tropical Pacific first warmed then cooled substantially towards the end of the epoch, and these trends are matched by isotopic temperatures in other parts of the globe. Haq et al. (1977) and McGowran (1978) document cooling in the middle Eocene, although at slightly different times (46-43, and about 49-46 million years, respectively), on the basis of foraminifera paleobiogeography. Keller (1983), using planktic foraminifera from all oceans, suggests cooling at the end of the middle Eocene and in the latest Eocene, the latter in agreement with the isotope determinations (Douglas and Savin, 1973; Shackleton and Kenett, 1975; Keigwin, 1980), but Corliss (1981) observed relatively little change in benthic foraminifera. The latest Eocene event has been given largely tectonic explanations (e.g. Frakes and Kemp, 1972; Kennett et al., 1972), having to do with interference with or establishment of oceanic currents. The latest Eocene event is important not so much because it indicates a substantial cooling, as because it represents the transition to a stratified ocean as we know it today (Benson, 1975).

The latest Eocene cooling coincides with the time of greatest abundance of hiatuses in the world ocean, possibly as a result of extensive erosion rather than corrosion (Moore et al., 1978). This may have been due to initiation of strong bottom currents through thermal isolation of Antarctica and formation of sea ice and/or ice shelves (Watkins and Kennett, 1972). In contrast, there is little correlation of the climatic record with the sea level curve; the climatically variable middle Eocene is a time of uninterrupted high stands, while the late Eocene cooling coincides with rising sea level (except for a short sharp fall and recovery in the early late Eocene).

A significant feature of the Eocene is the occurrence of subtropical floras to relatively high latitudes, a feature which continues on from the Cretaceous. Generally, this has been taken to signify low equator-to-pole temperature gradients and abundant precipitation in polar and high latitudes. However, Eocene floras which display sub-tropical to warm-temperate characteristics, while they come from paleolatitudes of 60-75°, have not been reported from true polar positions. Matthews and Poore (1980) in fact have suggested that accumulation of Antarctic ice may have begun in the Eocene, or earlier. It is not difficult to imagine steep temperature gradients beyond 75° paleolatitude or alternatively, orographic effects which permitted at least local development of ice bodies near to forests of the types indicated. The fact is that the nature of Eocene climates at and very near the poles remains unknown.

Oligocene

The major cooling at the end of the Eocene is seen not only in heavier isotopes but also in a major increase in the vertical temperature gradient in the tropics. Further, there are marked changes in planktic foraminifera assemblages (Haq et al., 1977; Keller, 1983). The latter study examined faunal characteristics in detail and established a faunal climatic curve with 1) cooling from the late Eocene into the early Oligocene; 2) warming from 34-31 million years ago in the early Oligocene; 3) pronounced cooling from 31 to 28 million years ago; and finally; 4) marked warming through most of the late Oligocene. These trends are in rough agreement with trends in the isotope record.

Oligocene climates described above agree fairly well with the sea level curve (Fig. 2), except that the pronounced cooling (3, above) took place in the late early Oligocene, while the possibly related major fall in sea level is dated as beginning early in the late Oligocene. If this age difference is real, the lag time (1-2 million years) would appear to represent the time necessary for ice accumulation to respond to global cooling.

Hiatuses are less common in the cool Oligocene than in the warm late Eocene (Moore et al., 1978), a fact which suggests that strong currents and/or corrosive bottom waters are not characteristic of all cool intervals.

Miocene

Detailed isotopic information on Miocene marine strata is available from the equatorial Pacific (Keigwin, 1979; Woodruff et al., 1981) and the mid-latitude South Atlantic (Boersma and Shackleton, 1979). These data, and hiatus distributions, have been summarized by Barron and Keller (1982) and Keller and Barron (1983). The generalized curve shows warming through the early Miocene and a severe cooling beginning in the early middle Miocene through to a cold late Miocene. Microfossil data are in good agreement with this record.

Seven Miocene hiatuses are correlated by the latter workers with cool times, as indicated by oxygen isotope and faunal and floral assemblages. Relationships of cool intervals to changes in sea level are less successful, with the main eustatic fall occurring well into the late Miocene, some 6 million years after the initial mid-Miocene cooling, and approximately coincident with initiation of abundant circum-Antarctic ice rafting. Again, a substantial lag in ice build-up (or imprecise dating of coastal onlap sequences) is suggested.

The end of the Miocene saw a deepening of the cooling trough in the isotope curve; there was a penecontemporaneous rise in sea level, expansion of Antarctic ice to the limits of the continental shelf and a severe expansion of the cold-water ring around Antarctica (Hayes and Frakes, 1975). Although both the middle and the end Miocene coolings have been attributed to large-scale growth of Antarctic ice, only the end Miocene event seems to have coincided in time with ice build-up, as indicated by falling sea level. Schnitker (1980) suggests the initial (middle Miocene) formation of North Atlantic Deep Water led to ice growth in Antarctica, by providing a moisture source for evaporation at high southern latitudes. According to recent interpretations of the carbon isotope record, NADW may have been in existence since the late Oligocene (Miller and Fairbanks, 1983).

The vertical temperature gradient, as derived from isotopic measurements, continued to steepen in the Miocene, with particularly large changes in the middle and late Miocene.

Pliocene-Pleistocene

The broad picture of climates in the Pliocene and Pleistocene is one of intensifying fluctuations, as revealed in isotope records. The present volume of Antarctic ice had already been achieved (or possibly exceeded) by the early Pliocene and some $\delta^{18}O$ variations, even in the Miocene (Woodruff et al., 1981), may have resulted from fluctuations in ice volume there, as controlled by orbital (Milankovitch) cycles (but see Moore et al., 1982). The initiation of northern hemisphere glaciation at around 2.5 million years ago was followed at about 700,000 years ago by establishment of regular cycles of about 100,000 year frequency.

Pronounced late Pliocene cooling of the globe between about 3.6 and 2.5 million years ago seems well documented, although quality of the dating is variable. This would seem to correspond quite well with an irregular though large fall in sea level illustrated by Vail and Hardenbol (1979).

Causes of the Glaciation

Changes in paleogeography leading to altered circulation and density configurations in the ocean are the most common type of explanation for the climatic deterioration and glaciation in the Mesozoic-Cenozoic. For rapid climate changes these have the advantage that "on-off" situations can be invoked. However, they are unsatisfactory in the broader sense because through the time considered, their effects would have necessarily always been cumulative toward cooling. The model of Berner et al. (1983) relating late Cretaceous-Cenozoic cooling to falling levels of atmosphere CO_2 due to decreasing rates of ocean crust generation, has attraction in that decreased spreading coincides neatly in time with the long-term cooling. The model has further value in that it provides an explanation for the very high temperature regime of the earth in the Mesozoic. Rapid climatic changes, on the other hand, might be related to global volcanic pulses which result in marked changes in the rate of CO_2 degassing and volcanic ash production.

CO_2 Budget

The role of CO_2 in controlling climate lies in its capability to generate a greenhouse effect with increasing concentration in the atmosphere. To determine the effect of CO_2 variation on past climates it is necessary to establish past CO_2 concentrations. There then remains the major problem of establishing whether changing levels of CO_2 have caused known climate changes, or alternatively, whether changes of climate have led to increases and/or decreases in CO_2 concentration. That is, does high CO_2 concentration, for example, contribute to warming in a primary sense, or is it merely functioning as a positive feedback mechanism? Answers to such questions are deeply embedded in the network of the CO_2 cycle, and they thus involve considerations of the CO_2 budget, particularly carbonate sediments.

Understanding of the complexity of the CO_2 cycle has been aided by summary works by Garrels and Mackenzie (1974), Milliman (1974), Berger (1977), Holland (1978), Arthur (1982), and Berner et al. (1983), and therefore only a short review is presented here. A balance is presumed to prevail between atmospheric and surface oceanic CO_2 (Broecker, 1974), such that any imbalances are compensated over geologically short periods (10^3-10^4 years). Increasing CO_2 in the atmosphere leads to larger proportions of dissolved CO_2 in the oceans and thus, to increased carbonate growth through plankton fertility, and vice versa. The link to deep waters acts through carbonate dissolution; increased dissolution provides more dissolved CO_2 in deep layers, thus leading to enhanced dissolution and positive feedback. Depending on the prevailing degree of oceanic upwelling, variable amounts of CO_2 and other nutrients are pumped back to the surface to complete the cycle. Along the way CO_2 is taken out of the system by deposition of carbonate and organic carbon; CO_2 also is released in the formation of oceanic carbonate. Growth of the ocean sediment (carbonate) reservoir is governed by relationships between fertility and dissolution and therefore varies both geographically and temporally. To this network, greatly simplified here, must be added other influences on the oceanic cycles-weathering rates and river supply; the workings of the intricately connected phosphate cycle; supply of Ca and oceanic nutrients; extent of deep-water oxygenation and its effect on sedimentated carbon.

In a recent important analysis of the CO_2 problem by modelling, Berner et al. (1983) have determined that atmospheric levels are most sensitive to the extent of seafloor spreading and of continental land area but relatively insensitive to the relative proportions of calcite and dolomite available to release CO_2 on weathering. Regardless of which spreading rate (rate of generation of oceanic crust) is assumed, the CO_2 level was found to be higher in the late Cretaceous than at present in the model, in accordance with the higher observed paleotemperatures.

Sediment Reservoirs

One method of evaluating CO_2 levels in the past involves determining the amounts of CO_2 sequestered in carbonate reservoirs (Budyko and Ronov, 1979). However, since many factors exert controls on fluxes and thus on total contents of the sediment reservoirs, it is unlikely that pCO_2 in the atmosphere can be estimated directly by such means. In attempting to assess past CO_2 levels, we first consider the carbonate reservoir on land. A summary of CO_2 in both oceanic and land carbonates over the last 65 m.y. is provided in Table 1. Antarctica is not included.

The main difference between oceanic and continental-basin reservoirs lies in the absence of one complicating factor -- dissolution -- in epicontinental and marginal seas. As with the deep ocean, the CO_2 content of surface waters is in equilibrium with that in the atmosphere.

TABLE 1. CO_2 Mass, ($X10^{15}g/y$)

Time	Total Oceanic Carbonate[1]	Total Land Sediments[2]	Total Shelves Etc.	Totals
Pliocene	6.82	0.7	3.4	10.9
Miocene	5.62	1.0	2.6	9.3
Oligocene	7.06	0.8	2.6	10.5
Eocene	5.91	3.1	2.6	11.6
Paleocene	7.8[3]	2.0	2.6	12.4

[1]from Davies and Worsley, 1981 (Table 2)
[2]from Budyko and Ronov, 1979 (Table 1)
[3]estimated from interval 55-60 m.y. (Davies and Worsley, 1981)

Unlike the deep ocean, seas on the continents permanently deposit their carbonate. They further tend to trap many dissolved anions and cations derived from river waters, though for none of these is the trap impermeable. Other factors (nutrient supply, oxygenation, etc.) remain important, though probably operating at varying rates and to a greater or lesser extent than in the oceans.

Estimating total CO_2 consumption in these environments in the past can be a problem owing to the fact that some continental basins now lie wholly or partly on the submerged continental margins. These therefore are not included in data from Budyko and Ronov (1979) (Table 1), but a useful estimate of long-term (post-Miocene) accumulation (2.6×10^{15} g/y CO_2, in carbonates in shelf-slope-rise-epicontinental-sea regimes) has been given by Milliman (1974). Since C_{org} is also a sink for CO_2, the figures for total organic carbon (Budyko and Ronov, 1979) are important in calculating a measure of CO_2 abundance over time. However, the range of epoch values from this source (early Paleocene to Pliocene) is only $.04 - .08 \times 10^{15}$ g/y CO_2, a factor of about 500 less than CO_2 consumed in sediments, so this sink is considered unimportant.

Examination of Table 1 shows that deep ocean consumption of CO_2 as carbonate dominates over land accumulation in the Cenozoic. It appears that carbonate is sequestered on land to a larger extent during warm intervals (Eocene, Paleocene and possibly the late Cretaceous) but that deep oceanic accumulation increases during cold times (Pliocene, Oligocene). The Miocene appears to be a time of moderate sequestration, possibly owing to great climatic variability during the epoch.

Thus far the figures are incomplete in that rates are not included for carbonate sediments deposited on the continental rise, slope and shelf. The Pliocene figure is from Milliman (1974, Table 62). Uniform application of the estimated combined rate (2.6×10^{15} g/y CO_2) (Milliman, 1974; Hay and Southam, 1977) yields epoch carbonate totals ranging from 9.3 to 12.4 x 10^{15} g/y CO_2. However, the amount of erosion of carbonate sediments on land can be expected to increase linearly with increasing age (Garrels and Mackenzie, 1974), a circumstance which indicates that CO_2 consumption rates in carbonate sediments were slightly higher than indicated for the early Cenozoic. The recalculated range is $9.3 - 12.5 \times 10^{15}$ g/y CO_2. The total variation, 3.2×10^{15} g/y CO_2, is large enough to require explanation, despite uncertainties in the land carbonate figures due to variability in erosion and in shelf figures arising from the assumption of constant sedimentation rates. Possibly as much as half this amount of CO_2 might be accounted for by variation in rise/slope/shelf/reef consumption rates (Milliman, 1974), leaving about 1.6×10^{15} g/y to be accounted for. Since this is well above the amount produced by steady-state volcanic degassing ($\sim 0.9 \times 10^{15}$ g/y CO_2), there is reason to believe the degassing rate has varied over time.

Volcanic Degassing at Present

There are two unknowns in reckoning the historical contribution of degassed juvenile and metamorphic/magmatic CO_2 to the carbon cycle, and hence to climate evolution. First, the present rate of degassing is poorly known owing to the difficulties in determining the average proportion of CO_2 among diverse volcanic gases and in estimating the total volume of volcanic gases being released. Second, there is the problem of determining variability in CO_2 degassing over geologic time.

Analyses have been performed on volcanic gases from erupting craters and lava lakes in such diverse sources as mid-ocean ridges (the island of Surtsey), oceanic intraplate islands (Kilauea, Hawaii) and continental margins (Paricutin, Mexico; Japan; Italy; Katmai, Alaska). In part because of variable collecting and analytical techniques, the concentration of CO_2 in these gases varies enormously; variability no doubt also results from compositional differences in the magmas (and wall rocks) and their particular physical environments (temperature, pressure, phase of eruption). Observed compositions are thus so affected that calculation of averages is almost meaningless.

Sigvaldason (1974), however, favors one analysis from Surtsey as being representative of basaltic degassing. From this, he estimates CO_2 concentration in volatiles at 1250 ppm, about two orders of magnitude lower than the generally accepted average of 10% CO_2. He further estimates total production of CO_2 (assumed to be solely from basaltic sources) at 0.86×10^{15} g/yr, a figure in agreement with that determined by Plass (1956). The degree of uncertainty in these calculations can be judged from the fact that production has been estimated at more than one order of magnitude higher (10×10^{15} g/yr) by Puccetti and Buddemeier (1974) and Anderson (1975). The matter is further complicated by the fact that andesitic rocks are likely to degass more CO_2 per unit volume of rock than basalts (Anderson 1975). Just how much more is uncertain but maximum estimates in the literature are 0.2 wt.% for basalts vs 0.5 wt. % for andesites. However, the mass rate of production of andesites is only about one fifth that of oceanic basalts.

In indirectly approaching the problem of degassing rate (since the rate cannot be measured directly), some workers have used the "excess volatiles" technique of Rubey (1955). Assuming that the atmosphere reached its present state from steady degassing since an early stage of earth formation, it is possible to calculate the annual degassing rate from the concentration of materials which tend to remain in the atmosphere and/or hydrosphere. In the case of CO_2, however, this type of calculation is not possible because of the involvement of C in the rock and biologic cycles. Calculations can be made from consideration of the water cycle assuming that the H_2O/CO_2 ratio in volcanic gases is about 4 and that all water at or above the surface originated from steady degassing. The figure 0.86×10^{15} g/y CO_2 degassed was derived in this way (Sigvaldason, 1974).

The degassing rate for CO_2 can also be estimated by equating degassing to the amount of CO_2 required to maintain the atmospheric composition in the carbon cycle, assuming a steady-state system in which CO_2 changes are buffered in the oceans on time-scales of 10^3 - 10^4 years. Holland (1978) showed that this leads to a value of 0.9×10^5 g/y for degassing of C (3.3×10^{15} g/y for CO_2). When the degassing rate is compared to figures for all active CO reservoirs, as determined by Holland (totalling to 1.5×10^{18} g/y CO_2), it is seen to be very small; it is also small compared to the total of all other fluxes between reservoirs ($\sim 8.3 \times 10^{16}$ g/y CO_2).

Attempts to estimate the rate at which the juvenile CO_2 component is degassed must take account of CO_2 addition to emissions from non-primitive sources, such as wall-rocks assimilated in magma chambers and pervasive hydrothermal leaching of oceanic crust. The proportion of truly juvenile gasses in emissions cannot presently be estimated but it may attain as much as 60%.

Degassing in the Past

In his paper on early atmospheres, Rubey (1955) showed that calculations based on gradual degassing gave approximately the same composition for the atmosphere as did those for the "residual" hypothesis, in which the atmosphere was inherited from an earlier stage of earth development. In either scenario, the CO_2 concentration in the atmosphere would have decreased gradually from an initial value of about 91% to its present level; drawdown of atmospheric CO_2 would have resulted from growth of the marine biologic realm, from deposition of shelf carbonates and organic matter in the oceans, and from weathering of silicate rocks, all in increasing amounts over time. Factors working against this trend, to increase concentration of CO_2 in the atmosphere, include oxidation of elemental carbon and release of CO_2 as a consequence of carbonate precipitation. Large imbalances, possibly also involving the degassing rate, are inherent in this scheme. It is not possible to say when the ocean-atmosphere system attained the present condition and it is not an easy task to specify the magnitude of fluctuations.

One way to estimate past rates of degassing is through knowing volumes of ancient volcanic rocks. At best, these will provide semi-quantitative results owing to the unknown gas compositions and abundances associated with the rock volumes. However, it seems safe to say that CO_2 degassing was greater during times of increased volcanic activity than during times of low activity. A corollary of this is that if volcanism can be shown to have varied over time, then the degassing rate has also varied.

The rate of volcanic activity can be correlated with thermal activity in general, and a reliable summary of radiometric age dates accordingly should provide a measure of the variability of degassing. Since early efforts, however, a comprehensive summary of age dates has not been attempted. Alternative methods involve using volumes of eruptions from individual volcanic piles as indicators of global production. Vogt (1972, 1979) found that mantle-plume convection affected two major centres (Hawaiian chain, Iceland) and perhaps several others simultaneously through the Cenozoic; the volume of volcanic accumulations per unit time at both Hawaii and Iceland varied by up to an order of magnitude, and intervals of relatively low and high discharge persisted for periods as long as 15 million years (Fig. 4). Although these data

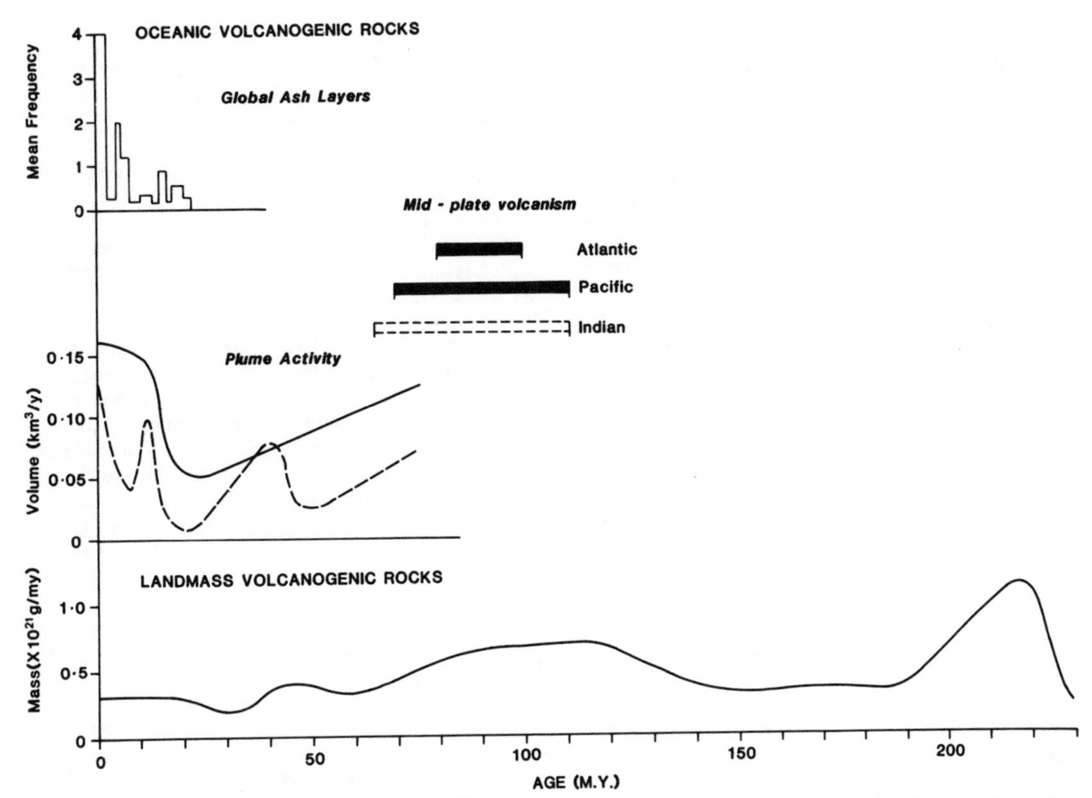

Fig. 4. Distribution of landmass and some oceanic volcanic products in the Mesozoic-Cenozoic. Global ash layers from Kennett and Thunell (1975); mid-plate volcanism from Schlanger et al. (1981); plume activity from Vogt (1972); landmass volcanogenic rocks from Budyko and Ronov (1979).

are suggestive of globally pulsed volcanic activity, much more comprehensive compilations are required before the history of degassing can be determined.

A compilation effort has now been completed by Ronov and his colleagues. In this work the volumes of several common types of sedimentary rock and of volcanogenic rocks from continental areas are tabulated. Budyko and Ronov (1979) utilize the volcanogenic summaries to construct a Phanerozoic curve for volcanic rocks generated per unit time (Fig. 4), in which it is seen that the rate of accumulation of volcanogenics varied between about 2.0 and 16.0 x 10^{14} g/y. Application to the above rates, of Sigvaldason's (1974) figure (0.86 x 10^{15} g/y) for degassing of CO_2, suggests that CO_2 would constitute up to 40% by weight of all volcanic materials during intervals of reduced output, but only about 5% during heightened discharge. The former figure is obviously much too high, particularly since the proportion of CO_2 in volcanic gases decreases as discharge slows. On the other hand, the suggested 5% during high discharge may be a reasonable estimate of total CO_2 output.

It must be remembered, however, that volcanic rocks from the oceans are not included in Ronov's figures and hence they under-represent the global rock totals. More recent estimates by Khain et al. (1984) similarly lack estimates of oceanic basalt volumes. Most basalts generated at mid-ocean ridges have probably released relatively small quantities of CO_2 directly, but over time leaching by hot fluids may have liberated practically all contained CO_2. Large subaerial edifices such as the Hawaiian chain, on the other hand, no doubt have contributed substantial amounts of CO_2 by direct degassing.

Mid-plate volcanism on a grand scale has been documented in the Pacific basin (Line Islands, Mid-Pacific Mountains, Nauru basin - Marshall Islands) by Menard (1964) and Schlanger et al. (1981). This occurred in the interval from the Aptian to the Campanian and may have had smaller counterparts in the south Atlantic (Walvis and Rio Grande rises) and Indian Oceans (Ninetyeast ridge, Kerguelen plateau, etc.). The Pacific edifice was totally subaqueous and probably liberated little CO_2, but the others may have contributed large quantities during the late Cretaceous.

From this discussion it is apparent that available totals are not totally reliable indicators of global degassing but instead represent a beginning point for estimation. This is important because Budyko and Ronov (1979) and

Budyko (1981) go on to compare trends in the volcanic data to calculated CO_2 levels and paleotemperatures, concluding that the rate of degassing of CO_2 has been the major control on Phanerozoic climates. However, Berner et al (1983) have shown that the relationship between CO_2 and temperature is not linear, as assumed by the Soviet workers, but instead non-linear and incorporates components due to runoff and rates of microbiological production of soil CO_2.

Although there is general correspondence of the calculated CO_2 levels with the Phanerozoic temperature record (high CO_2 during warm times, low CO_2 during cold ones), two intervals are strikingly aberrant. Both Late Cretaceous and the marked Miocene coolings are accompanied by higher CO_2 levels (and sequestration of carbonate in continental basins) than in immediately preceding times. It is noteworthy that during both these intervals, plume and other types of marine volcanism not related to mid-ocean ridge activity were widespread (Vogt, 1972; Schlanger et al., 1981). The effects of these on climate may have been substantial.

Volcanic Ash

Emissions of volcanic ash have been championed as a cause of short-term coolings as seen in historical records (Lamb, 1970; Bray, 1974). The component of ash eruptions which is of significance in climate studies is termed dust, that is, it is fine enough to remain suspended in air (generally, less than 4 m diameter) for long periods. Explosive volcanism is forceful enough to drive columns of dust and steam to levels within the stratosphere or mesosphere, up to 60 km or so, from where dust is dispersed by winds and, over periods of a few years, settles slowly back to the surface. At all levels volcanic dust, and newly-formed sulphate aerosol particles, absorb and back-scatter incoming solar radiation.

The extent of possible cooling by volcanic dust is unknown. Historical records suggest a maximum change of -1.6 to -1.8^{o}C over a few years at mid-latitude sites. The suggestion that the more voluminous the eruption, the greater the cooling has been shown to be not valid by Self et al (1981), and these same workers demonstrate that successive large eruptions do not afffect temperature to a greater extent than a single large eruption. Thus, long-term climatic effects would seem to be unlikely.

Episodicity of volcanism has been documented recently by many workers (Vogt, 1972, 1979; Kennett and Thunell, 1975; Kennett et al., 1977, 1979; Best et al., 1980; Axelrod, 1981) (Fig. 4). The evidence consists of histograms of age dates, calculated rock volumes, and frequency of ash layers, all of which are semi-quantitative methods. A limitation on all such studies is that they cover only part of the earth; there is no global summary. Further, many studies are restricted to late Cenozoic volcanism. The available information suggests that volcanic activity was heightened in a broad interval over the Cretaceous-Tertiary boundary, in the second half of the Eocene, and in the middle Miocene and the Plio-Pleistocene. All of these were times of substantial cooling.

In the case of ash emission it is not sufficient to add up all known occurrences or volumes of volcanic strata because the rate at which the critical product -- ash -- is generated, is highly dependent on the nature of the volcanism. Walker (1981) has provided a summary of this variation: the best ash producers are highly explosive eruptions of the ignimbritic, plinian, vulcanian and phreatomagmatic types, which also eject high into the stratosphere and mesosphere, thus permitting wide dispersal of their debris. Basaltic volcanism, on the other hand, generally emits relatively small quantities of ash. Thus, while the time-distribution of silicic to intermediate products obviously is of most importance, the data available mostly come from basaltic areas, and the time distribution of volcanic ash accordingly is poorly documented.

Importantly, since explosive volcanism generates the highest proportions of both volcanic ash and CO_2 the self-same eruptions carry the potential for both cooling and warming the earth's climate.

Discussion and Conclusions

The evolution of climate on the earth since at least the Santonian stage of the Cretaceous (that is, over about the last 80 million years) has been towards cooling. This resulted in substantial high latitude glaciation beginning in the Paleogene and culminated in the mid-latitude glaciers of the Pleistocene, although with warmer phases and irregularly spaced changes of large magnitude. These trends have been attributed largely to changes in both horizontal and vertical circulation patterns in the ocean, as arising from tectonic and/or regional salinity effects (the "major steps" of Berger et al., 1981). No doubt these have been important in climate evolution but why they should have led necessarily to a cooler earth has never been satisfactorily explained. The coincidence in time of decreased rates of ocean-crust generation now suggests a dependance of global climate on long-term variation in production of CO_2 and its decline in the atmosphere -- that is, an inverse greenhouse effect was in operation.

Before discussing CO_2 greenhousing further, some additional climatic aspects of decreased seafloor spreading must be considered. There are two major negative feedbacks and one important positive feedback associated with spreading-derived CO_2. Decreased spreading implies decreased tectonic activity globally, and therefore lowering of the mean elevation of continents. In itself, this further implies a warming rather than a cooling trend, although lag times might be lengthy. The decrease in atmosphere CO_2 outweighs mean lowered elevation as a factor in Mesozoic-Cenozoic climate history, as is evident in observed cooling trends. Effects of decreasing CO_2 must also be more important than the lessening in amounts of volcanic ash from collisional tectonics, which on the same time scale should have led to overall warming.

On the other hand, falls in sea level as a consequence of decreasing spreading would have tended to reinforce any cooling trend resulting from CO_2 reduction. Negative feedbacks obviously have not succeeded in reversing the CO_2 trend, except perhaps temporarily, but sealevel falls contributed sequentially to the ultimate cooling, although proportions are indeterminate. Model studies by Thompson and Barron (1981) indicate that changes in global albedo (land-sea proportions and distributions) may account for 25-50% off the total change in absorbed solar radiation between the Cretaceous and the present.

The evolution of Mesozoic-Cenozoic climate is broadly paralleled by the record of sea level change (Vail et al., 1977; Vail and Hardenbol, 1979; Vail et al., in press), which decreases from a maximum highstand in the late Cretaceous. If sea level and mean global temperature vary in tandem, both being controlled dominantly by spreading rate change, we should be able to match the curves in detail. Interference by other factors is suggested by Cenozoic divergence of the curves as follows:

1. General Paleocene warming is marked by gently rising sea level, but only in the second half of the epoch.
2. Variable climates of the Eocene are accompanied by fairly stable levels of the sea, except for sharp falls at the end of the early Eocene and the end of the middle Eocene.
3. The cooling at the end of the Eocene is accompanied by only a slight sea level drop.
4. A pronounced fall of sea level in the middle late Oligocene may have occurred during an interval of gentle warming.
5. The pronounced middle Miocene cooling is not reflected immediately in a fall of sea level (there is a lag of some 3 million years), and large late Miocene falls do not correlate well with the isotope curve, which shows gentle cooling.

The most serious of these discrepancies are those of the late Eocene, the late Oligocene and the late Miocene, when sea level apparently did not react in parallel with climate change. The sharp cooling at the end of the Eocene is among the best documented in the geologic record and the failure of sea level to respond strongly indicates that ice volume effects were insignificant; there was a cooling but expansion of ice volume was small.

The late Oligocene fall in sea level apparently is not firmly dated; it has recently been placed in the late early Oligocene and equated with cool planktic faunas by Keller (1983). There is indication of an ice volume increase in the late Oligocene; in Antarctica, Ross Sea ice expanded and reached sea level (Hayes and Frakes, 1975). Further, the arrival of NADW at the Antarctic perimeter at this time or slightly earlier (Miller and Fairbanks, 1983) could have provided the moisture source for growth of ice sheets. A large increase in surface $\delta^{18}O$ thus might be expected; however, available isotope records unfortunately are not closely spaced or are incomplete over this interval. Large $\delta^{18}O$ increases in bottom waters of the Pacific and North Atlantic in the late Oligocene (Douglas and Savin, 1975; Miller and Fairbanks, 1983), indicate cooler bottom waters and may reflect initial formation of ice shelves with attendant bottom freezing. The Late Miocene, in contrast, shows substantial falls in sea level but fairly flat isotope curves (~ 0.5 ‰ increase in $\delta^{18}O$). This suggests tectono-eustatic effects, possibly related to decreased spreading.

The three events described above can be attributed to three quite different scenarios as regards CO_2. First, the cooling with little sea level change (end Eocene) suggests only a minor contribution from changes in CO_2 production via spreading rate changes. That is, some other factor is likely to have served as the trigger for cooling. Second, adopting Keller's (1983) dating, the coincidence of a substantial fall in sea level with cooling (middle third of the Oligocene) is compatible with a causal connection between spreading/CO_2 production, and climate change. Third, the late Miocene scenario of sharp sea level fall and minor cooling is opposite to the end-Eocene conditions but allows similar conclusions -- possibly reduced CO_2 levels had relatively little effect on temperature. Again, other factors must be sought. Note that in none of these important times of change is CO_2 essential to the process of change, despite the fact that presumed CO_2 decrease due to decreased spreading coincided with the general cooling trend in the Cenozoic.

Tectonic events with effects on circulation patterns and/or density stratification in the oceans thus seem to be required for rapid changes in global climate. Among these might be included the construction of volcanic edifices, such as the Hawaiian chain, which are unrelated temporally to spreading rate changes. Plume-related volcanism appears to be concentrated near the Cretaeous-Tertiary boundary, in the second half of the Eocene and in the middle Miocene; reduction of incoming solar radiation by windborne ash from these structures may have played a role in latest Cretaeous and middle Miocene coolings. For these the timing is coincident, but this is not the case for the end-Eocene cooling, where a lag of several million years is indicated.

In conclusion, the CO_2 model of climate change fits very well the major cooling of the Mesozoic-Cenozoic earth and there seems little doubt that CO_2 concentrations related to rates of ocean-crust generation have played a significant role in climate deterioration. On the other hand, the more easily observable, rapid, changes are more likely the result of tectonic events which brought about reorganization of paleogeography and ocean dynamics. Among the latter may be included plume-related and mid-plate volcanism and its derived ash.

References

Ager, D.V. 1981. Major marine cycles in the Mesozoic. J. Geol. Soc. London, v. 138, p. 159-166.

Anderson, A.T. 1975. Some basaltic and andesitic gases. Rev. Geophys. Space Phys., v. 13, p. 37-55.

Arthur, M.A. 1982. The carbon cycle-controls on atospheric CO_2 and climate in the geologic past. In Climate in Earth History (Berger, W.H. and Crowell, J.C. eds.) National Academy Press, Washington, D.C., p. 55-67.

Axelrod, D.I. 1981. Role of Volcanism in Climate and Evolution. Geol. Soc. Amer. Spec. Paper 185, 59pp.

Barron, E.J. and Washington, W.M., 1982. Cretaceous climate: a comparison of atmospheric simulations with the geologic record. Palaeogeog., Palaeoclim., Palaeoecol., v. 40, p. 103-133.

Barron, E.J., Thompson, S.L. and Schneider, S.H. 1981. An ice-free Cretaceous? Results from climate model simulations. Science, v. 212, p. 501-508.

Barron, J.A. and Keller, G. 1982. Widespread Miocene deep-sea hiatuses: coincidence with periods of global cooling. Geology, v. 10, p. 577-581.

Benson, R.H. 1975. The origin of the psychrosphere as recorded in changes of deep-sea ostracode assemblages. Lethaia, v. 8, p. 69-83.

Berger, W.H. 1977. Carbon dioxide excursions and the deep-sea record: Aspects of the problem. In The Fate of Fossil Fuel CO in the Oceans, (Anderson, N.R. and Malahoff, A., eds.), Plenum, New York, p. 505-542.

Berger, W.H. Vicent, E. and Thierstein, H.R. 1981. The deep-sea record: Major steps inn Cenozoic ocean evolution. In The Deep Sea Drilling Project : A Decade of Progress (J. Warme, R.G. Douglas, and Winterer, E.L., eds.), Soc. Econ. Paleont. Mineral. Spec. Publ. 32, p. 489-504.

Berner, R.A., Lasaga, A.C. and Garrels, R.M. 1983. The carbonate-silicate geochemical cycle and its effect on atmospheric carbon dioxide over the past 100 million years. Amer. J. Sci., v. 283, p. 641-683.

Best, M.G. McKee, E.H. and Damon, P.E. 1980. Space-time-composition patterns of late Cenozoic mafic volcanism, southwestern Utah and adjoining areas. Amer. J. Sci., v. 280, p. 1035-1050.

Boersma, A. and Shackleton, N.J. 1977. Tertiary oxygen and carbon isotope stratigraphies, Site 357 (mid-latitude South Atlantic) Init. Reps. Deep Sea Drilling Project, v. 39, p. 911-924.

Bray, J.R. 1974. Glacial advance relative to volcanic activity since 1500 AD. Nature, v. 248, p. 42-43.

Broecker, W.S. 1974. Chemical Oceanography. Harcourt, Brace, Jovanovich, New York, 214pp.

Broecker, W.S. 1982. Ocean chemistry during glacial time. Geochim., Cosmochim. Acta., v. 46, p. 1689-1705.

Budyko, M.I. 1981. Variations in the atmosphere's thermal regime in the Phanerozoic. Meteorol. i. Gidrol., No. 10, p. 5-10.

Budyko, M.I. and Ronov, A.B. 1979. Chemical evolution of the atmosphere in the Phanerozoic. Geochim. Internat., v. 16, p. 1-9.

Davies, T.A. Worsley, T.R. 1981. Paleoenvironmental implications of oceanic carbonate sedimentation rates. In The Deep Sea Drilling Project : A Decade of Progress, (Warme, J., Douglas, R.G. and Winterer, E.L. eds.), Soc. Econ. Paleont. Mineral. Spec. Publ. 32, p. 169-179.

Donn, W.L. and Shaw, D.M. 1977. Model of climate evolution based on continental drift and polar wandering. Geol. Soc. Amer. Bull., v. 88, p. 390-396.

Douglas, R.G. and Savin, S.M. 1975 Oxygen and carbon isotopes of Tertiary and Cretaceous microfossils rom Shatsky Rise and other sites in the North Pacific Ocean. Init. Reps. Deep Sea Drilling Project, v. 32, p. 509-520.

Douglas, R.G. and Woodruff, F. 1981. Deep-sea benthic foraminifera. In The Sea, v. 7, (C. Emiliani, ed.), John Wiley, New York, p. 1233-1327.

Frakes, L.A. 1979. Climates Throughout Geologic Time. Elsevier, Amsterdam, 310pp.

Frakes, L.A. and Kemp, E.M. 1972. Influence of continental positions on early Tertiary climates. Nature, v. 240, p. 97-100.

Garrels, R.M. and Mackenzie, F.T. 1974. Evolution of Sedimentary Rocks, Norton, New York, 397 pp.

Hallam, A. 1982. The Jurassic Climate. In Climate in Earth History (Berger, W.H. and Crowell, J.C. eds.), National Academy Press, Washington, D.C., p. 159-163.

Hancock, J.M. and Kaufman, E.G. 1979. The great transgressions of the Late Cretaceous. J. Geol. Soc. London, v. 136, p. 175-186.

Haq, B.U., Premoli-Silva, I., and Lohmann, G.P.

1977. Calcareous plankton paleobiogeographic evidence for major climatic fluctuations in the early Cenozoic Atlantic Ocean. J. Geophys. Res., v. 82, p. 3861-3876.

Hay, W.W. and Southam, J.R. 1977. Modulation of marine sedimentation by the continental shelves. In The Fate of Fossil Fuel CO in the Oceans (Anderson, N.R. and Malahoff, A., eds.), Plenum, New York, p. 569-604.

Hayes, D.E. and Frakes, L.A. 1975. General synthesis. Init. Reps. Deep Sea Drilling Project, v. 28, p. 919-942.

Hays, J.D. and Pitman, W.C. III, 1973. Lithospheric plate motion, sea level changes and climatic and ecologic consequences. Nature, v. 246, p. 18-22.

Holland, H.D. 1978. The Chemistry of the Atmospheres and Oceans. Wiley Interscience, New York, 351 pp.

Keigwin, L.D. Jr. 1979. Late Cenozoic stable isotope stratigraphy and paleoceanography of Deep Sea Drilling Project sites from the east equatorial and central north Pacific Oceans. Earth Plan. Sci. Letters, v. 45, p. 361-382.

Keigwin, L.D. Jr. 1980. Paleoceanographic change in the Pacific at the Eocene-Oligocene boundary. Nature, v. 287, p. 722-725.

Keller, G. 1983. Paleoclimatic analyses of middle Eocene through Oligocene plankton foraminiferal faunas. Palaeogeog., Palaeoclim. Palaeoecol., v. 43, p. 73-94.

Keller, G. and Barron, J.A. 1983. Paleoceanographic implications of Miocene deep-sea hiatuses. Geol. Soc. Amer. Bull., v. 94, p. 590-613.

Kennett, J.P. and Thunell, R.C. 1975. Global increase in Quaternary explosive volcanism. Science, v. 187, p. 497-503.

Kennett, J.P., Burns, R.E., Andrews, J.E., Churkin, M. Jr., Davies, T.A., Dumitrica, P., Edwards, A.R., Gatehouse, J.S., Packham, G.H. and Van der Linden, W.J.M. 1972. Australian-Antarctic continental drift, palaeocirculation changes and Oligocene deep-sea erosion. Nature, v. 239, p. 51-55.

Kennett, J.P., Shackleton, N.J., Margolis, S.V., Goodney, D.E., Dudley, W.C. and Kroopnick, P.M., 1979. Late Cenozoic oxygen and carbon isotope history and volcanic ash stratigraphy : DSDP site 284, South Pacific. Amer. J. Sci., v. 279, p. 52-69.

Khain, V.Ye., Ronov, A.B. and Balukhovskiy, 1984. The Upper Mesozoic and Cenozoic lithological associatons of the continents and oceans (Early and Late Cretaceous). Internat. Geol. Rev., v. 26, p. 369-393.

Kumar, N. 1979. Origins of paired aseismic ridges : Ceara and Sierra Leone Rises in the equatorial, and the Rio Grande Rise and Walvis Ridge in the south Atlantic. Mar. Geol., v. 30, p. 175-191.

Lamb, H.H., 1970. Volcanic dust in the atmosphere : With a chronology and assessment of its meteorological significance. Philos. Trans. R. Soc. London, Ser. A., v. 266, p. 425-533.

Manabe, S. and Hahn, D.G. 1977. Simulation of the tropical climate of an ice age. J. Geophys. Res., v. 82, p. 3889-3911.

Matthews, R.K. and Poore, R.Z., 1980. Tertiary ^{18}O record and glacioenstatic sea-level fluctuation. Geology, v. 8, p. 501-504.

McGowran, B., 1978. Stratigraphic record of early Tertiary oceanic and continental events in the Indian Ocean region. Marine Geol., v. 26, p. 1-39.

Menard, H.N., 1964. Marine Geology of the Pacific. McGraw-Hill, New York, 214pp.

Miller, K.G. and Fairbanks, R.G. 1983. Evidence for Oligocene-middle Miocene abyssal circulation changes in the western North Atlantic. Nature, v. 306, p. 250-253.

Milliman J.D. 1974. Marine carbonates. In Recent Sedimentary Carbonates (Milliman, J.D., Mueller, G. and Foerstuer, U., eds.). Springer-Verlag, New York, 375pp.

Moore, T.C. Jr., van Andel, Tj.H., Sancetta, C. and Pisias, N., 1978. Cenozoic hiatuses in pelagic sediments. Micropal., v. 24, p. 113-138.

Moore, T.C. Jr., Pisias, N.G. and Keigwin, L.D. Jr., 1982. Cenozoic variability of oxygen isotopes in benthic foraminifera. In Climate in Earth History (Berger, W.H. and Crowell, J.C., eds.). National Academy Press, Washington D.C., p. 172-183.

Plass, G.N., 1956. The carbon dioxide theory of climatic change. Tellus, v. 8, p. 140-154.

Puccetti, A. and Buddemeier, R.W. 1974. Volcanic CO_2 emissions : A ^{13}C study of Kilauea volcano (Abst.), EOS, Trans. Amer. Geophys. Union, v. 55, p. 488.

Robinson, P. 1973. Palaeoclimatology and continental drift. In Implications of Continental Drift to the Earth Sciences (Tarling, D.H. Runcorn, S.K., eds.). Academic Press, London, v. 1, p. 451-476.

Ronov, A.B., Khain, V.E., Balukhovsky, A.N. and Seslavinsky, K.B., 1980. Quantitative analysis of Phanerozoic sedimentation. Sed. Geol., v. 25, p. 311-325.

Roth, P.H. and Bowdler, J.C. 1981. Middle Cretaceous calcareous nannoplankton biogeography and oceanography of the Atlantic Ocean. In The Deep Sea Drilling Project: A Decade of Progress, (Warme, J., Douglas, R.G. and Winterer, E.L., eds.). Soc. Econ. Paleont. Mineral. Spec. Publ. 32, p. 517-546.

Rubey, W.W. 1955. Development of the hydrosphere and atmosphere, with reference to probable composition of the early atmosphere. In Crust of the Earth, (Poldervaart, A., ed.). Geol. Soc. Amer. Spec. Pap. 62, p. 631-650.

Savin, S.M. 1977. The history of the earth's surface temperature during the past 100 million years. Ann. Rev. Earth Plan. Sci., v. 5, p. 319-355.

Savin, S.M., Douglas, R.G. and Stehli, F.G. 1975. Tertiary marine paleotemperatures. Geol. Soc. Amer. Bull., v. 86, p. 1499-1510.

Schlanger, S.O. and Jenkyns, H.C. 1976. Cretaceous oceanic anoxic events : causes and consequences. Geol. Mijnbouw, v. 55, p. 179-184.

Schlanger, S.O., Jenkyns, H.C. and Premoli-Silva, I. 1981. Volcanism and vertical tectonics in the Pacific basin related to global Cretaceous transgressions. Earth Plan. Sci. Letters, v. 52, p. 435-449.

Schnitker, D. 1980. Global paleoceanography and its deep water linkage to the Antarctic glaciation. Earth-Sci. Rev., v. 16, p. 1-20.

Schwarzbach, M. 1974. Das Klima der Vorzeit. Ferdinand Enke, Stuttgart, 3rd ed., 380 pp.

Self, S., Rampino, M.R. and Barbera, J.J. 1981. The possible effects of large 19th and 20th century volcanic eruptions on zonal and hemispheric surface temperatures. J. Volcan. Geotherm. Res., v. 11, p. 41-60.

Shackleton, N.J. and Kennett, J.P. 1975. Paleotemperature history of the Cenozoic and the initiation of Antarctic glaciation : oxygen and carbon isotope analyses in DSDP sites 277, 279 and 281. Init. Reps. Deep Sea Drilling Project, v. 29, p. 743-755.

Sigvaldason, G.E. 1974. Chemical composition of volcanic gases. In Physical Volcanology (Civetta, L. et al., eds.). Elsevier, Amsterdam, p. 215-240.

Stevens, G.R. 1971. The relationship of isotopic temperatures and faunal realms to Jurassic-Cretaceous palaeogeography, particularly the S.W. Pacific. J. Roy. Soc. N.Z., v. 1, p. 145-158.

Thompson, S.L. and Barron, E.J. 1981. Comparison of Cretaceous and present earth albedos : Implications for the causes of paleoclimates. J. Geol., v. 89, p. 143-167.

Vail, P.R. and Hardenbol, J. 1979. Sea-level changes during the Tertiary. Oceanus, v. 22, p. 71-79.

Vail, P.R., Mitchum, R.M. Jr. and Thompson, S. III, 1977. Seismic stratigraphy and global changes of sea level, 4. Global cycles of relative changes of sea level. In Seismic Stratigraphy - Applications to Hydrocarbon Exploration (Payton, C.E., ed.). Amer. Assoc. Petrol. Geol. Mem. 26, p. 83-97.

Vail, P.R., Hardenbol, J. and Todd, R.G. in press. Jurassic unconformities, chronostratigraphy, and sea-level changes from seismic and biostratigraphy. Proc. Geophy. Explor. Petrol., Beijing.

Vakhrameev, V.A. 1964. Jurassic and early Cretaceous floras of Eurasia and the paleofloristic provinces of this period. Trans. Geol. Inst. Moscow, v. 102, p. 1-263 (in Russian).

Vakhrameev, V.A. 1982. Classification and correlation of continental deposits based on paleobotanic data. Internat. Geol. Rev., v. 25, p. 1-8.

Vogt, P.R. 1972. Evidence for global synchronism in mantle plume convection, and possible significance for geology. Nature, v. 240, p. 338-342.

Vogt, P.R. 1979. Global magmatic episodes : New evidence and implications for the steady-state mid-oceanic ridge. Geology, v. 7, p. 93-98.

Walker, G.P.L. 1981. Generation and dispersal of fine ash and dust by volcanic eruptions. J. Volcan. Geotherm. Res., v. 11, p. 81-92.

Watkins, N.D. and Kennett, J.P. 1972. Regional sedimentary disconformities and upper Cenozoic changes in bottom water velocities between Australia and Antarctica. Amer. Geophys. Union, Antarctic Res. Sers., v. 19, p. 273-293.

Woodruff, F., Savin, S.M. and Douglas, R.G. 1981. Miocene stable isotope record: a detailed Pacific deep ocean study and its paleoclimatic implications. Science, V. 212, p. 665-668.

Yasamanov, N.A. 1981. Paleothermometry of Jurassic, Cretaceous and Paleogene periods of some regions of the USSR. Internat. Geol. Rev., v. 23, p. 700-706.

DEEP OCEAN CIRCULATION, PREFORMED NUTRIENTS, AND ATMOSPHERIC CARBON DIOXIDE: THEORIES AND EVIDENCE FROM OCEANIC SEDIMENTS

Edward A. Boyle

Massachusetts Institute of Technology, Cambridge MA 02139 U.S.A.

Abstract. The discovery of significant atmospheric carbon dioxide changes during the past 20,000 years (Neftel et. al., 1982) has stimulated observational and theoretical studies of the mechanisms controlling atmospheric carbon dioxide on the 10^3-10^5 year time scale. A number of recent papers have considered the role of the physical and chemical oceanic circulation in controlling p_{CO2}. Broecker (1982) proposed that glacial oceans have more phophorus, thereby decreasing atmospheric p_{CO2}. Oeschger et. al. (1984) noted that decreased pre-formed nutrients could result in greater storage of CO_2 in the deep sea. Wenk and Siegenthaler (1985) have suggested that a decrease in polar ocean temperatures increases the solubility of carbon dioxide in the polar ocean, thereby driving more CO_2 into the deep ocean. Several papers have noted that reduction of deep ocean ventilation rates would also reduce atmospheric carbon dioxide. Although low latitude upwelling has been proposed as a factor governing p_{CO2} previously, the suggested mechanisms have been criticized. In this review, new model is presented which illustrates how an increase in wind-driven upwelling could deplete the upper thermocline of nutrients and hence decrease pre-formed nutrients in polar waters. These processes are described in the context of a simple 5-box model of the ocean/atmosphere system. Presently-available data are inadequate to prove that which mechanisms are responsible for changes in atmospheric carbon dioxide. Some preliminary evidence concerning these processes (particularly the cadmium content of foraminifera fossils) is presented and some testable consequences of the models proposed.

Introduction

Neftel, Oeschger, Schwander, Stauffer, and Zumbrunn (1982) obtained measurements of carbon dioxide in bubbles trapped in glacial ice which showed atmospheric carbon dioxide lower by about 75 ppmV during the last glacial maximum. Subsequent work on ice cores from other locations has confirmed these observations (Barnola, Raynaud, Neftel, and Oeschger, 1983). Attempts to explain changes in p_{CO2} on time scales of several thousand years center on ocean chemistry and circulation, since no other potential controlling factor could have changed significantly on such a short time scale. Broecker (1982) proposed that increases in deep ocean nutrients during glacial time (caused by a stripping of the continental shelves of nutrient-rich organic material) would increase the storage of carbon in the deep sea, decrease the amount of carbon dioxide in ocean surface waters and thereby decrease atmospheric carbon dioxide. He utilized carbon isotope and sedimentologic data to support the hypothesis. Some recent ice-core data suggests that significant carbon dioxide changes can occur on time scales of a few hundred years (Stauffer et. al., 1983). If this data proves correct, it would be necessary to abandon Broecker's hypothesis because it cannot produce such rapid changes. Alternate explanations of atmospheric carbon dioxide have centered on changes in the pre-formed nutrient concentration of the deep ocean. Oeschger et. al. (1984) first pointed out that a decrease in preformed phosphorus (phosphorus which is present in an ocean water mass at the time it sinks from the ocean surface) would result in a greater depletion of carbon dioxide from the surface ocean. This depletion results from the greater proportion of phosphorus in the deep sea which would then be regenerated from falling particulate matter and accompanied by carbon dioxide in a ratio of approximately 106 C:1 P (Redfield, Ketchum, and Richards, 1963). Most subsequent examinations of deep ocean data and models have concentrated on high-latitude processes which might deplete the polar oceans of phosphorus (Broecker and Takahashi, 1984; Sarmiento and Toggweiler, 1984; Knox and McElroy, 1984; Siegenthaler and Wenk, 1984). Shackleton, Hall, Line and Shuxi (1983) presented comparative $\delta^{13}C$ data from benthic and planktonic foraminifera and estimated p_{CO2} changes over the last 140,000 years which were consistent with the ice core p_{CO2} measurements.

Newell et. al. (1978) proposed that increased low latitude upwelling could lower atmospheric p_{CO2} during glacial periods due to intensified upwelling and carbon removal by increased biological activity. As evidence, correlations of the El Nino/Southern Oscillation (ENSO) with changes in the rate of atmospheric CO_2 increase (Bacastow and Keeling, 1980) were extrapolated to glacial conditions. Broecker (personal communication) criticizes this idea because upwelling waters also supply excess carbon dioxide which almost exactly balances the CO_2 removal due to biological activity. Viecelli (1984) describes a model in which increased upwelling appears to reduce p_{CO2}. The model is a complex one-dimensional advection - diffusion system, however, and it is not particularly clear just what causes this sensitivity and how model-dependent the result is. Nonetheless, it has become clear that atmospheric p_{CO2} is sensitive to perturbations in ocean circulation and biological activity, even if it is not entirely clear which processes are most significant.

This manuscript will review low-latitude and polar ocean circulation mechanisms that might change atmospheric carbon dioxide. It will be shown that changes in low-latitude upwelling can contribute to changes in p_{CO2}, although for a different reason than previously supposed. Deep water formation rates and temperature may also be significant in glacial/interglacial carbon dioxide variability, and evidence for such changes will be reviewed.

Mechanisms for Changing the Preformed Nutrient Content of the Deep Ocean

Oeschger et. al. (1984) first pointed out that changes in pre-formed phosphorus could change atmospheric p_{CO2}. The same scenario could be presented using nitrate as the limiting variable; in this mansucript phosphorus will be used for reasons outlined by Broecker (1982). The reasons for this sensitivity and a simple estimate for its magnitude can be derived by examining the definition of preformed phosphorus and preformed total carbon dioxide, and by considering consequences of the equilibration of carbon dioxide between the polar ocean and atmosphere.

The preformed phosphorus content is conceptually simple: it is the phosphorus content of water (saturated with atmospheric oxygen) that sinks from the surface during deep water formation. It is difficult to make direct measurements, so pre-formed phosphorus (PFP) is estimated indirectly in deep water samples by analysis of total phosphorus (P), oxygen, and temperature:

$$PFP = P - r\,(AOU\,(T)) \qquad (1)$$

where r is the ΔO_2: ΔP 'Redfield Ratio' relating the amount of oxygen consumption during the decay of biogenic debris to the release of phosphorus, and AOU is the apparent oxygen utilization, the oxygen content at atmospheric saturation minus the observed oxygen content temperature. Redfield, Ketchum, and Richards (1963) considered that the oxidation state of average organic matter was approximately that of carbohydrates and proteins and estimated that $\Delta O_2 : \Delta P = 106 + 2*16 = 138$. More recently Takahashi, Broecker and Langer (1985) have used oceanographic observations to estimate that the actual ratio is $\Delta O_2 : \Delta P = 172$.

The preformed CO_2 content of polar surface water is then estimated as

$$\text{pre-formed } \sum CO_2 = \text{deep-ocean } \sum CO_2 - 127\,(PFP) \qquad (2$$

where the number 127 = 106 x 1.2 reflects the C:P ratio of decomposing particulate organic carbon augmented by recycled carbonate. The carbon dioxide pressure of this water can then be calculated from this preformed carbon dioxide and a pre-formed alkalinity. Here, the pre-formed alkalinity is computed in the same fashion as the preformed carbon dioxide assuming that the regeneration of alkalinity and carbon dioxide occur in the same ratio as in the falling particulate debris:

$$\text{preformed ALK} = \text{deep-ocean ALK} - 42\,(PFP). \qquad (3$$

Where 42 = (106 X 2)/5 reflects (a) the conversio of moles to equivalents (106 X 2) and and (b) the denominator (5) reflects the ratio of organic carbon to calcium carbonate. This assumption is not exactly correct (Edmond, 1974), but for the purposes of this discussion it is sufficiently close. The partial pressure of carbon dioxide of the polar water is then a function of the ratio o performed $\sum CO_2$/ALK, T, and S. Assuming T = 0°C and S = 34.7°/°°, then

$$\Delta P_{CO_2} = \left[\frac{\Delta P_{CO2}}{\Delta\,(ALK/\sum CO2}\right] \left[\frac{\Delta\ ALK/\sum CO2)}{\Delta\quad PFP}\right] \left[\Delta PFP\right] \qquad (4$$

Expanding about the present-day preformed $\sum CO_2 \approx$ 2158, ALK $\approx$ 2331:

$$\frac{\Delta P_{CO_2}}{\Delta\,(ALK/\sum CO_2)} \approx -30 \text{ ppmV/\%}$$

$$\frac{\Delta\,(ALK/\sum CO_2)}{\Delta\ PFP} \approx +4\%/(\mu mol/kg)\ .$$

So for a -0.5 μmol/kg change in PFP, we expect about -60 ppmV in P . Since the relation between P_{CO2} and AlK/$\sum CO_2$ is not linear, this estimate only applies to small perturbations about the present-day values. Recent calculations (Brewer, 1978; Chen and Millero, 1979) have shown that back-calculations lead to an estimates of atmospheric CO2 within an error of a few tens of ppmV (Broecker et. al., 1985) Based on these assumptions, there is a direct link between pre-formed phosphate and atmospheric CO_2, since polar P (during CO_2 periods of bottom water formation) is a function of preformed $\sum CO_2$ and ALK.

Sarmiento and Toggweiler (1984), Siegenthaler and Wenk (1984), Wenk and Siegenthaler (1984), and Knox and McElroy (1984) explored high-latitude mechanisms for changing pre-formed phosphate (changes in high latitude productivity and exchange with the nutrient-depleted warm surface ocean). Siegenthaler and Wenk (1984a) considered these processes and also argued that low-latitude upwelling could decrease pre-formed phosphate and p_{CO2}. In their original model, lowered p_{CO2} occurs because they assume that increased low-latitude upwellng leads to an increased flux of nutrient-depleted waters into the polar regions. In a later discussion (1984b) these authors minimize this effect as a possible artifact of the structure of their model, in which low latitude upwelling is forced to be equal to deep water formation rates.

In the following section an alternative mechanism for a decrease in high-latitude nutrients will be explored, in which a PFP decrease is caused by an increase in low-latitude upwelling, which therefore does not require a direct link between upwelling and deep water formation rates.

A Simple Box Model Including Upper Thermocline Processes

Present ideas about or ocean circulation do not envision a direct link between low latitude oceanic upwelling and deep water formation rates. Low latitude upwelling is a direct consequence of local wind velocity and the Coriolis force. Equatorial upwelling occurs because tropical winds flow from east to west and because the Coriolis force changes sign at the equator: due to the Ekman transport, the westerly winds move near-surface waters to the north in the Northern Hemisphere and to the south in the Southern Hemisphere. Deeper waters (from the upper few hundred meters) upwell to replace these divergent surface waters. The nutrient content of this upwelling water is enhanced by mixing processes and the thermal structure associated with the equator-bound equatorial undercurrent (Smith, 1968; Gill, 1974). Coastal upwelling occurs because winds are often equatorward along the eastern boundaries of the continents; the Coriolis force then moves the near-surface waters offshore, which are replaced by deeper water masses. The nutrient content of eastern boundary intermediate waters may be enhanced by a dynamically mandated eastern boundary stagnation (Rhines and Young, 1982). In both cases, changes in the upwelling flux are expected to be roughly proportional to changes in the local wind intensity.

Deep water formation processes are not understood as well as the low-latitude upwelling processes. Kilworth (1983) has reviewed deep water formation mechanisms. At present the deep ocean is filled by waters which form in a few locations (Warren, 1981). It is hard to specify how deep water formation would respond to changes in atmospheric forcing, although some of the physical limitations on deep water formation are known (Warren, 1981, 1983). It seems unlikely, however, that deep water formation rates are directly linked to low-latitude wind velocities. Although one could imagine global 'teleconnection' processes which might result in synchronous changes in low-latitude upwelling and deep water formation, I suspect that few physical oceanographers would pretend to know the sign of such a correlation. So in this discussion I will assume that low latitude upwelling and deep water formation are independent processes.

In order to model uncoupled changes in high latitude processes and low latitude upwelling, it is necessary to specify at least 4 oceanic reservoirs: warm surface, upper thermocline, deep ocean, and polar. The following discussion will be framed around the 5 reservoir model illustrated in figure 1 (adding another reservoir for the atmosphere). Although specific estimates for atmospheric carbon dioxide will be model-dependent, the basic process that leads to changes in PFP and p_{CO2} is not. As will be seen from the box model, increasing the upwelling rate leads to a depletion of nutrients in the upper thermocline. This depletion results from the depth-dependent remineralization of falling particulate debris: each time 10 atoms of organic carbon escape through the base of the mixed layer, 8 of these are respired back into inorganic carbon dioxide in the 100-600 m depth range while 2 of these are regenerated in the deep ocean (Jenkins, 1982; Suess, 1980). If the upwelling rate is increased, then the chance that a biogenic carbon detritus will 'escape' into the deep ocean reservoir increases; because of the increasing frequency with which the 1 in 5 loss to the deep sea occurs. Once in the deep box the carbon atom returns to the shallow reservoirs at a rate which is determined by deep ocean renewal, i.e. deep water formation. An increase in low-latitude wind-driven upwelling leads to a decrease of phosphorus in the thermocline. Since the thermocline is a major source of the high-latitude waters that contribute to deep-water formation, this decrease in upper thermocline nutrient levels leads to a decrease in preformed phosphorus in polar waters. This PFP decrease produces a decrease in atmospheric carbon dioxide concentration, as discussed previously.

A simple analogy can illustrate why increased low latitude upwelling rates can reduce the phosphate content of the upper thermocline. Consider a gambler with a steady source of income. This gambler always bets a fixed percentage of his current bankroll on black on each turn at the roulette wheel. Since the odds demand that the house will always win on the average, for each

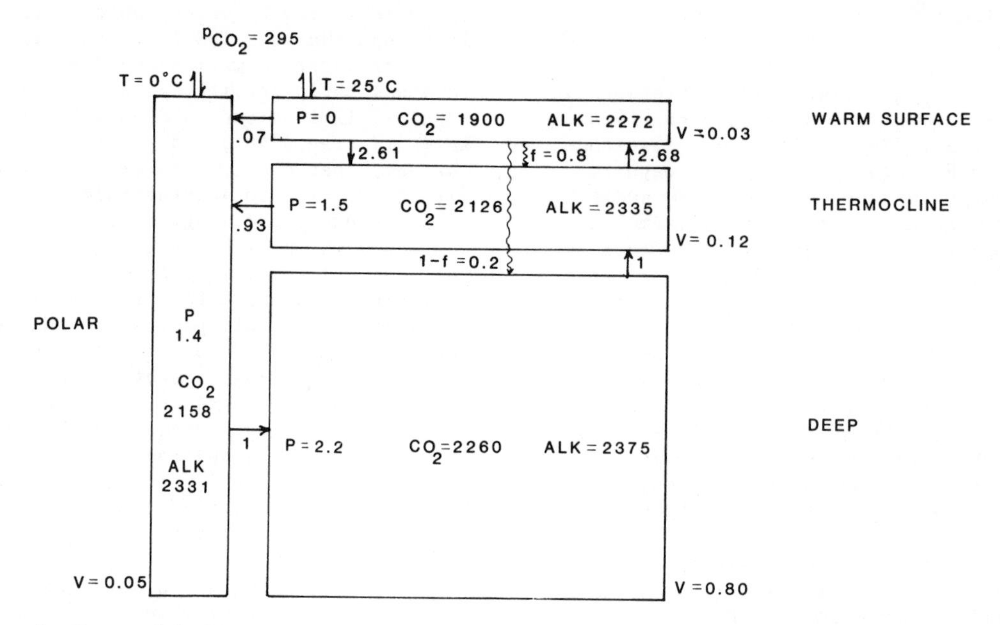

Fig. 1. Box model for atmospheric P_{CO2} during the present. Water fluxes are given as multiples of present-day deep water formation rates. Nutrient values are given in μmol/kg; P_{CO2} is reported in ppmV. Underlined quantities are fixed by present-day observations. All other values are determined from the steady-state solution.

turn a fixed percentage of the bet will be lost to the house (on the average). The faster the turns come up, the faster his money leaks into the coffers of the house. However, his bankroll will decline until a steady state is reached between his fixed income and loss as to the house. The fixed income is analogous to the steady renewal of deep water with a nearly fixed phosphate content; the rate at which the turn come is analogous to wind-driven upwelling; and the odds that the house will win are analogous to the probability that an atom of phosphorus that upwells into the warm surface box will escape into the deep ocean.

The box model has been streamlined to be the simplest and most determinate system which illustrates the behaviour described above. The exact numerical estimates which result are model-dependent, and slightly different results are to be expected from more realistic models. Nonetheless, any model which incorporates a depth-dependent particle regeneration rate and decoupled deep water formation and low latitude upwelling will display similar behavior to this simple box model, and the magnitude of the resulting P_{CO2} change should be comparable. The concentrations of phosphorus, total carbon dioxide, and alkalinity in the deep reservoir is based on averages obtained from GEOSECS data (Takahashi, Broecker, and Bainbridge, 1981). The concentration of phosphorus in the thermocline box is based on similar data for the 100-600 m interval. The concentration of preformed phosphate is as estimated by Broecker and Takahashi (in press) for the modern ocean. It is assumed that low-latitude organisms completely strip phosphorus out of the warm surface box, producing particulate matter with a composition 106 org C : 21 $CaCO_3$: 1 P. Finally, it is assumed that 80% of this falling debris is remineralized in the thermocline box and that 20% of it escapes into the deep ocean. The effect of nitrate regeneration on alkalinity (Brewer and Goldman, 1976) is neglected here only for simplicity and should be included in a comprehensive model.

The flow patterns have been specified to be consistent with the previous discussion. Polar waters are formed as a mixture of warm surface an upper thermocline waters; the proportion attributed to the two sources is determined by th phosphorus balance for the polar box. In this model, high latitude productivity is not included although several recent papers consider it important (Sarmiento and Toggweiler, 1984; Knox and McElroy, 1984; Siegenthaler and Wenk, 1984a,b,). This exclusion is mainly for the sake of simplicity, since this factor has already been explored in the previous studies.

In this model, fluxes of water are specified i units of multiples of present-day deep water formation. The ratio of wind-driven upwelling to deep water formation is computed from the hosphorus balance for the upper thermocline box. he P_{CO2} of the atmosphere is computed from the re-formed carbon dioxide and alkalinity values of he polar box computed as described previously sing the Mehrbach regression equations given by illero (1979), the CO_2 solubilities of Weiss 1974) and the calculation technique described by dmond and Gieskes (1970), assuming that this box s at 0°C. Other calculation schemes for stimating P_{CO2} will be offset by a constant; P_{CO2} hanges due to variations in $\sum CO_2$ and ALK will be argely parallel. This value of the atmospheric arbon dioxide pressure is then used to compute he carbon dioxide content of the warm surface box assumed to be at 25°C); it is assumed that both he warm surface and polar boxes are in as-exchange equilibrium with the atmosphere. All ther concentrations can then be computed from ass balance and the assumed stoichiometry. The eservoirs are assigned relative sizes based on easonable estimates for the modern ocean.

This model for the carbon dioxide cycle in the cean is oversimplified but contains a number of eatures that illustrate real-ocean processes: 1) In particular, there is a net flux of carbon ioxide from out of the warm surface box into the old polar box via the atmophere. Such a flux eems inescapable based on the observed oncentrations of carbon dioxide and alkalinity in he upper thermocline waters (GEOSECS) and the ear-equilibration of central gyre waters with tmospheric carbon dioxide and the observed attern of oceanic surface P_{CO2} (Keeling, 1968). (2) The model also indicates that the total low latitude upwelling rate is several times that of the bottom water formation. Warren (1982) reports that the total bottom water formation rate (prior to entrainment) is unlikely to be more than 15 x 10^6 m^3/sec. Quay, Stuiver, and Broecker (1983) estimate that upwelling in the equatorial Pacific alone is 47 x 10^6 m^3/sec. It seems likely that the global total wind-driven upwelling is several times higher than the bottom water formation rate.

Given this 'modern' analogue, we can then perturb the system in various ways and determine the response of the other variables. The initial change is specified, and flux parameters other than these assumed changes are held constant. The other parameters are computed from their new steady-state values subject to the constraint of mass balance. For the mass balance portion of the calculation, it is assumed that the mass of the ocean is 1.4 x 10^{21} kg and that the atmosphere contains 1.7 x 10^{20} moles of atmospheric gases. The calculation of the new steady state is done iteratively, starting with the assumption of no change in the deep ocean reservoir. Phosphorus and alkalinity in the upper thermocline box are then computed by mass balance. The phosphorus content of this box determines the pre-formed phosphorus content of the polar ocean, and from this the pre-formed carbon dioxide and alkalinity values are computed. The P_{CO2} of the polar box is computed, and from this value the total CO_2 of the warm surface box can be estimated. Then the carbon dioxide content of the upper thermocline can be computed from mass balance. At this point total quantities of phosphorus, alkalinity, and carbon dioxide are calculated, and the excess or deficiency is assigned to the deep ocean box. The computations are then repeated until mass is conserved in the whole system.

Figure 2 illustrates the steady state solution for this system under the assumption that the wind-driven upwelling doubles while the bottom water formation rate remains constant. The more rapid cycling of the upper waters results in a higher proportion of the phosphorus escaping into the deep ocean. This process depletes the upper thermocline of phosphorus and reduces the polar preformed phosphorus from 1.4 μmol/kg to 1.16 μmol/kg. This reduction in the preformed phosphorus content reduces atmospheric carbon dioxide by 25 ppmV, about one-third of the total observed glacial/interglacial range.

One rather surprising consequence of looking at the ocean in this way is that an increase in wind driven upwelling does not lead to a proportionate increase in productivity. For the case given in figure 2, productivity increases only 33% for a doubling of wind-driven upwelling. The reason for this diminished response is the depletion of nutrients in the upper thermocline by increased leakage into the deep sea.

Other physical changes can also alter atmospheric carbon dioxide, as shown in Table 1. A reduction of the deep water formation rate by a factor of two also decreases P_{CO2} by about 25 ppmV, which has been pointed out previously (Toggweiler and Sarmiento, 1984; Siegenthaler and Wenk, 1984). In fact, an examination of the structure of this steady state model shows that the polar preformed phosphorus and the atmospheric carbon dioxide concentration are functions of the ratio of the fluxes due to wind-driven upwelling and deepwater formation. If deepwater formation rates increased while upwelling also increased, there would be no net change in preformed phosphorus or atmospheric carbon dioxide. Lowering the temperature of the polar box by 2°C also decreases atmospheric carbon dioxide by 25 ppmV; this change occurs because carbon dioxide is more soluble at lower temperatures which then draws more CO_2 into the polar and deep ocean boxes. Wenk and Siegenthaler (1985) have also remarked on this sensitivity of P_{CO2} to polar ocean temperatures. One advantage of this mechanism is that it would increase the oxygen content of the deep ocean (due to higher gas solubilities at lower temperatures), since the lack of observed zero-oxygen glacial deep water is a problem in some models (Toggweiler and Sarmiento, 1984).

It should be noted here that in this context 'polar ocean temperatures' represent the average

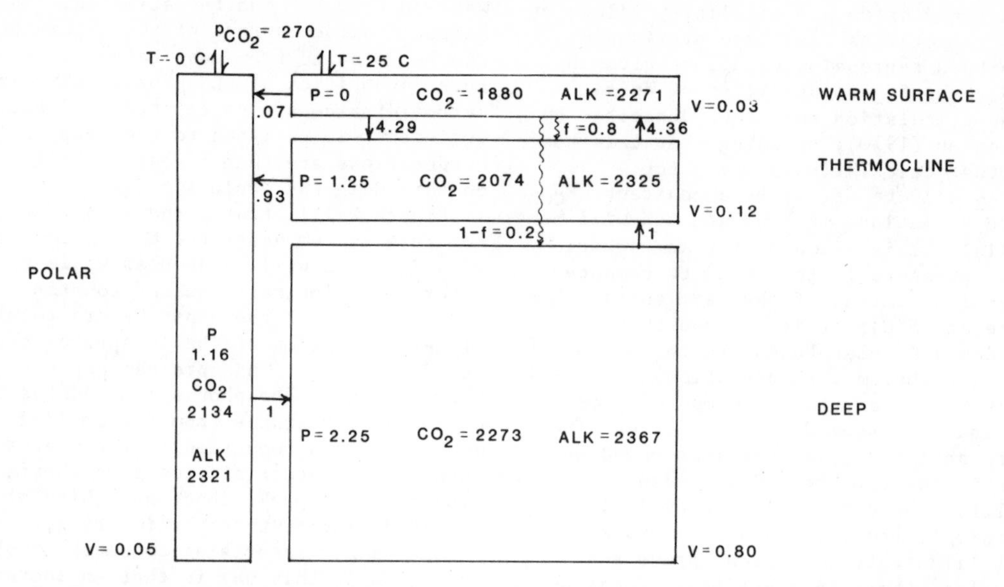

Fig. 2. Box model for p_{CO2} during the 'glacial' period, assuming a doubling of the wind-driven upwelling holding all other water fluxes constant. All other values are computed from the steady-state solution subject to mass conservation.

temperature of the primary bottom water formation and that of the entrained water that augments the flux. Since bottom water formation may occur near the freezing point (e.g. Foster and Middleton, 1980), it may be impossible to decrease the temperature of the primary flux. But it is quite possible that the entrained water is cooler.

It also should be noted that in the Antarctic, deep water formation draws water from deeper in the thermocline (1000m) than assumed in this model, which is more similar to bottom water formation in the North Atlantic. A realistic model should include two deep water sources, and allow for differences in the source water depth.

The next section considers evidence bearing on the significance of these mechanisms for changing atmospheric carbon dioxide. Before moving on, however, one important unverified assumption of the model should be emphasized. The model assumes that increases in wind strengths would not increase the rate of transfer of nutrients from the deep ocean into the upper thermocline, since the model has an upward nutrient flux only driven by deepwater ventilation. In reality there must be some eddy-diffusive transfer of water between boxes as well as this advective transfer, and it is possible that an increase in wind strength might increase this vertical eddy diffusivity. However, in the modern ocean the rate of vertical eddy diffusivity within the thermocline is very low (Jenkins, 1980) so that the upward flux of nutrients is dominated by general deep-ocean ventilation. If this situation were to change significantly in the past, the assumptions of this

TABLE 1. Results of Models for PFP(pre-formed phosphate) and p_{CO2} changes

Scenario	Result			
	p_{CO2}		PFP	
Double Wind-Driven upwelling	p_{CO2}	decreases 25 ppmV	decreases	0.25 μmol/kg
Decrease Deep Ocean Temperature 2°C	"	25 ppmV	invariant	
Half Deep Water Formation	"	25 ppmV	decreases	0.25 μmol/kg

odel would be violated. This possibility seems emote, however.

Evidence for Changes in Low Latitude Upwelling, Polar Ocean Temperatures, and Upper Thermocline Nutrient Levels

There is considerable evidence to support the otion that low latitude wind-strengths and pwelling were significantly greater in the past. here is evidence that the grain size of ind-blown sediments was greater during glacial eriods which implies a more vigorous atmospheric irculation (Parkin and Shackleton, 1973; Parkin, 974; Kolla, Biscaye and Hanley, 1979; Boyle, 983; Sarnthein, Tetzlaff, Koupman, Wolter, and flaumann, 1981). The CLIMAP (1976) reconstruction f ocean surface temperatures from temperature-ensitive fossil abundance data shows colder quatorial and eastern boundary temperatures contrasting with relatively stable central gyre emperatures, as does the CLIMAP (1981) winter easonal reconstruction. This evidence indicates n increased upwelling of cooler waters. Manabe nd Hahn (1977) used the CLIMAP boundary conditions to compute the atmospheric circulation patterns during the most recent glacial maximum. They found that while the response was spatially variable, in general wind strengths were higher by about a factor of two in many parts of the ocean. Since upwelling is roughly linear with the wind stress, this evidence indicates that the assumption of a factor of two increase in low-latitude upwelling is reasonable. Considered in isolation of other effects, it seems almost certain that there was increased low-latitude upwelling during the glacial period which would tend to reduce atmospheric carbon dioxide.

There is less certain evidence for changes in deep ocean temperatures, although it seems quite plausible that deep ocean temperatures were lower during the glacial period. The oxygen isotope ratio of benthic foraminifera responds to deep ocean temperature and to changes in the isotopic composition of seawater (Shackleton, 1968). The relative influence of these two variables is subject to debate at present, although everyone agrees that the ice volume effect is dominant. Shackleton (1968) estimated that no more than 2/3 of the 1.7°/°° increase in benthic forammiteral $\delta^{18}O$ during glacial time is due to ice volume. If so, then the residual 1/3 would correspond to a lowering of deep ocean temperature of about 2°C. More recently, Shackelton and Chappell (1985) and Duplessy and Shackleton (1985) have made stronger arguments for a 2°C decrease in deep ocean temperature during glacial periods. Although this number is more uncertain than the increase in wind-driven upwelling, it is plausible to expect that changes in polar temperatures may have decreased atmospheric carbon dioxide by 25 ppmV.

The rate of deep water formation during glacial time is unknown at the present. There is considerable evidence that deep ocean circulation patterns were different during glacial time (e.g., Streeter and Shackleton, 1977; Curry and Lohmann, 1983; Shackleton, Imbrie, and Hall, 1983) from studies of deep sea benthic fauna and benthic foram carbon isotope variability. Variations in the cadmium content of benthic foraminifera has also been useful in this regard (Boyle and Keigwin, 1982). Since this technique may be unfamiliar to many readers of this report, it is worthwhile to review the principles of this method. The oceanic distribution of cadmium is quite similar to that of phosphorus, being depleted to very low levels in surface waters through uptake by organisms and returned to the deep water by the decomposition of sinking biogenic debris. The crystal chemistry of cadmium and calcium are very similar, so foraminifera incorporate cadmium into their shells in proportion to the content of the water they inhabit:

$$\left[\frac{Cd}{Ca}\right]_{foram} = D \left[\frac{Cd}{Ca}\right]_{water}$$

where D is an empirically-determined distribution coefficient approximately equal to 2.0 ± 0.4 (Hester and Boyle, 1982). The deep ocean distributions of cadmium and phosphorus reflect the input of nutrient-depleted North Atlantic Deep water which intrudes into nutrient-enriched Antarctic deep water sources. Hence changes in the cadmium content of benthic foraminifera reflect variations in the relative intensity of these deep water sources. Boyle and Keigwin (1982) showed that the cadmium concentration of the deep North Atlantic was a factor of two higher 18,000 years ago, but was not so high as to indicate a cessation of a nutrient-depleted source of deep water to the North Atlantic. Further data is presented by Boyle (1984) and Boyle and Keigwin (1985), from which the North Atlantic Cd/Ca record for the last 225,000 years is shown in figure 3. Chemical and isotopic data cannot determine absolute rates of deep ocean ventilation, and one must be brave even to guess whether total deep water formation rates increased or decreased. Eventually, we may be able to ascertain this from studies of the carbon-14 content of planktonic and benthic foraminifera from deep ocean sediment cores (Andree et. al., 1985). If it should prove that deep water formation rates decreased, then the whole of the glacial/interglacial p_{CO2} change could be accounted for within the framework of this model. On the other hand, if it should turn out that deepwater fluxes increased significantly, the magnitude of change would be reduced and other mechanisms would be needed to account for the change in atmospheric carbon dioxide.

The advantage of the mechanisms proposed in this paper is that they all are capable of being fully effective on time scales of several hundred years and could account for the rapid changes in atmospheric carbon dioxide observed in ice cores

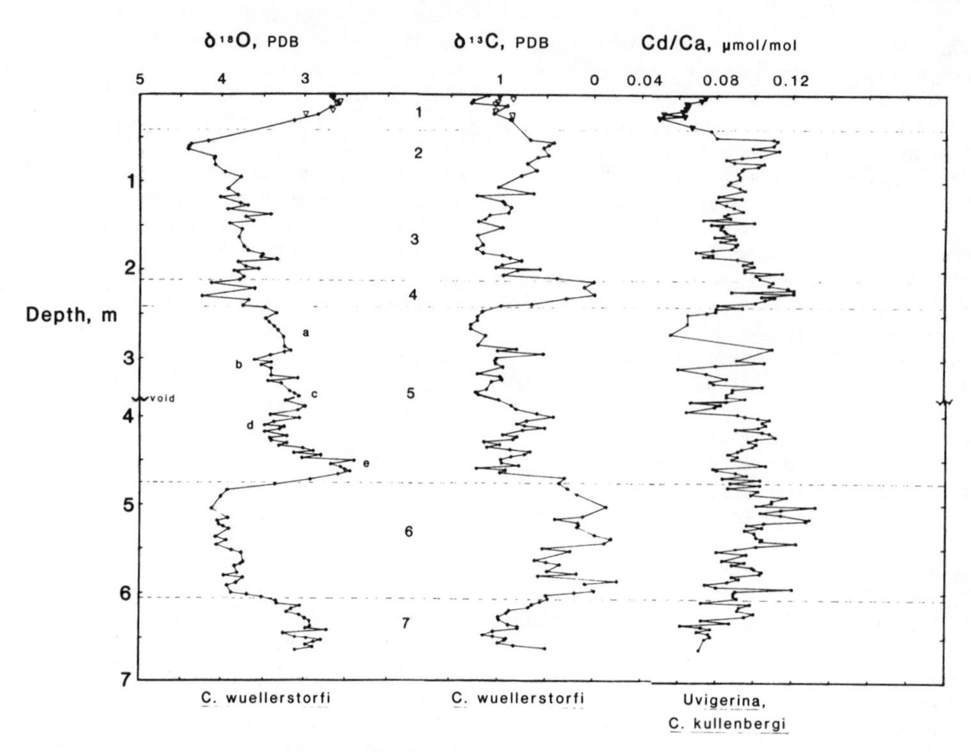

Fig. 3. Cd/Ca, $\delta^{13}C$ and, $\delta^{18}O$ data from core CHN 82 Sta 24 core 4PC on the western flank of the mid-Atlantic ridge at 42°N, 33°W; 3427 m depth.

(Stauffer et. al., in press). The most important difficulty with PFP mechanism is that preliminary evidence on the pre-formed nutrient content of polar waters suggests that there have not been large changes in PFP. This evidence is based the $\delta^{13}C$ of planktonic foraminifera in regions near bottom-water formation; it does not appear that high-latitude surface $\delta^{13}C$ changes significantly relative to the global mean of $\delta^{13}C$, which it should if PFP changes. (Mix and Fairbanks, 1985; Fairbanks, personal communication). This issue is far from settled as this review goes to press; there is a great need for data on the $\delta^{13}C$ and Cd content of high-latitude and upper-thermocline forammifera to resolve this issue.

Finally, I would like to present some new evidence which provides modest support for the hypothesis that increases in low-latitude upwelling decreased the nutrient content of the upper thermocline. The model described here predicts that upper thermocline nutrient contents should decrease in response to intensified low-latitude upwelling. If an indicator of upper thermocline nutrient levels could be found, then this prediction could be tested. An ideal tracer of such a depletion is the cadmium content of foraminifera tests.

Ideally, one could look at benthic foraminifera from a transect of cores through the thermocline to see if such a nutrient decrease occured. Alternatively, the cadmium content of deep-dwelling planktonic foraminifera may reflect nutrient concentrations in the upper thermocline. Although deep-dwelling planktonic foraminifera will not be affected by sea-level change, they are subject to artifacts related to vertical migration within the water column caused by changes in temperature, density, and nutrient structure.

Finally, it may be worthwhile to remark on what might be a significant implication of response of atmospheric carbon dioxide to increased low-latitude upwelling, namely, that it provides a mechanism whereby orbitally-driven changes in the global distribution of incoming sunlight could be translated into global climatic change. There is overwhelming evidence that climate responds to these orbital changes (Imbrie, Hays, and Shackleton, 1976; Ruddiman and McIntyre, 1981), but the mechanisms whereby the relatively slight insolation changes (Hansen et. al., 1984) are translated into climate change are not yet resolved. Kutzbach and Otto-Bleischer (1982) utilized a low-resolution global circulation model to show that the African-Asian monsoon would be expected to be strongly influences by precession-induced changes in incoming radiation. Prell (1984) showed that there are significant increases in the percentages of upwelling-correlated oraminifera in sediment cores off the Arabian oast at the time predicted by their model. eigwin and Boyle (1985) observed a significant 23 yr global carbon isotope periodicity which is oherent with precession; they argued that this ignal was due to low-latitude biomass control by recession-driven humidity cycles. Other portions f the global climate system may prove to be imilarly sensitive. If insolation drives wind trengths and low-latitude upwelling, the esultant changes in atmospheric carbon dioxide ould induce global temperature changes which ould then be amplified by the response of the ocean and cryosphere (Hansen et. al., 1984).

Summary

The influence of low-latitude upwelling and olar temperatures on atmospheric p_{CO2} has been nvestigated through the use of a simple -reservoir model of the ocean. Since each ecycling of a nutrient atom into the ocean surface results in a 20% chance of driving the nutrient into the deep ocean, an increase in low-latitude upwelling (upper ocean recycling rate) depletes the upper thermocline of nutrients. This depletion of the upper thermocline lowers the nutrient content of polar waters and hence also decreases the preformed nutrient content of the deep ocean and atmospheric carbon dioxide. The model predicts a 25 ppmV decrease in p_{CO2} for the doubling of low latitude upwelling which appears to be likely based on CLIMAP data and modelling studies. A 2°C decrease in polar ocean temperatures would decrease p_{CO2} by the same amount. Such a temperature decrease is plausible and within the limits set by oxygen isotope studies.

Acknowledgements. I thank Erwin Suess and Rob Toggweiler for thoughtful readings of the first version of this manuscript. This research was supported by NSF grant # 8209362-OCE; my attendance at the IGC was partially supported by travel grants arranged by Ken Hsu.

References

Bacastow, R.B., J.A. Adams, C.D. Keeling, D.J. Moss and T.P. Whorf, Atmospheric carbon dioxide, the Southern Oscillation and the weak 1975 El Nino, Science, 210, 66-68, 1980.

Barnola, R., A. Neftel, and H. Oeschger, Comparison of CO_2 measurements by two laboratories on air from bubbles in polar ice, Nature, 303, 410-412, 1983.

Barnola, J.M., D. Raynaud, A. Neffel, and H. Oeschger Comparison of CO_2 measurements by two laboratories on air from bu bbles in polar ice, Nature, 303, 410-413, 1983.

Boyle, E.A. and L.D. Keigwin, Deep circulation of the North Atlantic over the last 200,000 years: geochemical evidence, Science, 218, 784-787, 1982.

Boyle, E.A., Chemical accumulation variations under the Peru current during the past 130,000 years, J. Geophys. Res., 88, 7667-7680, 1983.

Boyle, E.A., Cadmium in benthic foraminifera and abyssal hydrography: evidence for a 41 kyr. obliquity cycle, in Climate Processes and Climate Sensitivity (eds. J.E. Hansen and T. Takahashi) Am. Geophys. Union Monograph 29, 1984.

Boyle, E.A. and L.D. Keigwin, Comparison of Atlantic and Pacific paleochemical records for the last 215,000 years: changes in deep ocean circulation and chemical inventories, Earth Planet. Sci. Lett., in press.

Brewer, P. and J.C. Goldman, Alkalinity changes generated by phytoplankton growth, Limnol. Oceanogr., 21, 108-117, 1976.

Brewer, P.G. Direct observation of the oceanic CO_2 increase, Geophys. Res. Lett., 5, 997-1000, 1978.

Broecker, W.S., Ocean chemistry during glacial time, Geochem. Cosmochim. Acta, 46, 1689-1705, 1982.

Broecker, W.S. and T. Takahashi, Is there a tie between atmospheric CO_2 content and ocean circulation? in Climate Processes and Climate Sensitivity, (eds. J.E. Hansen and T. Takahashi, pp. 314-326, Am. Geophys. Union, 1984.

Chen, C.-T. and F. Millero, Gradual increase of oceanic CO_2, Nature, 277, 205-206, 1979.

CLIMAP, The surface of the ice-age earth, Science, 191, 1131-1137, 1976. CLIMAP, Project Members, A. McIntyre, Leader of LGM Project, Seasonal reconstructions of the earth's surface at the last glacial maximum, Geological Society of America Map Series, 1981.

Curry, W. and G.P. Lohmann, Reduced advection into Atlantic Ocean deep eastern basin during last glacial maximum, Nature, 306, 577-579, 1983.

Duplessy, J.C. and N.J. Shackleton, The oxygen isotope record of benthic foraminifera: effect of deep water temperature and ice volume changes, Trans. Am. Geophys. Union, 66, 292, 1985.

Edmond, J.M., On the dissolution of carbonate and silicate in the deep ocean, Deep-Sea Res., 21, 455-480, 1974.

Edmond, J.M. and J.M. Gieskes, On the calculation of the degree of saturation of sea water with respect to calcium carbonate under in situ conditions, Geochim. Cosmochim. Acta, 34, 1261-1291, 1970.

Emiliani, C., Pleistocene temperatures, J. Geol., 63, 538-578.

Foster, T.D. and J.H. Middleton, Bottom water formation in the western Weddell Sea, Deep-Sea Res., 27A, 367-381, 1980.

Gill, A.E., Models of equatorial currents, in Numerical Methods of Ocean Circulation, pp. 181-203, National Academy of Sciences, 1975.

Haigs, J.D., J. Imbrie and N.J. Shackleton, Variations in the earth's orbit: pacemaker of the ice ages, Science, 194, 1121-1132, 1976.

Hansen, J., A. Lacis, D. Rind, G. Russell, T. P.

Stone, I. Funy, R. Rued, and J. Levrer, Climate sensitivity: analysis of feedback mechanisms, in Climate Processes and Climate Sensitivity (eds. J.E. Hansen and T. Takahashi) Am. Geophys. Union, Geophysical Monograph 29, pp. 130-163, 1984.

Hester, K. and E., Boyle Water chemistry control of cadmium content in recent recent benthic foraminifera, Nature, 298, 260-262, 1982.

Jenkins, W., Tritium and He-3 in the Sargasso Sea, J. Mar. Res., 38, 533-569, 1980.

Jenkins. W., Oxygen utilization rates in North Atlantic subtropical gyre and primary production in oligotrophic systems, Nature, 300, 246-248, 1982.

Keeling, C.D., Carbon dioxide in surface ocean waters, 4 global distribution, J. Geophys. Res., 73, 4543-4553, 1968.

Keigwin, L.D. and E.A. Boyle, Carbon isotopes in deep-sea benthic foraminifera: precession and changes in low-latitude biomass, in The carbon cycle and atmospheric CO_2: Natural Variations Archean to Present, (eds. E.T. Sundquist and W.S. Broecker) Am. Geophys. Union, Chapman Series, pp. 319-328, 1985.

Kilworth, P., Mechanisms of Deep Water Formation, Rev. Geophys. Space Phys., 21, 1-26, 1983.

Kolla, V., P.E. Biscaye and A. Hanley, Distribution of quartz in late Quaternary sediments in relation to climate, Quat. Res., 11, 261-277, 1979.

Knox, F. and M. McElroy, Changes in atmospheric CO_2: influence of biota at high latitude, J. Geophys. Res., 89, 4629-4637, 1984.

Kutzbach, J. and B.L. Otto-Bleischer, The sensitivity of the African-Asian monsoonal climate to orbital parameter changes for 9000 years B.P. in a low-resolution climate model, J. Atm. Sci., 39, 1177-1188, 1982.

Manabe, S. and D.G. Hahn, Simulation of the Tropical Climate of an Ice Age, J. Geophys. Res.,82, 3889-3912, 1977.

Mix, A. and R.G. Fairbanks, North Atlantic surface-ocean control of Pleistocene deep-ocean circulation, Earth Planet. Sci. Lett., 73, 231-243, 1985.

Neftel, A.H., J. Oeschger, J. Schwander, B. Stauffer, and R. Zumbrumn, Ice core sample measurements give atmospheric CO_2 content during the past 40,000 years, Nature, 295, 220-223, 1973.

Newell, R.E., A.R. Navato, and J. Hsiung, Long-term global sea-surface temperature fluctuations and their possible influence on atmospheric CO_2 concentrations, Pure Appl. Geophys., 116, 351-371, 1973.

Oeschger, H., J. Beer, U. Siegenthaler, B. Stauffer, W. Dansgaard, and C.C. Langway, Late glacial climate history from ice cores, in Climate Process and Climate Sensitivity (eds. J.E. Hansen and T. Takahashi), pp. 299-306, Am. Geophys. Union, 1984.

Oeschger, H., B. Stauffer, R. Finkel, and C.C. Langway Jr., Variations of the CO_2 concentration of occluded air and of anions and dust in polar ice, in: The Carbon Cycle and Atmospheric CO_2: Natural Variations, Archean to Present, (eds, E.T. Sundquist and W.S. Broecker) Am. Geophys. Union, Wash. D.C., pp. 132-142, 1985.

Parkin, D.W., Trade winds during the glacial cycles, Proc. R. Soc. Lond. A., 346, 245-260, 1974.

Parkin, D.W. and N.J. Shackleton, Trade wind and temperature correlations down a deep-sea core off the Saharan Coast, Nature, 245, 455-457, 1973.

Prell, W., Variation of monsoonal upwelling: a response to changing solar radiation, in Climate Processes and Climate Sensitivity (eds. J.E. Hansen and T. Takahashi) pp. 48-57, Am. Geophys. Union, 1985.

Quay, P., M. Stuiver, and W.S. Broecker, Upwellin rates for the equatorial Pacific Ocean derived from the bomb ^{14}C distribution, J. Mar. Res., 41, 769-792, 1983.

Redfield, A., B. Ketchum, and F.R. Richards, The influence of organisms on the chemical composition of seawater, in The Sea, Vol. 2 (ed. M.N. Hill), pp. 26-77, Interscience, New York, 1963.

Rhines, P. and W. Young, A theory of wind-driven circulation I, Mid-ocean gyres, J. Mar. Res., 40, (supp.), 559-596, 1982.

Ruddiman, W. and A. McIntyre, Oceanic mechanism for amplification of the 23,000 year ice volume cycle, Science, 212, 617-627, 1981.

Sarmiento, J.L. and J.R. Toggweiler, A new model for the role of the oceans in determining atmospheric p_{CO2}, Nature, 308, 621-623, 1984.

Sarnthein, M., G. Tetzlaff, B. Koopman, K. Wolter, and U. Pflaumann, Glacial and interglacial wind regimes over the eastern subtropical Atlantic and North-west Africa, Nature, 293, 193-196, 1981.

Shackleton, N.J., Oxygen isotope analyses and Pleistocene temperatures re-assessed, Nature, 215, 15-17, 1967.

Shackleton, N.J., J. Imbrie, and M. Hall, Oxygen and carbon isotope record of East Pacific core V19-30: implications for the formation of deep water in the late Pleistocene North Atlantic, Earth Planet. Sci. Lett., 65, 233-244, 1983.

Shackleton, N.J., M.A. Hall, J. Line and C. Shuxi, Carbon isotope data in core V19-30 confirm reduced carbon dioxide concentration in the ice age atmosphere, Nature, 306, 319-322, 1983.

Shackleton, N.J. and J. Chappell, The ocean deep-water isotope record and the New Guinea sea-level, Trans. Am. Geophys. Union, 66, 293, 1985.

Siegenthaler, U. and T. Wenk, Rapid atmospheric CO_2 variations and ocean circulation, Nature, 308, 624-625, 1984.

Siegenthaler, U. and Th. Wenk, Rapid atmospheric CO_2 variations and ocean circulation, Nature, 308, 624-627, 1984.

Smith, R.C., Upwelling, Oceanogr. Mar. Biol. Ann. Rev., 6, 11-46, 1968.

Stauffer, B., H. Hofer, H. Oeschger, J. Schwander, and U. Siegenthaler, Atmospheric CO_2 concentration during the last glaciation, Ann. Glacial., 5, 1984.
Streeter, S.S. and N.J. Shackleton, Paleocirculation of the deep North Atlantic; 150,000 year record of benthic foraminifera and oxygen-18, Science, 203, 168-171, 1979.
Suess, E., Particulate organic carbon flux in the oceans -surface productivity and oxygen utilization, Nature, 288: 260-263, 1980.
Takahashi, T., W.S. Broecker, and A.E. Bainbridge, Supplement to the alkalinity and total carbon dioxide concentration in the world oceans, in Carbon Cycle Modelling, Scope 16, (ed. B. Bolin) pp. 159-200, John Wiley, New York, 1981.
Takahashi, T., W.S. Broecker, and S. Langer, Redfield Ratio based on chemical data from isopycnal surfaces, J. Geophys. Res., (in press), 1985.
Viecelli, J.A., The atmospheric carbon dioxide response to oceanic primary productivity fluctuations, Climatic Change, 1984.
Warren, B., Deep circulation of the world ocean, in The Evolution of Physical Oceanography (eds. B.A. Warren and C. Wunsch) MIT press, Cambridge, MA, pp. 6-41, 1981.
Warren, B., Why is no deep water formed in the North Pacific? J. Mar. Res., 41, 327-347, 1983.
Wenk, T.H. and U. Siegenthaler, The high-latitude ocean as a control of atmospheric CO_2, in The Carbon Cycle and Atmospheric CO_2: Natural variations Archean to Present (eds E. Sundquist and W.S. Broecker) Am. Geoph. Union., Wash. D.C., pp. 185-194, 1985.

HIGH FREQUENCY SEA-LEVEL FLUCTUATIONS IN CRETACEOUS TIME: AN EMERGING GEOPHYSICAL PROBLEM

Seymour O. Schlanger

Department of Geological Sciences, Northwestern University, Evanston, Illinois, U.S.A. 60201

Abstract. Refinements in dating and correlation of Cretaceous strata coupled with the mapping of shoreline migrations and determinations of water depths that obtained during deposition of these strata show that from Albian through Maastrichtian time cratonic platforms experienced 6 major marine transgressive pulses. These pulses show a periodicity of $\sim$5.5$\pm$1 Ma. During these marine transgressions water depths on the cratons increased by 100 to 250 m. The global synchroneity of the transgressions implies rates of relative sea-level rise and fall of from 50 to 100 m/my. Recent revisions of geologic time scales show that while changes in ridge-crest spreading rates and global ridge volume and ridge-crest length may account for the general long-term flooding of major cratons from $\sim$110 to $\sim$85 Ma, such changes cannot account for the $\sim$5.5 Ma transgressive-regressive cycles seen in Cretaceous strata. The possibility exists that ridge geometries were highly variable in Mesozoic time but the lack of significant magnetic reversals between chrons 34 and M0 makes it impossible to discern short-term changes in these rates. Other mechanisms such as continent-continent collisions, sediment accumulation in oceanic basins, dessication of marginal seas, hot spot-induced rejuvenation of older sea floor, and stress variations in the lithosphere related to major plate reorganizations also cannot account for the frequency of the Cretaceous transgressive-regressive cycles. If we accept the premise of an ice-free Cretaceous world, and the preponderance of evidence points to the lack of significant glaciation from Albian through Maastrichtian time, we are faced with the challenge of finding a geologically reasonable mechanism capable of affecting major, globally synchronous, short-term changes in the freeboard of the major cratons. The linkage between transgressions and oceanic anoxic events in the Cretaceous ocean adds an economic impulse to the geophysical-geological problem. During oceanic anoxic events large amounts of organic carbon were preserved in marine sediments, sedimentary manganese deposits formed in shallow basins, and sulfide mineral formation was enhanced in axial ridge vent settings.

Introduction

The observation that the oceans have periodically inundated the major continental land masses is one of the oldest in the earth sciences. Lyell's [1873, p. 7] discussion of Indian and Egyptian writings emphasizes that

> No point of the Eastern cosmogony...is more interesting to the geologist than the doctrine, so frequently alluded to, of the reiterated submersion of the land beneath the waters of an universal ocean.

H. Suess's [1906] studies of global Cretaceous epicontinental transgressions led to his proposal of the term eustasy for describing widespread, synchronous relative sea-level rises and falls. Over the past 10 to 15 years the study of sea-level fluctuations has been particularly intensive and fruitful due to the steadily increasing interaction between stratigraphers who apply paleontological and paleobathymetric analyses to refine both the dating and amplitude of transgressions and regressions, geophysicists and geochemists who develop magnetostratigraphic and radiometric time scales, and exploration geologists and marine geophysicists who interpret and relate continental margin stratigraphy and oceanic bathymetry to sea floor processes. These integrated studies have resulted in a flood of papers on the subject of relative sea-level changes and their causes and consequences.

A discussion of Cretaceous sea-level history necessarily involves use of the terms eustasy, relative sea level, and transgression and regression. Inherent in the concept of eustasy proposed by Suess [1906] was the implication that the water depth in oceanic basins, marginal seas, and epicontinental seaways could increase or decrease in a globally synchronous manner. If the volume of water remained constant these global, synchronous changes in water depth could only mean that large portions of the cratons were moving towards or away from the center of the earth. Suess [1906, v. 2., p. 538] favored suboceanic sinking in his scheme when he stated that "the formation of the sea basins produces spasmodic eustatic negative movements." In order to account for transgressions Suess [1906] called upon the accumulation of sediments in the oceanic basins. In recent years fluctuations in the elevation and length of the global mid-ocean ridge system have been invoked to account for the simultaneous flooding of widely separated cratons. Further complications arise from studies that relate earth rheology to isostasy and eustasy. According to Clark [1980, p. 525], "ustatic sea level cannot be measured anywhere." This statement is based on the argument that a transfer of water from the global ocean to geographically restricted ice caps causes both deformation of the crust and perturbation of the geoid due to the redistribution of mass on the surface of the earth. Geoidal deflections are such that during a glacial stage sea level can apparently rise in certain areas and fall in other areas; a glacial event can cause both transgressions and regressions simultaneously. In an ice-free world the problem of the effect of geoidal deflections on deciphering or documenting Suessian eustatic sea-level changes still exists. Geoidal deflections due to major changes in the global ridge-crest system would be significant. Further, the long wave length highs and lows seen on modern geoid maps are caused by mass inhomogeneities deep within the earth that are not spherically symmetrical. It could be argued that the motions of continents through geoid highs and lows could result in relative sea-level changes at the edges of moving continents.

It is now generally accepted that true eustatic sea-level changes in the sense of Suess, assuming that the volume of water remains constant, do not take place. However, we can perceive a Cretaceous history of global relative sea-level changes in the stratigraphic record of transgressions and regressions. In this paper a transgression is said to have taken place when the shoreline migrates inland over a land surface. The extent of the transgression can be mapped by correlating the shoreline position with the stratal age and the local increase in water depth can be deduced from considerations of vertical facies and faunal changes at a given point. The linkage between transgressions and regressions and global sea-level changes is in itself complex. As Sloss [1962] pointed out, in an analysis of shelf sequences, stratigraphically recorded transgressions and regressions can be the result of variations in the rate of basin subsidence and sediment input; no global sea-level change need be invoked. Pitman [1978] analyzed Sloss's argument in terms of global sea-level rises and falls related to considerations of passive margin subsidence and sedimentation rates and concluded that "...transgressive and regressive events may not be indicative of eustatic sea-level rises and falls respectively, but may be caused by changes in the rate of sea-level change."

That there is a problem in relating stratigraphically recorded rates and magnitudes of relative sea levels in Cretaceous time to known mechanisms for causing such changes has been clearly pointed out by Pitman and Golovchenko [1983]. Their review of causes, magnitudes and rates of sea-level change is shown as Table 1. These authors conclude [1983, p. 57] "...that our inability at present to propose a mechanism that will cause lowerings of sea level of the rate and magnitude suggested by Vail and others does not mean that such events did not occur. We can only conclude that either there is a sufficient, but as yet unknown, mechanism or that the interpretation of the sedimentary patterns in terms of sea-level changes is incorrect."

It is the purpose of this paper to argue that the geological interpretation of the stratigraphic record and sedimentary patterns of Albian through Maastrichtian sequences in terms of high-frequency relative sea-level rises and falls is correct and to reinforce Pitman and Golovchenko's argument that known mechanisms cannot account for the recorded transgressive-regressive cycles in strata of these ages.

Review of Previous Work

A seminal paper, taken here as the starting gun of the recent surge of research, was that of Hays and Pitman [1973] in which an analysis of lithospheric plate motion related to sea-level changes and climatic and ecological consequences was made. Their analysis led to the conclusion that Cretaceous sea levels of up to 521 m above present sea level were the result of a pulse of rapid ridge-crest spreading between 110 and 85 my. Following this paper a series of analyses using refined time scales and plate motion models and hypsometric approaches [Larson and Pitman, 1975; Berggren et al., 1975; Pitman, 1978; Bond, 1978; Parsons, 1982; Harrison et al., 1981; Kominz, 1984; Kent and Gradstein, 1985] resulted in the reduction of earlier estimates of Cretaceous sea-level rises to the range of $\sim$100 to $\sim$300 m. It has been proposed by a number of workers, following the Hays and Pitman paper, that the major controls on long-term changes in sea level include temporal variations in the volume and length of the global mid-ocean ridge system [Hallam, 1977] and the thermal subsidence of the aging plates [Parsons and Sclater, 1977] modulated by other processes such as hot spot epierogeny [Heestand and Crough, 1981; Crough, 1983], mid-plate volcanism [Schlanger et al., 1981], sedimentation rates [Harrison et al., 1981], orogenic collisions, basin dessication, and of course glaciation [Donovan and Jones, 1979]. Cloetingh et al. [1985] and Cloetingh [1986] have raised the interesting point that apparent sealevel changes of up to 100 m can be produced at the edges of sedimentary basins through the interaction between variations in the horizontal stress fields in lithospheric plates and sediment loading. Major recent reviews include those by Hallam [1984], Steckler [1984], and Worsley et al. [1984]. Pitman and Golovchenko [1983] defined the magnitudes and rates of sea-level changes attributable to these several mechanisms as shown in Table 1. The application of seismic stratigraphic analysis to passive margin sequences developed by Vail et al. [1977] led to the development of detailed coastal onlap-offlap curves interpretable in terms of relative sea level. This development was an outgrowth of the sequence concept of Sloss [1963]. A further series of papers analyzed sea-level changes in terms of margin subsidence and sediment loading [Steckler and Watts, 1978; Watts and Steckler, 1979; Pitman, 1978; Guidish et al., 1984].

Was the Cretaceous World Ice-Free?

One of the major problems in analyses of Cretaceous sea-level history centers on the question of whether or not there was significant glaciation during the Cretaceous; in an ice-free Cretaceous world the most obvious mechanism for producing short term, high amplitude, sea-level changes would be denied us. For the purposes of this paper it is taken that there was no significant glaciation during Cretaceous time. This assertion is based on the lack of strong, direct geological evidence for major continental glaciation in the Cretaceous [Crowell, 1982] and the existence of strong evidence in the oxygen isotope record of high, 10-15°C, bottom water temperatures in the Cretaceous oceans [see Savin, 1982, for review]. The validity of the oxygen isotope record has recently been questioned by Killingsley [1983] who proposed that diagenesis could strongly bias the isotope record due to recrystallization of pelagic carbonates upon burial. However, the consistency of the Cretaceous isotope data set argues against the processes of diagenesis as a major factor in altering the isotope signal. Further, studies of the global distribution of faunas and floras argue for a warm and equable Cretaceous climate [Barron, 1984; Barron and Washington, 1984; Barron et al., 1984; Morris, 1985]. While high rates of sea-level change can be ascribed to the melting and growth of small mountain glaciers [Meier, 1984] it is doubtful if the total reservoir of these could have produced the amplitude of sea-level changes observed in Cretaceous strata.

Transgressive-Regressive Cycles in Albian through Maastrichtian Time: A Search for Periodicity

Since the middle 1970's a number of workers, based on several cratons, have attempted to document transgressive-regressive cycles in Cretaceous sequences, particularly those of Albian through Maastrichtian age, in a search for transgressive pulses that appear to be due to global, synchronous sea-level changes. Cooper [1977] argued that there were 13 transgressive episodes between Valanginian and early Maastrichtian time, 11 from late Aptian through Maastrichtian time (Figure 1), and proposed that if these transgressions were in fact linked to volumetric changes in the mid-oceanic ridge system then plate motion must have been highly episodic. Cooper's transgression count compares, considering the various time scales used, with the plot of Kauffman [1977] based on the Western Interior Basin of the U.S. where 10 transgressive-regressive cycles are shown between Valanginian through Maas-

TABLE 1. Causes, magnitudes, and rates of sea level change [from Pitman and Golovchenko, 1983]

Mechanism	Probable Maximum		Maximum Maximum		Time Interval
	Mag. m.	Rate cm/1000/yr	Mag. m.	Rate cm/1000/yr	(m.y.)
Glaciation	150	1000	250	1000	0.1
Ridge Volume	350	0.75	500	1.2	70
Orogeny	70	0.10	150	0.20	70
Sediment	60	0.11	85	0.25	70
Hot Spots	50	0.08	100	0.14	70
Flooding of Ocean Basins			Instantaneous		

trichtian time (Figure 2). Studies of the paleobathymetric history of western France by Juignet [1980] revealed transgressive-regressive cycles similar in form and frequency to those previously proposed (Figure 3).

Recognizing that local basin tectonics complicate the deciphering of global transgressive-regressive cycles, Matsumoto [1980] attempted a global stacking of paleobathymetric curves (Figure 4) and came to the conclusion that the periodicity in the records of transgression-regression is a feature which is clearly recognized in every area studied. But he added the observation that detecting a world-wide synchronous periodicity is difficult. He also related transgression-regression phenomena to movements in the interior of the earth. Hancock and Kauffman [1979] sifted through the various cycles and proposed that 5 major transgressive peaks seen in Cretaceous strata are so marked that they could be considered as globally synchronous. These transgressive peaks are: early Late Albian, Earliest Turonian, Coniacian, Middle Santonian, and Late Campanian-Early Maastrichtian. That there may actually be 6 peaks was recognized by Hancock and Kauffman [1979] in that separate Late Campanian and early or middle Maastrichtian peaks appear in the studies of Matsumoto [1980] and Hancock [1975]. Weimer's studies [1984] of the Cretaceous of the Western Interior of the U.S. are summarized on Figure 5. I propose that the Weimer curve can be reconciled with the curves of other workers by considering the following.

1) The Greenhorn cycle [G of Weimer, 1984] can be accepted as a strong tie point with other workers. As discussed below in detail, this cycle is the best documented of all the proposed transgression-regression pulses; it is the equivalent of the T6 cycle of Kauffman [1977] and is clear on the Hancock [1975] curve as well as Matsumoto's compilation [1980].

2) The Coniacian transgression is shown (un-named) by Weimer as correlative with the north Europe spike and is equivalent to the T7a-b transgressions of Kauffman [1977] and the Coniacian transgression of Hancock and Kauffman [1979].

3) The Niobrara (N) transgression of Weimer is seen in the northern Europe record, is equivalent to the T7c-l peak of Kauffman [1977] and the Santonian transgression of Hancock and Kauffman [1979].

4) The Clagget and Bearpaw transgression (Cl and Be of Figure 5) are controversial in that Hancock and Kauffman [1979] recognize only a Late Campanian-Early Maastrichtian single major transgression. However, these workers do recognize that both Matsumoto [1980] and Hancock [1975] argued for 2 peaks as shown by Weimer. I propose that these separate peaks are real global transgressive pulses.

5) At the older end of the scale is the problem of the definition of the resolution of the Albian T5 cycle of Kauffman [1977] and Hancock and Kauffman [1979]. Weimer shows Skull Creek (SC) and Mowry (M) transgressions separated by a Julesberg (J) regression. Hancock [1975] also shows a minor regression correlative with the Julesberg event. I propose that this minor regression is within the major T5 transgression of Kauffman [1977] and should be considered a single transgressive pulse as proposed by both Matsumoto [1980] and Hancock and Kauffman [1979].

If these interpretations are accepted a pattern of 6 major transgressions emerges shown as I through VI on Figure 5.

I. Late Albian; ~100-95 Ma
II. Cenomanian-Turonian; 95-90 Ma

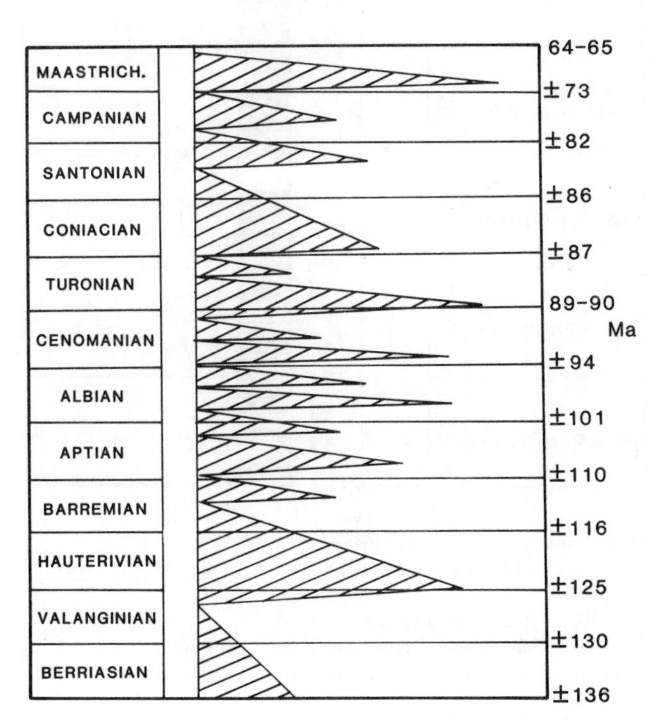

Fig. 1. Schematic representation of eustatic fluctuations in mean sea level during the Cretaceous. Oblique hatching designates transgressions [from Cooper, 1977].

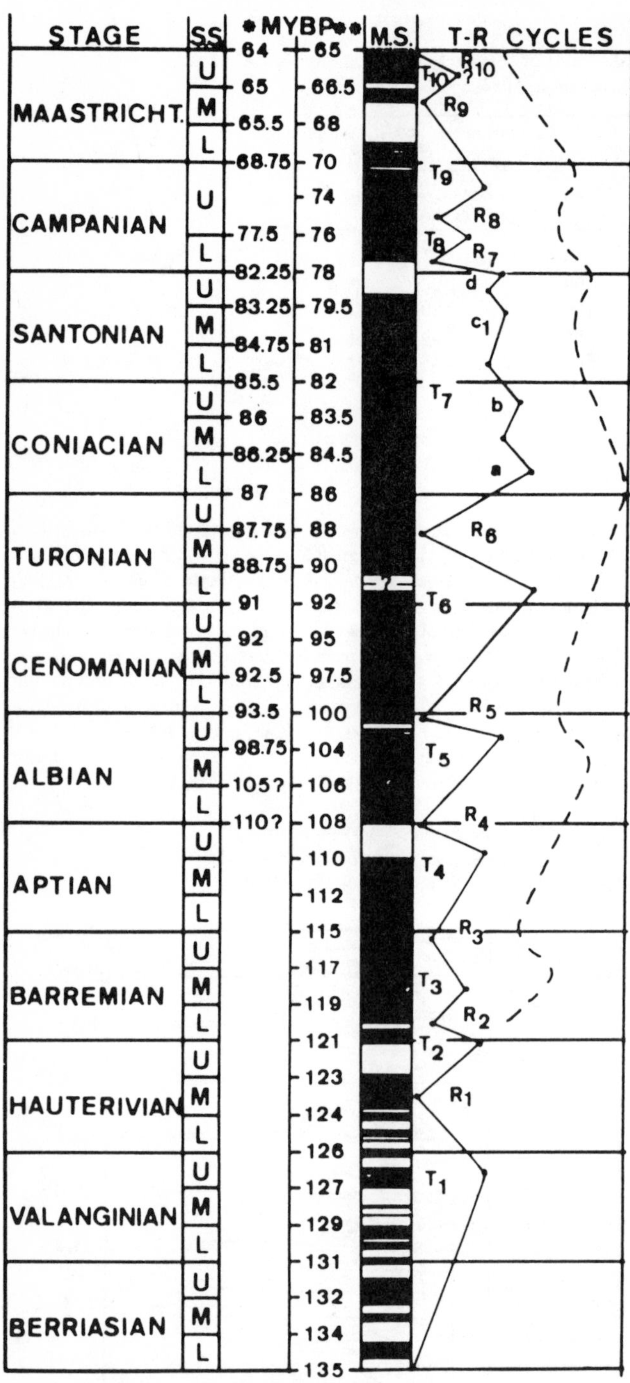

Fig. 2. Transgressive-regressive cycles in the Cretaceous of the U.S. Western Interior Basin [from Kauffman, 1977]; the time scales are from Obradovich and Cobban* [1975] and van Hinte** [1976].

III. Coniacian; 90-87 Ma
IV. Santonian-Early Campanian; 87-80 Ma
V. Late Campanian; 80-73 Ma
VI. Early Maastrichtian; 73-67 Ma

These six show trough-peak-trough time spans of 5, 5, 3, 7, 7, 6 Ma, respectively. Given the uncertainties of dating I propose that transgressive-regressive cycles with an average length of ~5.5±1 Ma characterize Cretaceous sections on a global basis and that these 6 cycles over-ride local tectonic and margin subsidence factors.

The Late Cenomanian-Early Turonian Transgression: Duration and Magnitude

Of the 6 Cretaceous transgressions, which are considered as true global sea-level rises, as evidenced from the stratigraphic record, the Cenomanian-Turonian transgression, Cycle II in Figure 5, is perhaps the best documented on a global basis in terms of evidence for its synchroneity on various continents, its duration and its amplitude. As pointed out by Reyment [1980] this short-lived transgression is so marked that H. Suess [1906] used this particular event in the development of the concept of eustasy. To the eye of the field stratigrapher the sharp onset of the transgression is obvious in European terranes as a cliff of white Turonian chalk that surmounts gentler slopes developed on flaggy-bedded glauconitic sands of the Cenomanian shelf sea; this morphological expression is the "alaise Turonnienne" of North Africa. As Juignet [1980] has clearly illustrated, the Cenomanian-Turonian contact in western France is marked by a rapid vertical lithologic transition from highly glauconitic sands and silts containing a shallow shelf fauna upward into more massive, slightly glauconitic chalks which in turn are overlain by pure open marine chalks containing bands of nodular black chert. Russian workers [Naidin et al., 1980] note an identical vertical succession across the Cenomanian-Turonian boundary in the Moldavian region where the latest Cenomanian and basal Turonian nannofossil and foraminiferal chalks contain beds of diatomaceous and radiolarian-rich sediments. The vertical transition from detrital, glauconitic shelf sands to open marine chalks certainly implies a sharply deepening sea.

The following section is an attempt to define the duration of the Cenomanian-Turonian transgressive cycle and its amplitude in order to establish boundary conditions necessary in a discussion of the efficacy of spreading rate and ridge-crest volumetric requirements in accounting for such a transgressive cycle. This transgression is the transgressive peak phase of Matsumoto [1980] shown on Figure 4, the T6 peak (Figure 2) of the Greenhorn cycle [Kauffman, 1977] in the Cretaceous North American western Interior Basin, oscillation 5 (Figure 3) of Juignet [1980], and Cycle II of this paper (Figure 5). According to Kauffman's [1977] depiction of the T6 cycle the transgressive pulse was relatively long lasting beginning in Early Cenomanian time and ending by the end of Middle Turonian time; using the van Hinte time scale [1976] the total pulse time would have been ~12 Ma (100 Ma to 88 Ma). However, according to the Obradovich and Cobban time scale [1975] the transgressive pulse took place between 94 and 88 Ma. Barron, Arthur and Kauffman [1986] show the T6 cycle as lasting from 92 to 88 Ma—a total of 4 Ma. Weimer [1984] brackets the entire T6 cycle between 95 and 90 Ma (Figure 5). Reyment [1980] correlates the T6 transgression with the Ezeaku cycle of Nigeria and points out that at the time of maximum transgression the basal Early Turonian epicontinental sea extended from the Tethys across the Sahara to the Atlantic in the Nigerian basin. The time-stratigraphic bounds of the Ezeaku transgressive pulse are rather well-defined as 92 to 87 Ma [on the time scales of van Hinte, 1976], i.e., latest Cenomanian through early Turonian. Even allowing for variations between time scales the correlation with the T6 cycle is clear and the length of the transgressive pulse is ~5 Ma. Other estimates of the duration of the transgressive pulse can be derived. In western France (Figure 3) the number 5 oscillation of Juignet [1980] took place as a complete

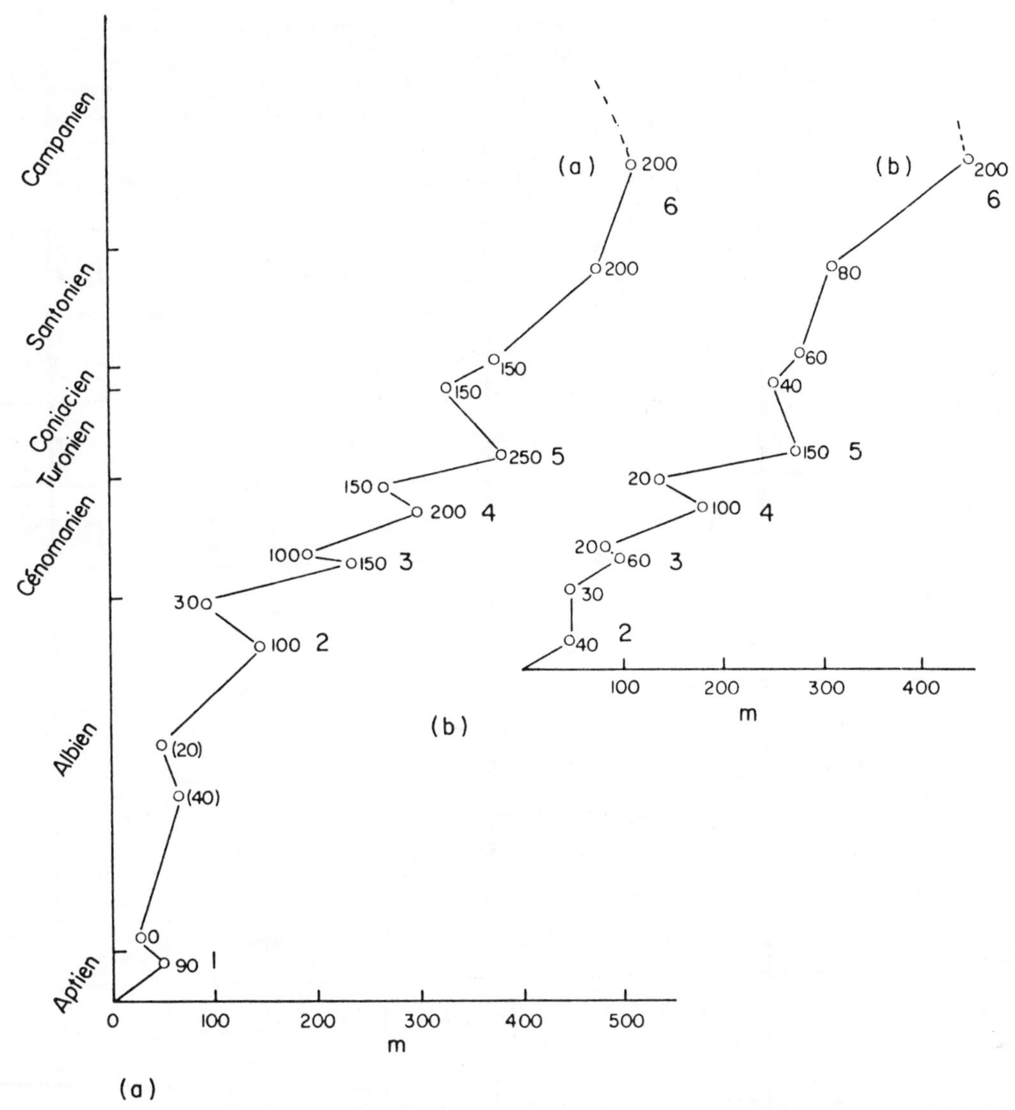

Fig. 3. Bathymetric oscillations related to Cretaceous stages and water depth in the Normandy Basin (a) and the Amoricain region (b) of France. Water depth scale in meters [from Juignet, 1980].

transgressive-regressive cycle from preceding low sea level in latest Cenomanian time to succeeding low sea level in Late Turonian time in a period of only 2.5 Ma (using the Harland et al. [1982] time scale of 88.5 and 91 Ma as the bounds of the Turonian stage). The duration of the transgressive pulse is also seen in the paleobathymetric curves of Naidin et al. [1980] for the East European sea and Central Asian area where the entire transgressive cycle took place in the Turonian stage. On the Russian Platform the pulse is broader and less well-defined (with rare local shallowing rather than deepening of the sea) and extends from Late Cenomanian through Middle Turonian time. The slope of the Russian sea-level curve resembles that of the Greenhorn cycle and the trans-Sahara Ezeaku cycle with a time span of ~95-90 Ma. On the east coast of Brazil this transgression is limited to Early Turonian time [Matsumoto, 1980]. The transgression shown by Naidin et al. [1980] began in Late Cenomanian time, peaked in the middle or late Early Turonian, and ended by Late Turonian time. Here again the cycle lasted a maximum of 5 Ma. Thus on many major continental terranes (the record for Australia and India not withstanding, as seen on Figure 4) this clearly defined transgressive pulse lies within a time envelope of at most 5 Ma. If the time span of the total transgressive pulse, which peaked in earliest Turonian time, from the preceding Cenomanian low to the succeeding Middle or Late Turonian low, is accepted to be ~5 Ma then the amplitude of the transgression must be fixed so that the rate of sea-level rise and fall can be determined.

The present intensification of interest in this amplitude problem can be taken as beginning with the paper of Hays and Pitman [1973] in which the relationship between sea level and ridge-crest volumes and spreading rates was analytically attacked. Their analyses revealed a long term rise in sea level between 110 Ma and 85 Ma of from +200 to +521 meters above present sea level; by 90 Ma sea level was proposed to be at +462 m. The stratigraphic resolving power of the method is insufficient to decipher short-term changes in sea level. More recently, Pitman

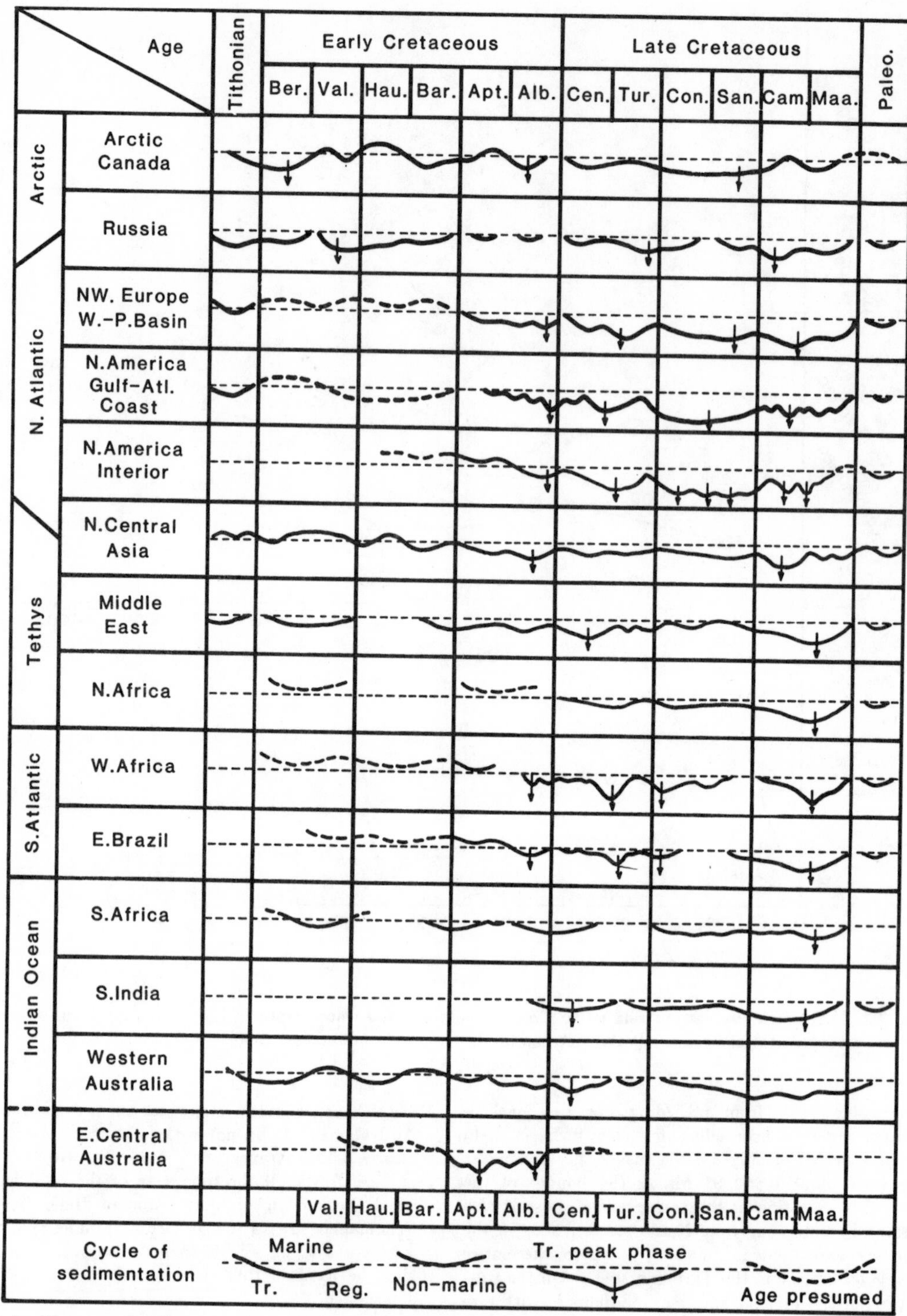

Fig. 4. Diagram showing records of transgressions and regressions during Cretaceous time on major platforms [from Matsumoto, 1980].

[1978] revised the +521 m level down to ~+350 m. Watts and Steckler [1979], based on passive margin subsidence modeling, revised the maximum Cretaceous sea-level peak downward to 150 m (Figure 6). Since then, Kominz [1984] has further revised the 85 Ma sea-level estimate, upon review of the uncertainties and various time scales used, to between a minimum of 45 m, a maximum of 365 m and a most probable height of 230 m (Figure 7). As more studies are done the amplitude of the long term Cretaceous sea-level rise apparently decreases.

With reference to the amplitude of the Cenomanian-Turonian

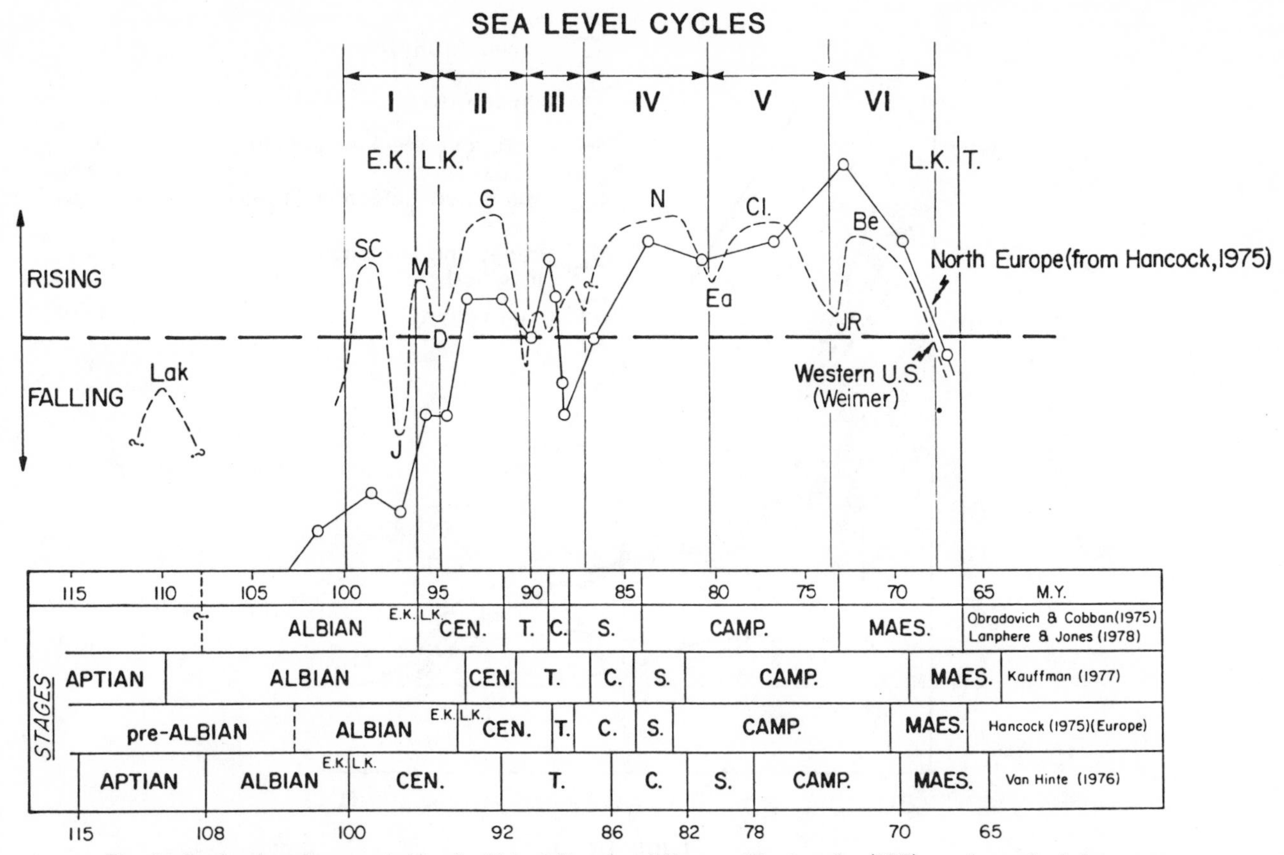

Fig. 5. Sea-level cycles recorded in the United States and Europe. The 6 cycles (I-VI) are the author's interpretation [modified from Weimer 1984].

transgressive pulse evidence has been presented by several authors based on work in various basins. The absolute minimum sea-level rise can be derived from the studies of Reyment [1980] on the Saharan epicontinental transgressions. At the time of the union of the Tethyan and Nigerian arms of the trans-Saharan sea in Early Turonian the water depth in northwest Nigeria, where topographic highs were submerged, was 20-30 m. Considering the stability and elevation of Africa the sea-level rise (correlated with

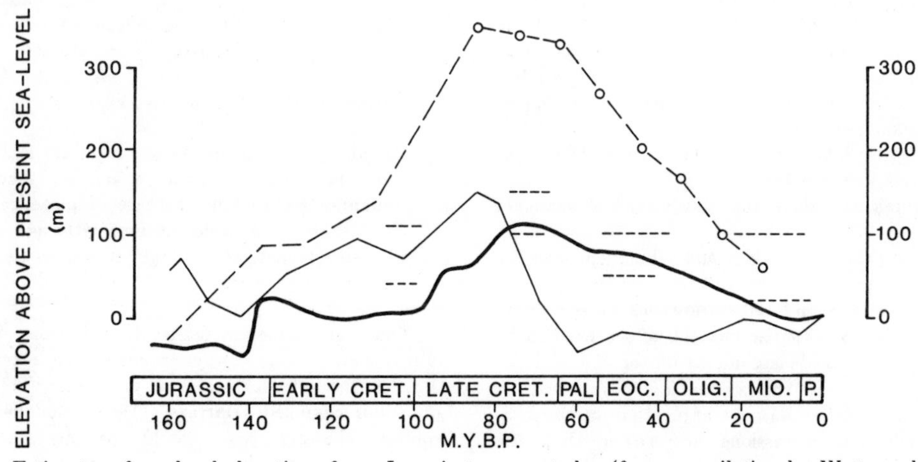

Fig. 6. Estimates of sea-level elevations from Jurassic to present day (from compilation by Watts and Steckler, [1979]). The solid line is the estimate of Watts and Steckler [1979], the dashed line [from Vail et al., 1977] is marked with circles from Pitman [1978]. The fine solid line is from Wise [1974]. The horizontal dashed lines are from Bond [1978].

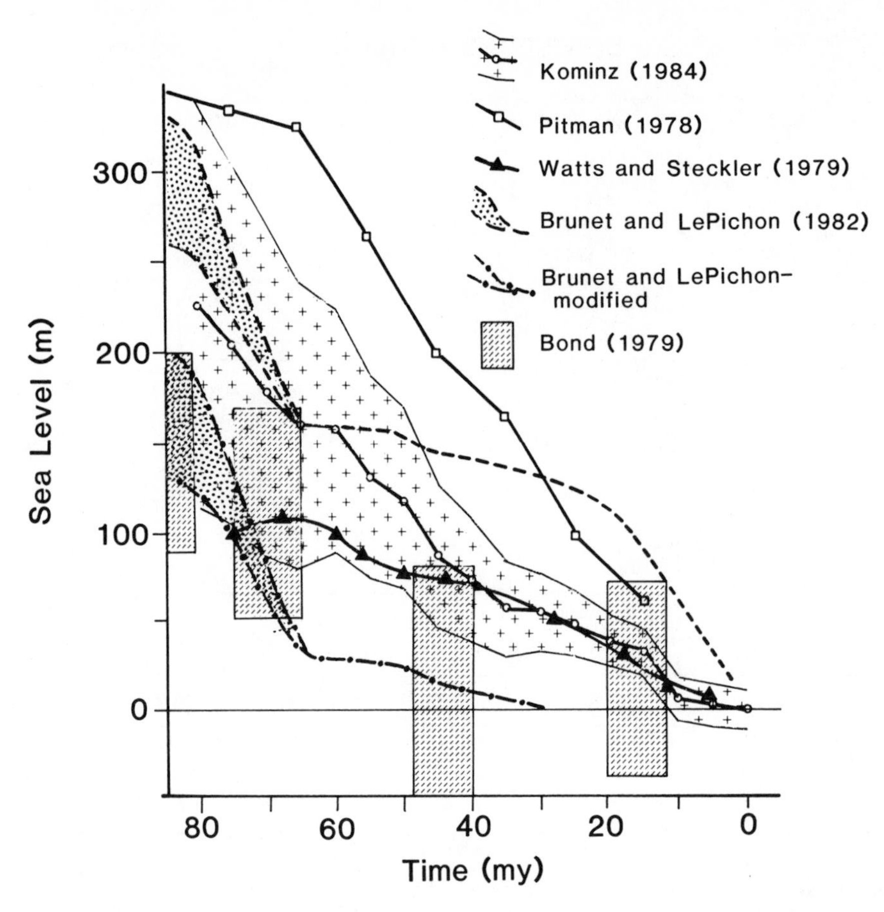

Fig. 7. Sea level estimates for the period 85 Ma to the present [from Kominz, 1984]. The solid line marked by circles shows the most probable height of sea level at 230 m above present sealevel at 80 Ma. Other estimates are from the authors cited by Kominz [1984].

the T6 cycle in the U.S.) could not have been less than ~150 meters. Juignet's [1980] paleobathymetric curves of this pulse, (oscillation number 5 on Figure 3) show that the maximum amplitude of this transgression in western France was ~150 meters. Hancock and Kauffman [1979] show the amplitude of the T6 at ~200 m; a later discussion of the Greenhorn cycle transgressive pulse by Barron et al. [1986] shows a rise and fall during the T6 cycle of from 150 meters water depth at 92 Ma to ~400 m at the Cenomanian-Turonian boundary 90 Ma with a fall back to 150 meters depth 88 Ma—a bathymetric range of 250 m. Given the uncertainties in these estimates and the general tendency of recent workers to reduce the amplitudes of sea-level fluctuation estimates it is nonetheless reasonable to accept an amplitude of ~200 meters above previous and subsequent lows for the Cenomanian-Turonian sea-level rise and transgressive pulse.

The rates of sea-level rise and fall during this pulse can be approximated by considering both rise and fall of sea level to be linear. This assumption sets aside arguments on the shape of the sea-level curve; the Vail curve shows slow onlap followed by instantaneous offlap while other authors argue that in the U.S. western Interior Basin the regressions are slower than the transgressions. With a cycle length of 5 Ma and a half width of 2.5 Ma we see that a 200 m rise of sea level above the preceding low implies a rise and a similar fall rate of 80 m/my. Hancock and Kauffman [1979] show a rise rate of 27 m/my leading up to the T6 cycle peak followed by a fall rate of 97 m/my. Rise and fall rates of from 50 to 100 m/my appear reasonable for the T6 transgression-regression cycle.

Can Sea Floor Processes Account for High-Frequency Sea Level Fluctuations?

Parsons [1984, p. 289] pointed out that

> If long-term changes in sea level are less than 150 m, as suggested by recent estimates, then only small changes (~12%) in the rate of plate generation and distribution of consumption with age are required. Sea-level changes as large as 300 m require about a 50% change in the rate of crustal generation.

The high-frequency sea-level oscillations discussed here certainly then would require previously undetected major fluctuations in the rates of sea floor processes. This problem has not gone unrecognized. Harrison [1985] modelled the ridge crest volume changes that would be necessary to cause the transgressive-regressive cycles seen in Cretaceous strata of the U.S. Western Interior Seaway and came to the conclusion that "impossibly large changes in sea-floor spreading would have to be postulated to obtain eustatic signals of this magnitude."

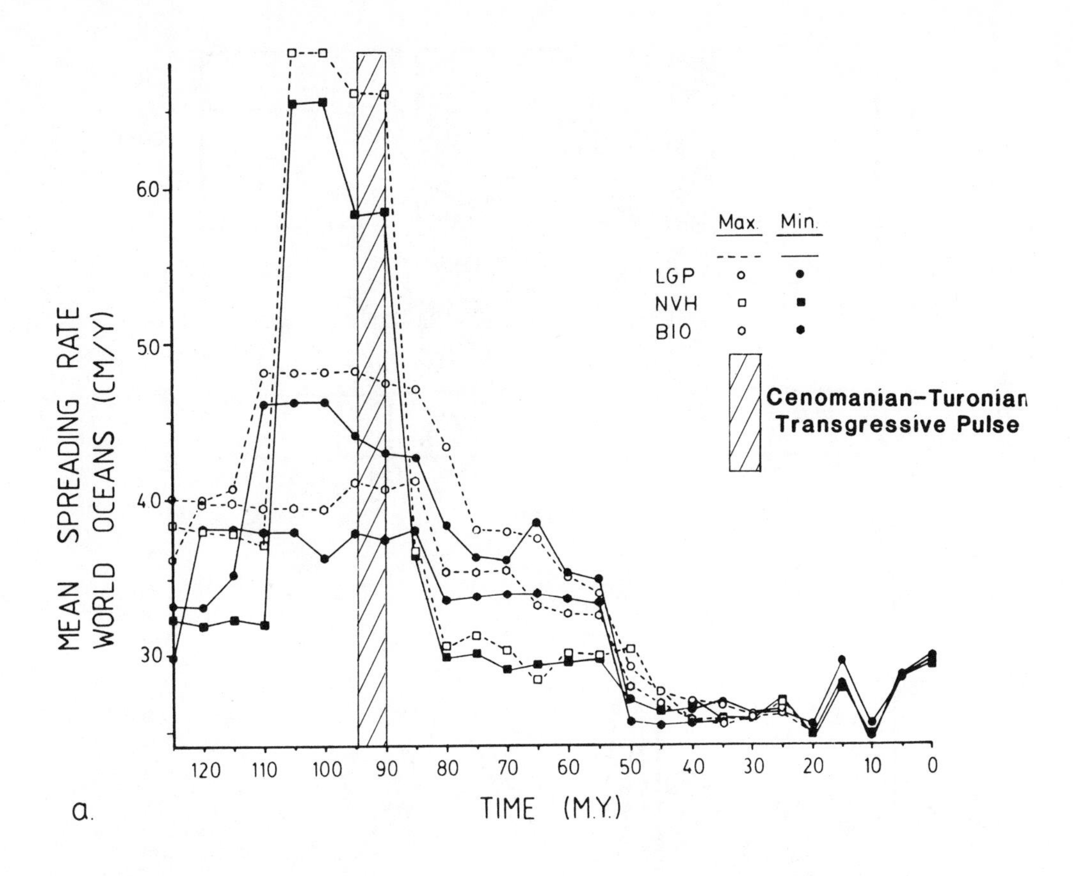

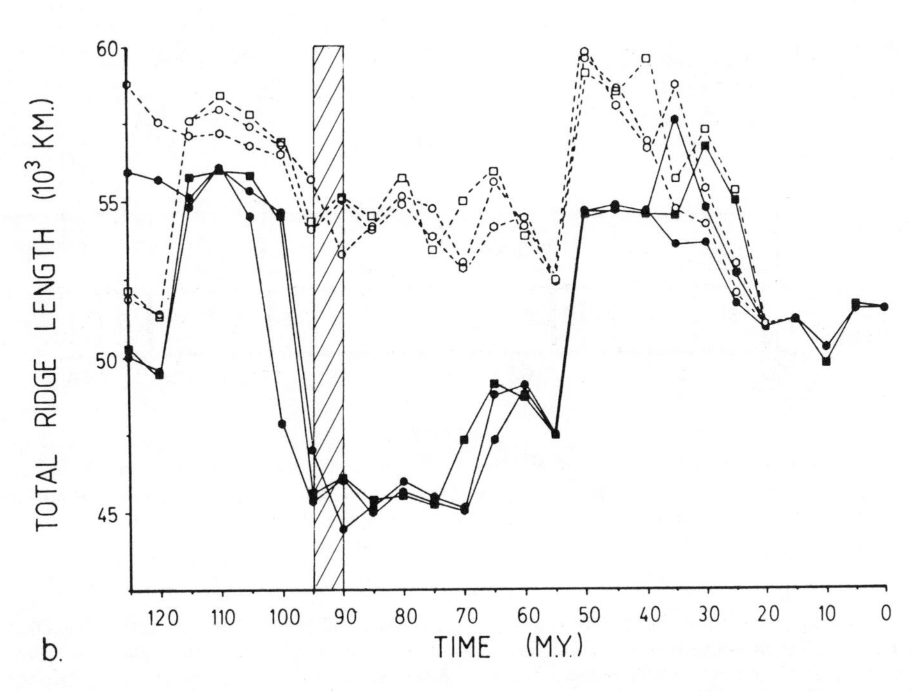

Fig. 8. Mean spreading rates and total ridge lengths for the period 120 Ma to the present [from Kominz, 1984]. The 6 curves represent maximum and minimum estimates based on 3 time scales: LGP [Larson, Golovchenko, and Pitman, 1982]; NVH [Ness, Levi, and Couch, 1980]; BIO [a combined time scale, see Kominz, 1984]. The shaded bar at 90-95 Ma shows the timing of the Cenomanian-Turonian transgressive cycles.

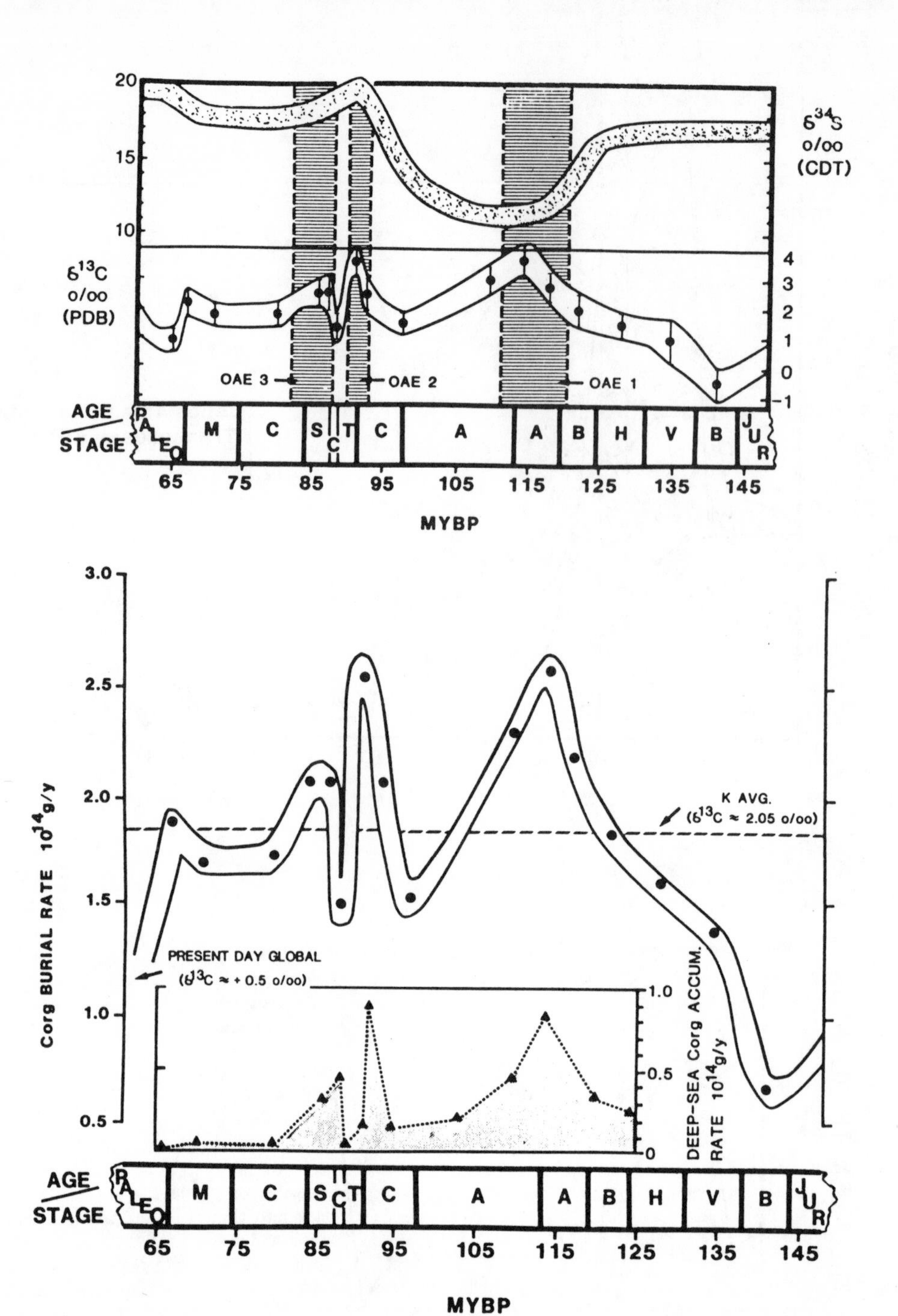

Fig. 9. The upper diagram shows variations in $\delta^{13}C$ and $\delta^{34}S$ values in pelagic limestones through Cretaceous time correlated with oceanic anoxic events. The lower diagram shows organic carbon burial rates calculated from the $\delta^{13}C$ values. The inset shows the burial rate of organic carbon determined from DSDP data, Legs 1-48 [from Arthur et al., 1985].

In this paper I wish to emphasize that the 6 Cretaceous transgressions, particularly the T6 transgression, discussed are synchronous and that the magnitude and rates of these transgressions are not compatible with known sea floor process rates. The "maximum-maximum" rate of sea-level change, as shown above in the tabulation by Pitman and Golovchenko [1983], ascribable to any mechanism except glaciation, is 12 m/my, at least 1/4 that shown above for the T6 cycle. The T6 cycle is taken as the example and examined in the light of Kominz's [1984] analysis of ridge length and spreading-rate data. Kominz [1984] published an intensive error analysis of oceanic-ridge volumes and ridge lengths in terms of placing bounds on sea-level changes (Figure 8). Kominz showed that the use of different time scales results in widely varying estimates of both ridge volume and length. While

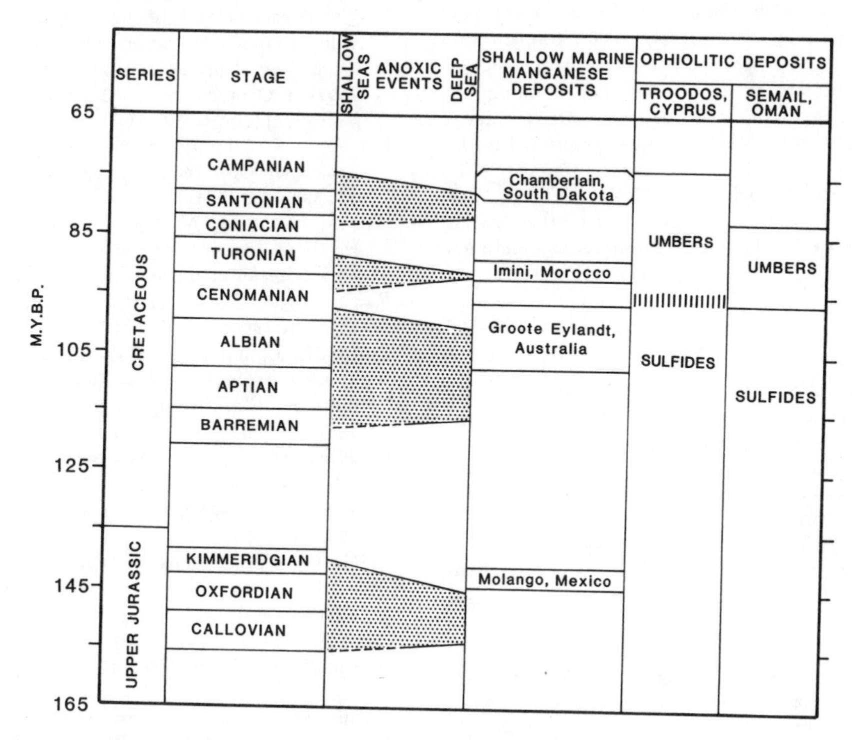

Fig. 10. Correlation of oceanic anoxic events, shallow marine manganese deposits and the sulfide-umber associations of ophiolites [from Force et al., 1983].

recognizing that the time averaging used and thus the determination of points on the curves tends to flatten trends it is nevertheless clear that the period in which the transgressive cycles took place (Figure 8), from ~100 to ~67 Ma, was one of decreasing spreading rate and ridge length. The 12% to 50% positive changes in the rate of crustal generation indicated by Parsons [1982] as necessary to account for sea-level changes of from 150 to 300 m are not seen. In particular, the Cenomanian-Turonian, or T6 cycle, which took place between 95 and 90 Ma, is not at all in synchronization with any ridge length or spreading rate increase. Of the 3 sets of curves shown by Kominz [1984] 2 sets show decreasing spreading rates while the set based on the B10 time scale is essentially flat during the 100 to 85 Ma period. All of the total ridge length curves show a decrease during the 100 to 85 Ma period.

Economic Implications of Transgressions

Concomitant with the explosive growth of interest, research and publication on the timing and magnitude of sea-level fluctuations and their relationship to geophysical and paleoceanographic factors has come the realization that both hydrocarbon accumulations and sedimentary metallic mineral deposits are genetically linked to transgressive events. [Donovan and Jones, 1979; Hallam and Bradshaw, 1979; Barron and Washington, 1984; Brass et al., 1981; Schlanger and Jenkyns, 1976; Arthur and Schlanger, 1979; Jenkyns, 1980; Arthur et al., 1985; Schlanger et al., 1986; Arthur et al., 1986; Fischer and Arthur, 1977].

Irving et al. [1974], impressed by statistics indicating that "...72% of all known oil was probably formed in the late Mesozoic, most of it (60%) in the mid-Cretaceous," related this stratigraphic clustering of oil-rich strata to the repeated flooding of cratonic embayments. The resulting warm, shallow seas were proposed to be the sites of increased organic HC-source production which led to the deposition of prolific Mesozoic source beds. Later studies of the stratigraphic distribution of source beds [see North, 1980; Tissot, 1979; Schlanger and Cita, 1982] reinforced the Irving et al. argument that mid-Cretaceous time was a period of enhanced carbon burial. The correlation of organic carbon-rich black shales penetrated during drilling operations of the DSDP-IPOD programs with sections exposed on land led to the development of the concept of Oceanic Anoxic Events (OAE's) by Schlanger and Jenkyns [1976]. They argued that during transgression globally expanded O_2-minimum zones were widespread in the world ocean and that sediments deposited in these O_2-depleted water masses sequestered much of the carbon that in normally oxygenated oceanic basins would have been recycled back into the oceanic reservoir. Schlanger and Jenkyns [1976], Arthur and Schlanger [1979] and Jenkyns [1980] defined distinct OAE's of Aptian-Albian, Cenomanian-Turonian and Coniacian-Santonian age [see also Ryan and Cita, 1977; Fischer and Arthur, 1977; Schlanger et al., 1986; Arthur et al., 1986]. Scholle and Arthur [1980] discovered that skeletal carbonate deposited during these OAE's was enriched in C^{13} as the result of the sequestering of C^{12} in the buried organic carbon. Thus, carbon burial rates during OAE's could be calculated (Figure 9). Hallam and Bradshaw [1979] noted the high degree of correlation between transgressive pulses in the Jurassic and the deposition of organic carbon-rich beds; the best known of these carbonaceous strata being the Jet Rock of England and the Toarcian black shales of the Paris Basin—the latter being the source beds for oil in that basin. Other examples of transgression-linked carbonaceous beds include those described by Heckel [1977] in Pennsylvanian cyclothems of the U.S. mid-continent and the Devonian black shales of the eastern U.S. and Europe [Roen, 1984].

The genetic link between sedimentary manganese deposits and OAE's has been explained by Force et al. [1983], Cannon and Force [1983], and Frakes and Bolton [1984]. These workers argue that the O_2-depleted waters in the widespread O_2-minimum layers present during an OAE serve as reservoirs for large amounts of dissolved manganese which is then precipitated in the oxygenated waters above the O_2-minimum zone forming "bathtub ring" manganese deposits such as the Groote Eylandt, Australia deposits, or in the case of manganese and other metals which issued from mid-ocean ridge crests within an O_2-minimum layer ophiolitic sulfide deposits of the Cyprus type (Figure 10).

Conclusions

Taking into consideration problems of intercontinental and local correlation and the absolute dating of particular strata the post-Aptian Cretaceous records 6 major transgressive-regressive cycles with an average trough-peak-trough period of $\sim$5.5$\pm$1 Ma. Sea-level change amplitudes of 150-250 meters imply rates of sea-level rise or fall between 50 and 100 m/my. That such periodicity is not necessarily confined to Cretaceous time is evidenced by similar cycles in Jurassic time [Hallam and Bradshaw, 1979; Hallam, 1981]. Further the Cretaceous and Jurassic transgressive-regressive cycles fall within the Zuni episode, one of 4 "submergent modes" of cratonic behavior in Phanerozoic time described by Sloss and Speed [1974] during which the worlds cratons are subjected to gradual submergence punctuated by regressions. In such an "submergent mode" relative sea level changes of 100-200 m/Ma take place involving transgressive pulses of a few millions of years to ten million years on globally distributed cratons.

The wave lengths and amplitudes of these sea-level oscillations are not ascribable to documented rates of sea-floor processes such as changes in ridge-crest length or ridge-crest volume. Continent-continent collisions, mid-plate volcanism, dessication of local basins, and variations in oceanic sedimentation rates also fail to explain the apparent geological record of cratonic behavior. The lack of evidence for major glacial control during the Cretaceous deprives us of the only known mechanism for producing such major short-term periodicity.

The challenge to tectonicians and stratigraphers is clear. In addition to the intellectual challenge we need to further explore the important economic implications of the apparent linkage between transgressions and periods of organic carbon burial and the formation of several types of mineral deposits.

Acknowledgements. I wish particularly to thank L. L. Sloss who has worked so effectively for so many years to convince us that sea-level changes are linked to global tectonic events. This work was supported by Grant no. EAR-8207387 from the U.S. National Science Foundation to Northwestern University.

References

Arthur, M. A., S. O. Schlanger and H. C. Jenkyns, The Cenomanian-Turonian oceanic anoxic event, II. Paleoceanographic controls on organic matter production and preservation, in *Marine Petroleum Source Rocks*, edited by J. Brooks and A. Fleet, Geological Society of London, in press, 1986.

Arthur, M. A., and S. O. Schlanger, Cretaceous "oceanic anoxic events" as causal factors in development of reef-reservoired giant oil fields, *Bull. Am. Assoc. Pet. Geol., 63*, 870-885, 1979.

Arthur, M. A., W. E. Dean, and S. O. Schlanger, Variations in the global carbon cycle during the Cretaceous related to climate, volcanism, and changes in atmospheric CO_2, in *The Carbon Cycle and Atmospheric CO_2: Natural Variations Archean to Present*, Geophys. Mon. 32, edited by E. T. Sunquist and W. S. Broeker, pp. 504-530, AGU, Washington, D.C., 1985.

Barron, E. J., Climatic implications of the variable obliquity explanation of Cretaceous-Paleogene high-latitude floras, *Geology, 12*, 595-598, 1984.

Barron, E. J., M. A. Arthur, and E. G. Kauffman, Cretaceous rhythmic bedding sequences: A plausible link between orbital variations and climate, *Earth Planet. Sci. Lett.*, in press, 1986.

Barron, E. J., W. W. Hay, and E. G. Kauffman, Cretaceous climates, *Geology, 12*, 377, 1984.

Barron, E. J., and W. M. Washington, The role of geographic variables in explaining paleoclimates: Results from Cretaceous climate model sensitivity studies, *J. Geophys. Res., 89*, 1267-1279, 1984.

Berggren, W. A., D. P. McKenzie, J. G. Sclater, and J. E. van Hinte, World-wide correlation of Mesozoic magnetic anomalies and its implications: Discussion, *Geol. Soc. Am. Bull., 86*, 267-269, 1975.

Bond, G., Speculations on real sea-level changes and vertical motions of continents at selected times in the Cretaceous and Tertiary periods, *Geology, 6*, 247-250, 1978.

Bond, G., Evidence for some uplifts of large magnitude in continental platforms, *Tectonophysics, 61*, 285-305, 1979.

Brass, G., E. Saltzman, J. Sloan, J. Southam, A. Hay, W. Holzer and W. Peterson, Ocean circulation, plate tectonics and climate, in *Climate in Earth History, Studies in Geophysics*, pp. 83-89, National Academy Press, Washington D. C., 1981.

Brunet, M. F., and X. LePichon, Subsidence of the Paris Basin, *J. Geophys. Res., 87*, 8547-8560, 1982.

Cannon, W. F., and E. R. Force, Potential for high-grade shallow-marine manganese deposits in North America, in *Unconventional Mineral Deposits*, edited by W. C. Shanks III, pp. 175-200, Soc. Econ. Geol., Am. Inst. Mining, Met., and Pet. Eng., New York, 1983.

Clark, J. A., A numerical model of worldwide sea-level changes in a viscoelastic earth, in *Earth Rheology, Isostasy and Eustasy*, edited by N.-A. Morner, pp. 525-553, Wiley, New York, 1980.

Cloetingh, S., Intraplate stresses: A new tectonic mechanism for relative sea-level fluctuations, *Geology*, in press, 1986.

Cloetingh, S., H. McQueen, and K. Lambeck, On a tectonic mechanism for regional sea-level variations, *Earth Planet. Sci. Lett., 75*, 157-166, 1985.

Cooper, M. R., Eustasy during the Cretaceous: Its implications and importance, *Palaeogeogr., Palaeoclimatol., Palaeoecol., 22*, 1-60, 1977.

Crough, S. T., Hotspot swells, *Annu. Rev. Earth Planet. Sci., 11*, 165-193, 1983.

Crowell, J. C., Continental glaciation through geologic time, in *Climate and Earth History*, pp. 77-82, National Academy Press, Washington, D.C., 1982.

Donovan, D. T., and J. W. Jones, Causes of world-wide changes in sea level, *J. Geol. Soc. London, 136*, 187-192, 1979.

Fischer, A. G., and M. A. Arthur, Secular variations in the pelagic realm, in *Deep Water Carbonate Environments*, Soc. Econ. Paleontol. Mineral. Spec. Pub. 25, edited by H. E. Cook and P. Enos, pp. 19-50, Society of Economic Paleontologists and Mineralogists, Tulsa, Okla., 1977.

Force, E. R., W. F. Cannon, R. A. Koski, K. T. Passmore, and B. R. Doe, Influences of oceanic anoxic events on manganese deposition and ophiolite-hosted sulfide preservation, in *Paleoclimate and Mineral Deposits*, U.S. Geol. Surv. Circ. 822, edited by R.

M. Cronin, W. F. Cannon, and R. Z. Poore, pp. 26-29, U.S. Geological Survey, Washington, D.C., 1983.

Frakes, L. A., and B. R. Bolton, Origin of manganese giants: Sea-level change and anoxic-oxic history, *Geology, 12*, 83-86, 1984.

Guidish, T. M., I. Lerche, C. G. St. C. Kendall and J. J. O'Brien, Relationship between eustatic sea level changes and basement subsidence, *Am. Assoc. Pet. Geol. Bull., 68*, 164-177, 1984.

Hallam, A., Secular changes in marine inundation of USSR and North America through the Phanerozoic, *Nature, 269*, 769-772, 1977.

Hallam, A., A revised sea-level curve for the early Jurassic, *J. Geol. Soc. London, 138*, 735-743, 1981.

Hallam, A., Pre-Quaternary sea-level changes, *Annu. Rev. Earth Planet. Sci., 12*, 205-243, 1984.

Hallam, A., and M. J. Bradshaw, Bituminous shales and oolitic ironstones as indicators of transgressions and regressions, *J. Geol. Soc. London, 136*, 157-164, 1979.

Hancock, J.M., The sequence of facies in the Upper Cretaceous of northern Europe compared with that in the Western Interior, in *The Cretaceous System in the Western Interior of North America*, Geol. Soc. Canada Spec. Paper 13, edited by W.G.E. Caldwell, pp. 83-119, Geological Association of Canada, Waterloo, Ontario, 1975.

Hancock, J. M., and E. G. Kauffman, The great transgressions of the Late Cretaceous, *J. Geol. Soc. London, 136*, 175-186, 1979.

Harland, W. B., et al., *A Geologic Time Scale*, Cambridge Univ. Press, England, 1 p., 1982.

Harrison, C. G. A., Modelling fluctuations in water depth during the Cretaceous, in *Fine-Grained Deposits and Biofacies of the Cretaceous Western Interior Seaway: Evidence of Cyclic Sedimentary Processes*, Field Trip Gdbk. No. 4, edited by L. M. Pratt, E. G. Kauffman, and F. B. Zelt, pp. 11-16, Society of Economic Paleontologists and Mineralogists, Tulsa, Okla., 1985.

Harrison, C. G. A., G. W. Brass, E. Saltzman, J. Sloan II, J. Southam, and J. M. Whitman, Sea level variations, global sedimentation rates and the hypsographic curve, *Earth Planet. Sci. Lett., 54*, 1-16, 1981.

Hays, J. D., and W. C. Pitman III, Lithospheric plate motion, sea level changes and climatic and ecological consequences, *Nature, 246*, 18-22, 1973.

Heckel, P. H., Origin of phosphatic black shale facies in Pennsylvanian cyclothems of mid-continent North America, *Am. Assoc. Pet. Geol. Bull., 61*, 1045-1068, 1977.

Heestand, R. L., and S. T. Crough, The effect of hot spots on the oceanic age-depth relation, *J. Geophys. Res., 86*, 6107-6114, 1981.

Irving, E., F. K. North, and R. Couillard, Oil, climate, and tectonics, *Can. J. Earth Sci., 11*, 1-17, 1974.

Jenkyns, H. C., Cretaceous anoxic events: From continents to oceans, *J. Geol. Soc. London, 137*, 171-188, 1980.

Juignet, P., Transgressions-regressions, variations eustatiques et influences tectoniques de l'Aptien au Maastrichtian dans le bassin de Paris Occidental et sur la bordure au Massif Amoricain, *Cretaceous Res., 1*, 341-357, 1980.

Kauffman, E. G., Geological and biological overview, Western Interior Cretaceous Basin, in *Cretaceous Facies, Faunas, and Paleoenvironments Across the Western Interior Basin*, edited by E. G. Kauffman, pp. 75-100, Rocky Mt. Assoc. Geol., 14, 1977.

Kent, D. V. and F. M. Gradstein, A Cretaceous and Jurassic geochronology, *Geol. Soc. Am. Bull., 96*, 1419-1427, 1985.

Killingsley, J. S., Effects of diagenetic recrystallization on 18O/16O values of deep-sea sediments, *Nature, 301*, 594-597, 1983.

Kominz, M. A., Oceanic ridge volumes and sea-level change—an error analysis, in *Interregional Unconformities and Hydrocarbon Accumulation*, Am. Assoc. Pet. Geol. Mem. 36, edited by J. S. Schlee, pp. 109-127, American Association of Petroleum Geologists, Tulsa, Okla., New York, 1984.

Larson, R. L., and W. C. Pitman III, World-wide correlation of Mesozoic magnetic anomalies, and its implications, *Geol. Soc. Am. Bull., 83*, 3645-3662, 1972.

Larson, R. L., X. Golovchenko, and W. C. Pitman III, Geomagnetic polarity time scale, Tectonic Map Circum-Pacific Region, Pacific Basin Sheet, *Am. Assoc. Pet. Geol. Plate*, 1982.

Lyell, C., *Principles of Geology*, 11th ed., vol. 1, Appleton and Company, New York, 671 p., 1873.

Matsumoto, T., Inter-regional correlation of transgressions and regressions in the Cretaceous period, *Cretaceous Res., 1*, 359-373, 1980.

Meier, M. F., Contribution of small glaciers to global sea level, *Science, 226*, 1418-1420, 1984.

Morris, S. C., Polar forests of the past, *Nature, 313*, 739, 1985.

Naidin, D. P., I. G. Sasonova, Z. N. Pojarkova, M. R. Djalilov, G. N. Papulov, Yu. Senkovsky, V. N. Benjamavsky, and L. F. Kopaevich, Cretaceous transgressions on the Russian Platform, in Crimea and Central Asia, *Cretaceous Res., 1,*, 373-387, 1980.

Ness, G., S. Levi, R. Couch, Marine magnetic anomaly timescales for the Cenozoic and Late Cretaceous; a precise critique, and synthesis, *Rev. Geophys. Space Phys., 18*, 753-770, 1980.

North, F. K., Episodes of source-sediment deposition: The episodes in individual close-up, *J. Petrol. Geology, 2-3*, 323-338, 1980.

Obradovich, J. D., and W. A. Cobban, A time scale for the Late Cretaceous of the western interior of North America, in *The Cretaceous System in the Western Interior of North America*, Geol. Assoc. Canada Spec. Paper 13,edited by W.G.E. Caldwell, pp. 31-54, Geological Association of Canada, Waterloo, Ontario, 1975.

Parsons, B., Causes and consequences of the relation between area and age of the ocean floor, *J. Geophys. Res., 87*, 289-301, 1982.

Parsons, B., and J. G. Sclater, An analysis of the variation of ocean floor bathymetry and heat flow with age, *J. Geophys. Res., 82*, 803-827, 1977.

Pitman, W. C., III, Relationship between eustasy and stratigraphic sequences of passive margins, *Geol. Soc. Am. Bull., 89*, 1389-1403, 1978.

Pitman, W. C., III and X. Golovchenko, The effect of sealevel changes on the shelf edge and slope of passive margins, *Soc. Econ. Paleontol. Mineral. Spec. Pub. 33*, 41-58, 1983.

Reyment, R. A., Biogeography of the Saharan Cretaceous and Paleocene epicontinental transgressions, *Cretaceous Res., 1*, 299-327, 1980.

Roen, J. B., Geology of the Devonian black shales of the Appalachian Basin, *Org. Geochem., 5*, 241-251, 1984.

Ryan, W. B. F., and M. B. Cita, Ignorance concerning episodes of ocean-wide stagnation, *Marine Geol., 23*, 197-215, 1977.

Savin, S., Stable isotopes in climatic reconstructions, in *Climate and Earth History*, pp.164-171, National Academy Press, Washington, D. C., 1982.

Schlanger, S. O., H. C. Jenkyns and I. Premoli-Silva, Volcanism and vertical tectonics in Pacific Basin related to global Cretaceous transgressions, *Earth Planet. Sci. Lett., 52*, 435-449, 1981.

Schlanger, S. O., M. A. Arthur, H. C. Jenkyns and P. A. Scholle, The Cenomanian-Turonian oceanic anoxic event, I. Stratigraphy and Distribution of organic carbon-rich beds and the marine δ13 excursion, in *Marine Petroleum Source Rocks*, edited by J. Brooks and A. Fleet, Geological Society of London, in press, 1986.

Schlanger, S. O., and H. C. Jenkyns, Cretaceous oceanic anoxic events: Causes and consequences, *Geol. en Minj., 55,* 179-184, 1976.

Schlanger, S. O., and M. B. Cita, Introduction, in *Nature and Origin of Cretaceous Carbon-Rich Facies,* edited by S. O. Schlanger and M. B. Cita, pp. 1-6, Academic Press, New York, 1982.

Scholle, P. A., and M. A. Arthur, Carbon isotope fluctuations in Cretaceous pelagic limestones: Potential stratigraphic and petroleum exploration tool, *Am. Assoc. Pet. Geol. Bull., 64,* 67-87, 1980.

Sloss, L. L., Stratigraphic models in exploration, *Am. Assoc. Pet. Geol. Bull., 46,,* 1050-1057, 1962.

Sloss, L. L., Sequences in the cratonic interior of North America, *Geol. Soc. Amer. Bull., 74,* 93-114, 1963.

Sloss, L. L., and R. C. Speed, Relationships of cratonic and continental-margin tectonic episodes, in *Tectonics and Sedimentation,* Soc. Econ. Paleontol. Mineral. Spec. Pub. 22, edited by W. R. Dickinson, pp. 98-119, Society of Economic Paleontologists and Mineralogists, Tulsa, Okla., 1974.

Steckler, M. S., and A. B. Watts, Subsidence history and tectonic evolution of Atlantic-type continental margins, in *Dynamics of Passive Margins,* vol. 6, edited by R. D. Scrutton, pp. 184-196, AGU, Washington, D.C., 1982.

Steckler, M., Changes in sea level, in *Patterns of Change in Earth Evolution,* edited by H. D. Holland and A. F. Trendall, pp. 103-121, Dahlem Konf., Springer-Verlag, New York, 1984.

Suess, H., *The Face of the Earth,* vol. 2, Clarendon Press, Oxford, 759 p., 1906.

Tissot, B., Effects on prolific petroleum source rocks and maj coal deposits caused by sea-level changes, *Nature, 277,* 46 4654, 1979.

Vail, P. R., R. M. Mitchum, Jr., and S. Thompson III, Glob cycles of relative changes of sea level, in *Seism —Stratigraphy—Applications to Hydrocarbon Exploration,* An Assoc. Pet. Geol. Mem. 26, edited by C. E. Payton, pp. 83-9 American Association of Petroleum Geologists, Tulsa, Okla., 197

Van Hinte, J. E., A Cretaceous time scale, *Am. Assoc. Pet. Geo Bull., 60,* 489-497, 1976.

Watts, A. B., and M. S. Steckler, Subsidence and eustasy at th continental margins of eastern North America, in *Deep Drillin Results in the Atlantic Ocean: Continental Margins an Paleoenvironments, Maurice Ewing Series,* vol. 3, edited by M Talwani, W. Hay, W. B. F. Ryan, pp. 218-234, AGU, Washington, D. C., 1979.

Weimer, R. J., Relation of unconformities to tectonics and sea level changes, Cretaceous of Western Interior, U.S.A., in *Interregional Unconformities and Hydrocarbon Accumulation* Am. Assoc. Pet. Geol. Mem. 36, edited by J. S. Schlee, pp. 7 36, American Association of Petroleum Geologists, Tulsa, Okla., 1984.

Wise, D. U., Continental margins, freeboard and the volumes of continents and oceans through time, in *The Geology of Continental Margins,* edited by C. A. Burk and C. L. Drake, pp. 45-68, Springer-Verlag, New York, 1974.

Worsley, T. R., D. Nance, and J. B. Moody, Global tectonics and eustasy for the past 2 billion years, *Marine Geol., 58,* 373-400, 1984.

CRETACEOUS/TERTIARY BOUNDARY EVENT

Kenneth J.Hsü

Geological Institute, Swiss Federal Institute of Technology
Zurich, Switzerland

Abstract. The biotic and sedimentologic records of detailly investigated sedimentary sequences across the Cretaceous/ Tertiary boundary give clear indication of a major catastrophe at the end of the Cretaceous. Mass mortality, followed by environmental deterioration, led to mass extinction of many groups of organisms, especially the floaters and swimmers in the oceans and the giant reptiles on land. The ocean was almost lifeless when plankton fertility was drastically reduced. Excessive influx of decaying organic matters led to oxygen-deficiency in seawater and widespread seafloor-dissolution of calcite. The siderophile-enrichment and the presence of microtektites and shock-quartz in the boundary clay indicate that the environmental catastrophe was triggered by the impact of a large meteorite.

Introduction

The gradual and slow changes of the ocean have been punctuated from time to time by perturbations. The crisis at the end of the Cretaceous was one of the most remarkable in the Earth history. Cuvier (1812) proposed that an earth revolution as the cause of sudden disappearance of dinosaurs on land and ammonites at sea. The evidence for this mass extinction troubled Charles Darwin, but he dismissed the fact, however, with a postulate of imperfect geological record, noting the common presence of an unconformity, - a gap in the record - , between the last Mesozoic and the first Cenozoic strata (Darwin, 1859). Lyell (1833) had concluded, having assumed constant rate of faunal changes, that the time interval represented by the gap in record is longer than the Cenozoic Era itself (Figure 1). Darwin was, therefore, convinced that catastrophes are artifacts of our imperfect record. This substanstive uniformitarianism has been deeply rooted in our science, but we now know that Darwin is wrong. Radiometric dating has proven that the time of the "lost record" cannot be longer than a million years. Nearly continuous record of deep sea sedimentation has indicated conclusively that mass extinction and catastrophic changes did take place at the end of the Cretaceous. This paper summarizes the data on the extent of this biotic crisis and reviews the evidence on environmental changes at that time.

The Biotic Record

Diversity reduction gives an expression of the magnitude of extinction. Russell (1979) made a count of the number of genera of fossil organisms which lived just before and after the end of the Cretaceous. As shown by Table 1, the diversity of floating marine microorganisms was reduced by 42%, of bottom-dwelling marine organisms by 49% and of swimming marine organisms by 70% across the C/T boundary, while the diversity-reduction of terrestrial organisms, except for reptiles, was small. Diversity data give, however, only a minimal measure, for if the extinct taxa have been quickly replaced by the new, the diversity reduction could be numerically insignificant. I shall, therefore, analyse the record of extinction in the following sections.

Shallow Marine Benthic Organisms

There is a common impression that the shallow marine benthic communities were little affected by the terminal Cretaceous event in view of the relatively small diversity-reduction of several groups.Birkelund and Hakansson (1982, p. 378), for example concluded that the evolution of cheilostome bryozoans reached a plateau, or "standstill" at the end of Cretaceous, because the number of families per stage is about the same before and after the boundary event. A closer look of the data tells quite a different story: The authors found diversified faunas, totalling 115 species in the boundary sequence of Denmark. Of those, 60 species have a restricted Maastrichtian range, 44 species are restricted to the Danian; only 11 species or about 10% "occur in both the Maastrichtian and the Danian" (p. 380). The so-called evolutionary standstill is, in fact, represented by a

TABLE 1. Number of Genera of Fossil Organism Currently Recognized as Having Lived in the Geologic Epochs Just Before and Just After the C/T Boundary Event (After Russell, 1979)

Taxa	before	after	% reduced
Terrestrial			
Fishes & amphibians	24	19	21
Reptiles	54	24	55
Other faunas	48	50	0
Plants	100	90	10
Total	226	183	19
Marine			
Floating micro-organisms:			
Calcareous	61	7	89
Noncalcareous	237	166	39
Total	298	173	42
Bottom-dwellers	1976	1012	49
Swimming organisms:			
Ammonites & belemnites	38	0	100
Others	294	99	49
Total	332	99	70

catastrophic extinction event, when 90% of the species became extinct within a very short time interval. The severity of the catastrophe is best illustrated by the own words of those authors (p. 380):

"All Maastrichtian chalk populations are highly diverse and very uniform, having as many as 60 species in individual samples. In marked contrast to this, the basal Danian chalk contains an extremely poor fauna increasing from a single species at the very base to no more than four species within the first metre. None of the species surviving the Maastrichtian-Danian boun-dary are present in this pioneer community. Following this extreme reduction, both diversity and density rise rapidly to a maximum of more than 40 species in the bryozoan limestone some 6 metres above the boundary".

Surlyk and Johansen (1984) studied the brachiopods of the same sequence and they were not misled by the "standstill" in diversity. They found 27 species in the uppermost Maastrichtian, and the mass extinction abrupt, coinciding in timing with the boundary event. "The basal few metres of the Danian are almost devoid of brachipods, and a Danian brachiopod fauna started almost as abruptly as the Maastrichtian fauna disappeared. The new fauna is similar to the Maastrichtian as regards diversity and density" (p. 1174), but only a maximum of six species are common to both faunas. In other words, the Maastrichtian fauna became 80% extinct, while the diversity reduction was practically nil.

In fact even Birkelund and Hakansson acknowledged that the record indicates a "collapse of the benthonic communities, involving a very high species extinctions" (p. 383) at the end of Cretaceous. But they followed blindly the Darwinian dogma and attributed the sudden faunal changes as an artifact of an imperfect geologic record (Figure 1); they stated unequivocally that "a single catastrophic event is improbable" (p. 383). Surlyk and Johansen (1984) came to an opposite conclusion: The terminal Cretaceous extinction is real and abrupt. No ecological crisis is discernable in the Maastrichtian record. Then the crisis came suddenly, and without any warning. Faunal groups which were specia-

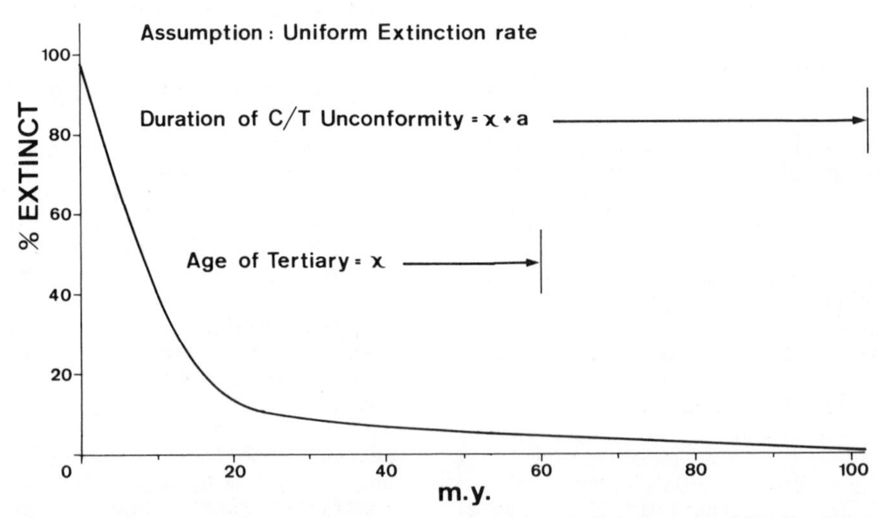

Fig.1. Lyell found no Maastrichtian taxa in the Paleocene mollusk faunas of his collection, but a few Paleocene taxa in the Holocene. He concluded, therefore, the lost time interval represented by the C/T boundary event was longer than the duration of the Cenozoic. He is wrong.

lized and restricted to the chalk substrate, such as the chalk brachiopods, all became extinct. The surviving species are forms with wider substrate tolerance. When conditions eventually became normal again, migration and adaptive radiation within surviving groups led to a rapid restoration of benthic faunas in the chalk macrohabitat, but the Danian shelly benthos is almost totally different on species level from that of the Maastrician.

Aside from bottom habitat, there seems to have been also a paleogeographical selection. Kauffman (1979) thought that virtually all Cretaceous organisms which became extinct near a major peak of their evolutionary development lived predominantly in or very near to the equatorial Tethys; Kauffman, an expert on rudists, was no doubt impressed by the total extinction of those reef-building organisms. Newell (1982, p. 262) also referred to the terminal Cretaceous event as being "highlighted by the dramatic exit of climate-sensitive reef-forming corals". This pattern of selectivity suggests that sudden changes in physical environment, rather than biotic interactions, were the main cause of the mass extinction.

Swimming Marine Organisms

Ocean swimmers as a group sustained the greatest loss, having been reduced from 332 to 99 genera, or some 70%, despite of the Danian replacement. The ammonites and belemnites became all, and the marine reptiles almost all, extinct at the end of Cretaceous. Other groups, especially bony fishes, suffered greatly. Those are indisputable facts. The favorite arguments among paleontologists are, however, centered on the issue if those swimming organisms were destined for oblivion before the final end came.

Birkelund (1979) found 7 ammonite genera in the youngest Maastrichtian rock at Stevns Klint, Denmark, and they all belonged to ancient lineages without much evolutionary changes before their last appearance. Birkelund thought the ammonites have lost their "evolutionary vigor" because no important new characters evolved toward the end. The same fact could, however, be interpreted as evidence that a catastrophe struck without forewarning.

The idea that the ammonites went into a decline prior to their final extinction may in fact be a reflection of the provincial outlook. The generalization was based upon studies in Europe and North America, where the Upper Cretaceous sequence is largely regressive. In a transgressive sequence on Antarctica, Macellari and Zinsmeister (1985) found an increase, not a decrease, in the ammonite diversity toward the end of the Cretaceous.

Floating Marine Organisms

The extinction records of two major groups of planktons differ remarkably. The calcareous micro- and nannoplanktons became almost all extinct, whereas the damage to the siliceous and other noncalcareous planktons was considerable less.

The last Cretaceous microfauna includes diversified assemblages, belonging to the families Globotruncanids, Rugoglobigerinids, and Heterohelicids. Those floaters had undergone little evolutionary change before a catastrophe came and wiped out all of them (Figure 2). There was no forewarning, no sign of decay till the very end. The only survivors, first identified above a boundary clay at Gubbio, Italy by Luterbacher and Premoli-Silva in 1962, are *Globogerina eugubina* and related dwarf species. Those "dwarfs" came into prominence when all the "giants" were wiped out. Foraminiferal diversity was very quickly restored by explosive radiation. The sickly *G.eugubinas* were soon replaced by a diversified fauna of robust floaters which are found in early Paleocene limestones 5 m above the boundary clay (Figure 2). The present planktonic foraminiferas of the oceans are almost all descendents of those early Tertiary "dwarfs".

The pattern of the nannoplankton evolution across the C/T boundary is similar to that of the microplankton. Thierstein (1981) found also a stasis in the evolution of the Late Cretaceous nannoplankton; all but a few Cretaceous taxa became extinct in a very brief transitional interval of les than 50,000 years (Perch-Nielsen et al., 1982). "The typical earliest Danian nannofossil taxa have been recorded in Late Cretaceous sediments in extremely low abundances and mainly at high latitude sites. Their survival into the Tertiary may have been due to their adaptation to higher seasonality of temperature, salinity, or light (Thierstein, 1981, p. 384). The earliest Paleocene nannoplankton communities are characterized by strongly fluctuating abundances, indicative of an ecologic instability.

In contrast to the crisis for the calcareous planktons, dinoflagellates were not much affected by the terminal Cretaceous event, nor were the siliceous planktons (see Russell, 1979). Their different response to the crisis suggests a higher tolerance by the noncalcareous planktons to the ocean pollution at the end of the Cretaceous (Ekdale and Bromley, 1984).

Deep marine benthic organisms

The Upper Cretaceous taxa differ only slightly from those in the Paleocene. Numerous species evolved across the C/T boundary with little or no changes (R. Wright in Hsü et al., 1984, p. 335).

Although the influence of the terminal Cretaceous event on the faunal composition of deep benthos was minimal, there was apparently a drastic reduction of the biomass at the time of biotic crisis. The uppermost Maastrichtian and

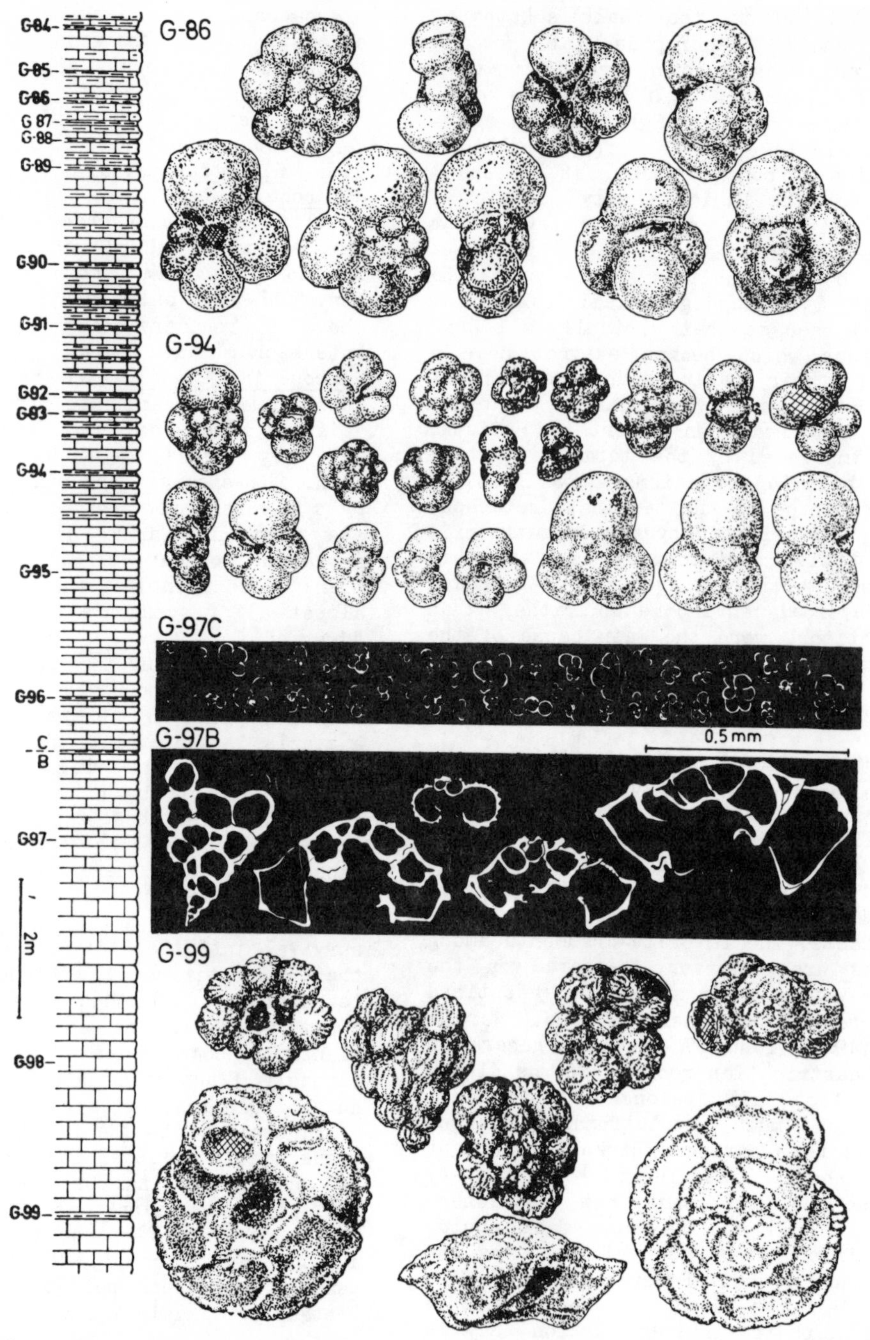

Fig. 2. The sudden extinction of the planktic foraminifers at the end of Cretaceous, replaced by "dwarfs" of the first Paleocene taxa, illustrated by this figure from Luterbacher and Premoli-Silva, 1962.

the lower Paleocene sediments are much bioturbated (Ekdale and Bromley, 1984), but the boundary clay between the two is commonly laminated. This lack of bioturbation suggests a mass mortality of burrowing organisms (e.g., Hsü et al., 1984, p. ii).

Terrestrial Faunas and Floras

On land, the extinction of dinosaurs defines the end of the Cretaceous. Precision stratigraphy has put the timing of the event within Chron C-29-R, synchronous to plankton extinction

(Butler and Lindsay, 1985). Other giant reptiles, with body weight more than 25 kg also became extinct, but the invertebraete faunas on land seemed to have suffered little damage (Russell, 1979).

It has been stated that the plant kingdom was little affected by the catastrophe at the end of the Cretaceous. Looking at the taxonomic turn-over, one might conclude that no great changes took place except for the extinction of the *Aquilapollenites* flora at the high northern latitudes (Hickey, 1981).

Pollen studies have revealed, however, a remarkable disruption of the terrestrial plant ecosystm at the end of the Cretaceous. The sudden disappearance of tree pollen, which are abundant in both Cretaceous and Paleocene floras, in the boundary indicates that forests were destroyed during a catastrophe, to be replaced by the growth of ferns. The crisis did not last long, and those first colonizers of a deforested landscsape were soon crowded out again, when the angiosperms reestablished their dominance (Tschudy et al., 1984).

Summary of the Bitoic Records

The biotic record indicates that a catastrophic event came suddenly and without forewarning. Many taxa were eliminated. Particularly hard hit were the floaters and the swimmers in the oceans, as well as the giant reptiles on land. Other groups suffered drastic population reduction, but many taxa recovered quickly and did not become extinct.

The Sedimentologic Record

The C/T boundary in the oceans is commonly defined by the first appearance of typically Tertiary taxa, characterized by the foraminiferal species *G. eugubina.* But the micro- and nannofossils of numerous typically Cretaceous taxa did not disappear altogether in the sediments intermediately above the boundary, and they are present in decreasing abundance in a transitional zone a few centimeters to a few meters thick (Perch-Nielsen et al., 1982). The base of this zone is commonly a pelagic clay, almost devoid of calcium carbonate. Sedimentological and geochemical studies during the last decade have revealed sharp anomalies indicative of catastrophic environmental perturbations.

Crisis in Plankton Production

The very first sediment deposited in the Tertiary ocean is a boundary clay, or a sediment with reduced calcite content. It has been postulated that the calcite-compensation-depth (CCD) of the ocean rose suddenly to the photic zone (Worsley, 1974). The postulate is wrong because the boundary sediment at several deepsea drilling sites is a marl or even a calcareous ooze (see Hsü, 1984).

One could explain the calcite-deficiency in the boundary sediments by assuming that they consist wholly or partly of ejecta-fallout detritus after a meteorite impact (Alvarez et al., 1980). This interpretation has been disputed (Officer and Drake, 1983). Although the boundary clay in Denmark may indeed have been altered from glassy impact detritus (Kastner et al., 1984), the low calcite content at numerous other sites, where the clay mineralogy of the boundary sediment is indistinguishable from that of those above and below, is more probably caused by reduced production or more intensive dissolution.

A negative perturbation of the carbon-isotope values across the C/T boundary was first found in the bulk-analysis data (Brennecke and Anderson, 1977); the bulk of ocean oozes consists of nannoplankton. It was found later that the planktic foraminifers in the boundary sediment also registers this isotope anomaly, but not the benthic foraminifers (Boersma and Shackleton, 1981; He et al., 1984). Normally the planktic skeletons have a delta carbon-13 value 2 parts per mil more positive than the benthic, because of the carbon-13 enrichment in surface waters after the preferential utilization of the light carbon by photosynthetic organisms (Figure 3). This carbon- isotope gradient is eliminated if and when the planktic production is significantly curtailed. This concept of a *strangelove ocean* after a terminal Cretaceous catastrophe can best explain the observed carbon-isotope perturbation across the C/T boundary (Hsü and McKenzie, 1985). Such a suppression of production can also explain, at least in part, the reduced calcite-content in the boundary sediments.

The reduced fertility most likely led to plankton extinction. Since the calcareous planktons were the hardest hit, one might postulate that their production was most curtailed, while those of non-calcareous planktons far less. Such a selectivity suggests further that the ocean has become more acid during the environmental crisis: Calcareous planktons would then cease to reproduce, but the same change did not much disturb siliceous planktons (Ekdale and Bromley, 1984).

The carbon-isotope perturbation at the boundary is very sharp (Figure 4). Blooms of single species of nannoplankton may have led to quick re-establishment of the carbon-isotope gradient. There seemed to have been a second depletion, coinciding in time approximately with the first appearance of *G. eugubina* (Perch-Nielsen et al., 1982). I am not certain if this represents a second "strangelove perturbation", or a general increase of the light carbon atoms in the oceans.

A 2 parts per mil increase of carbon-12 in

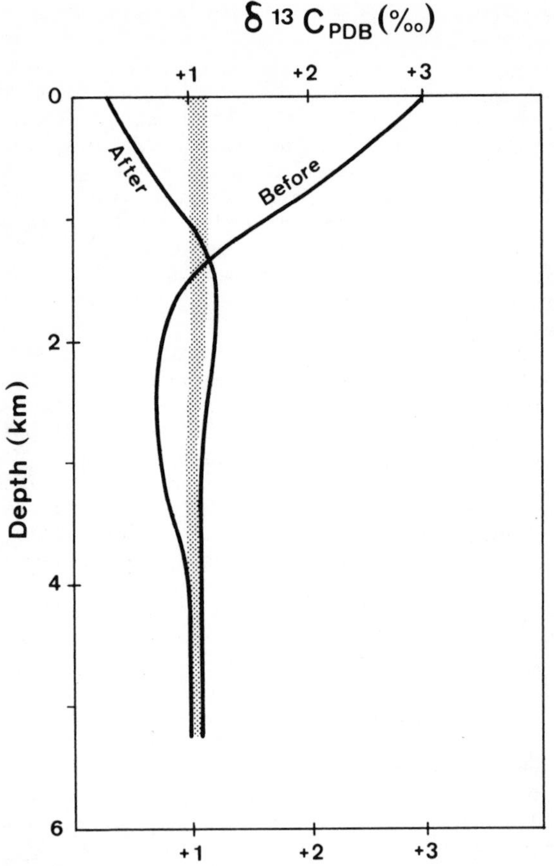

Fig. 3. Carbon-13 is enriched in surface waters normally because of the preferential utilization of carbon-12 by photosynthetic plants. This enrichment is registered by the skeletons of planktic organisms. The carbon-isotope profile in the oceans before the C/T boundary event is the same as today, indicative of normal plankton fertility. The profile in the boundary sediment after the event shows an enrichment of carbon-12 in surface water, because of the suppression of plankton production at that time.

the organic matter in the boundary sediments has also been detected. This perturbation has been correlated to the second carbon-isotope perturbation at Site 524 and considered evidence of a carbon-12 enrichment in the atmosphere, brought about by the ocean-atmosphere exchange during the first 50,000 years of the Tertiary (Shimmelmann and DiNero, 1984).

Calcite Dissolution

Although the low carbonate content in the boundary sediment is almost certainly related to a fertility anomaly, there is evidence of extensive calcite-dissolution at the end of the Cretaceous. The uppermost Cretaceous sediments have progressively less calcite toward the boun-dary horizon, indicative of subbottom removal where the bottom water was grossly under-saturated (Figure 4). Planktic foraminiferal tests have been largely dissolved in those uppermost sediments.

The presence of a boundary clay at many outcrop localities and its absence at several deepsea sites indicates an apparent inversion of calcite-compensation; more calcite was dissolved at shallower sites of deposition. This paradox may be explained if we recall that the dissolved carbon-dioxide in seawater is maximal in the zone of oxygen-minimum at intermediate depth. Deforestation on land at the end of Cretaceous should have supplied an excess of organic matters to the oceans and their decay caused an expansion of the oxygen-minimum. Outcrop localities in Denmark, Italy, Spain, and Tunisia was located within this zone of most corrosive water. At greater paleodepth of 2 or 3 km, the ocean water should have been less corrosive and thus more favorable for the preservation of calcitic sediments (Figure 5). At still greater paleo-depth, the CCD was reached, below which only pelagic clays were accumulating.

Oxygen Deficiency

A color change from red or white to grey, green, or black is observed at many sites across the C/T boundary. At the famous outcrop at Stevns Klint, for example, the Cretaceous chalk is white and the Tertiary chalk is white, but the boundary clay is black. One also sees paper-thin laminations in the clay. Both the color and the laminated structure indicate that the bottom water was devoid of oxygen at the time of boundary clay deposition. The black color owes its origin to very finely disseminated organic carbon. The laminations owe their origin to the settling of sedimentary particles on a bottom where no life existed to disturb the delicate sedimentary structure. There was no life, because there was no oxygen, and there was no oxygen because it had all been used up, having been consumed to oxidized the excessive decaying matter that was brought in.

The organic-carbon content of the boundary sediments is 4-5 times higher than the adjacent sediment. Assuming an ejecta-dust origin of the boundary clay and a sedimentation rate of 1 cm/yr, Wolbach et al. (1985) obtained a figure of a thousandfold increase of the organic-C flux after a terminal Cretaceous catastrophe. Their conclusion will have to be revised because the actual sedimentation rate was probably much slower. Significant is, however, the observation that the organic-carbon particles "show the characteristic morphology of carbon deposited from flames, such as soot or carbon black". Wolbach et al. postulated soot-production by forest fires which destroyed some 5 percent of the biomass at the time. This corroborates the

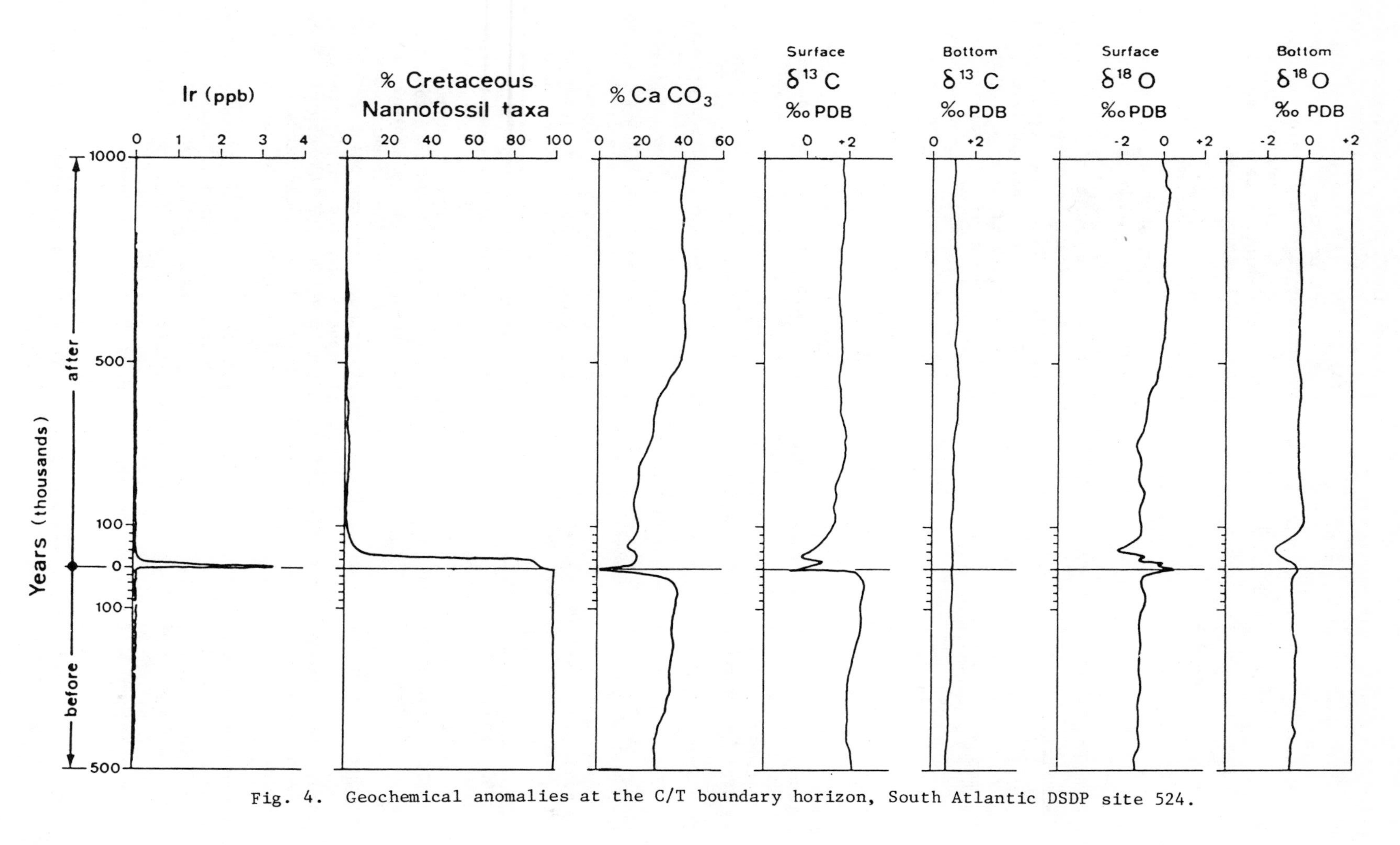

Fig. 4. Geochemical anomalies at the C/T boundary horizon, South Atlantic DSDP site 524.

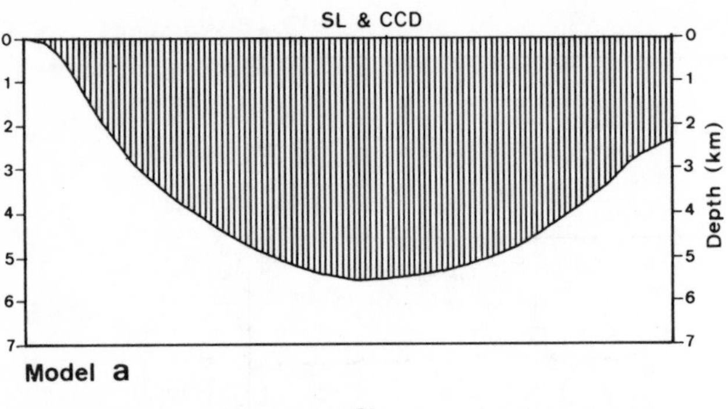

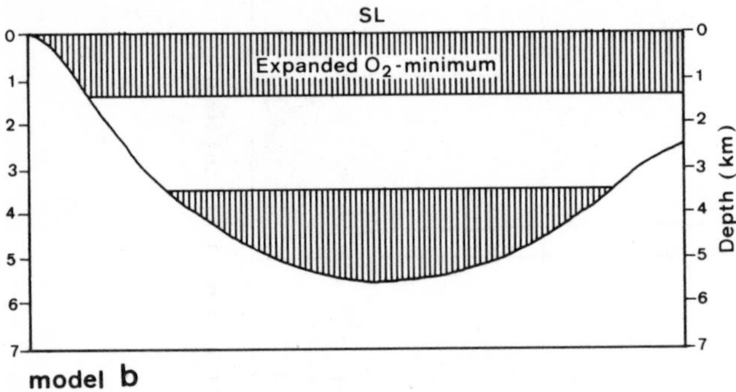

Fig. 5. Worsley (1974) postulated the rise of CCD to the photic zone (Model A). More probably, the deposition of calcite-free sediments in shallow depths was related to increased carbon-dioxide content in an expanded oxygen-minimum zone (Model B).

observation of regional deforestation by Tschudy et al. (1984). The partial oxidation of the soot and other land-derived organic matters should have caused an oxygen-depletion in seawater. That reducing conditions prevailed at that time is manifested by the common presence of pyrite and/or glauconite in the boundary clay at many sites (Smit and Kyte, 1984). Even in the Pacific region of red clay deposition, arsenic and antimony sulphides have been found in the boundary clay, and their precipitation is indicative of an oxygen-deficient condition, which lasted about 10,000 years (Kyte and Wasson, 1985).

Temperature changes

Abrupt temperature changes across the C/T boundary have been suggested by oxygen-isotope data. First analyses of closely spaced samples from the Atlantic indicated a warming of 5 degrees Celsius (Boersma et al., 1979). Our work on South Atlantic cores revealed that the temperature oscillated with an overall warming trend during the 40,000 years after the boundary event (Hsü et al., 1984). Temperature increase at such a rate may have been caused by the greenhouse effect of an increased carbon dioxide content in the Earth"s atmosphere.

Unfortunately, not all studies came to the same conclusion. Boersma and Shackleton (1981), for example, found no systematic change of bottom temperature, and perhaps even a slight drop of surface temperature, across the boundary at the Pacific DSDP Site 465. Considering that the temperature may have oscillated rapidly during the time of crisis, we could expect no reliable conclusion if the sampling interval is not closely enough spaced.

Summary of Sedimentologic Record

The geochemistry of sediments records an environmental catastrophe at the start of the Teritary:

The surface water of the ocean was pobably sufficiently acid to suppress the production of calcareous planktons. The intermediate and bottom waters were also unusually corrosive. An abnormal influx of organic carbon from land led to widespread oxygen-deficiency or even anoxic conditions. Meanwhile the ocean and atmospheric temperature may have fluctuated widely.

Cause of Environmental Catastrophe

The coincidence in timing of the biotic crisis and the environmental catastrophe leaves little doubt that one was caused by the other. But what triggered the the sudden environmental changes?

Of the numerous ideas proposed, only the hypothesis of meteorite-impact or that of explosive volcanism could provide an adequate scenario to explain the postulated enviromental changes (Hsü and McKenzie, 1985; McLean, 1985):

After a very large impact or an unusually explosive volcanism, ejecta dust rose to the stratosphere, enveloped and darkened the Earth for several months,when plankton productivity were but ceased. An ocean almost devoid of planktons would become more acid and this, in turn, should have led to further suppression of fertility. Meanwhile the destruction of forests on land by darkness should have provided an excess flux of organic carbon to the oceans, inducing oxygen-deficient conditions. Finally the sudden decrease of organic productivity on land and in the oceans may have caused disturbance of carbon-dioxide flux to the atmosphere and the rapidly changing greenhouse effect may have induced temperature oscillations.

Of the two possible mechanisms triggering the postulated environmental catastrophe, the volcanism scenario is practically eliminated by recent sedimentological investigations. The enrichment of siderophile elements in the boundary clay is best explained by a bolide-impact theory (Alvarez et al., 1980; Hsü et al., 1982; Jordan, in prep.). Now that microtektites and shock quartz grains, products of bolide impact, are found in the boundary clay (Smit and Kyte, 1985; Bohor et al., 1984), only a stubborn few still cling to the hypothesis of catastrophic volcanism.

References

Alvarez, L.W.,W. Alvarez, F. Asaro and H.V. Michel, Extraterrestrial cause for the Cretaceous-Tertiary extinction, Science 208, 1095-1108, 1980.

Birkelund, T., The last Maastrichtian ammonites, in: Birkelund,T. and Bromley, R.G., eds., Cretaceous-Tertiary Boundary Events, v. 1, Copenhagen, Univ. of Copenhagen, 51-57, 1979.

Birkelund,T. and E. Hakansson, The terminal Cretaceous extinction in Boreal shelf seas -a multicausal event, in: Silver, L.T. and Schultz, P.H., eds., Geological Implications of Impacts of Large Asteroids and Comets on Earth, Geol. Soc. America, Spec. Paper 190, 373-384,1982.

Bohor, B.F., E.E. Foord,P.J. Modreski and D.M. Triplehorn Mineralogic evidence for an impact event at the Cretaceous-Tertiary boundary, Science, 22, 867-869, 1984.

Boersma, A. and N. Shackleton, Oxygen and carbon isotope variations and planktonic depth habitats, late Cretaceous to Paleocene, central Pacific, DSDP sites 463 and 465, Initial Reports of DSDP,62, 513-526, 1981.

Boersma,A.,N. Shackleton, M.Hall and Q. Given, Carbon and oxygen records at DSDP site 384 (North Atlantic) and some Paleocene paleotemperatures and carbon isotope variations in the South Atlantic, Initial Reports of DSDP, 43, 695-718, 1979.

Brennecke, J.C. and T.F. Andrson, Carbon isotope variation in pelagic carbonates (abstract), EOS,Trans.AGU, 58, 415, 1977.

Butler, R.F. and E.H. Lindsay, Mineralogy of magnetic minerals and revised magnetic polarity stratigraphy of continentals sediments, San Juan Basin, New Mexico, Jour. Geology,93. 535.554,1985.

Darwin, C., On the Origin of Species, London: Murray, 513 pp., 1859.

Ekdale, A.A. and R.D. Bromley, Sedimentology and ichnology of the Crtaceous-Tertiary boundary in Denmark: implications for the causes of terminal Cretaceous extinction, Jour. Sed. Petrology, 54, 681-703, 1984.

He, Q., J.A.McKenzie and H.Oberhänsli, Stable-isotope and percentage-of-carbonate data for upper Cretaceous/lower Tertiary sediments from DSDP site 524,Cape Basin, South Atlantic, Initial Rept. DSDP,73, 749-754, 1984.

Hickey, L., Land plant evidence compatible with gradual, not catastrophic, change at the end of the Cretaceous, Science, 292, 529-531, 1981.

Hsü, K.J., Geochemical markers of impacts and their effects on environments, in: Holland,H.D. and A.F.Trendall, eds., Patterns and Changes in Earth Evolution, Berlin:Springer-Verlag, 63-76, 1984.

Hsü, K.J. and J.A. McKenzie, A "strangelove ocean" in the earliest Tertiary, in: Sundquist, E.T. and W.S.Broecker, eds., The Carbon Cycle and Atmospheric CO_2: Natural Variations Archean to Presnet,AGU Geoph. Monograph 32, 487-492, 1985.

Hsü,K.J., and J. L. LaBrecque, et al., The Initial Reports of the Deep Sea Drilling Project,v.73,Washington,DC:U.S.Govt.Printing Office,798pp.,1984.

Hsü, K.J., Q. He, J.A.McKenzie, H. Weissert, K. Perch-Nielsen, H. Oberhänsli, K.Kelts, J. La Brecque, L. Tauxe, U. Krähenbühl, S.F. Percival, Jr., R. Wright, A. M. Karpoff, N. Petersen, P. Tucker, R.Z.Poore, A.M.Gombos, K. Pisciotto, M. F. Carman, Jr. and E. Schrieber, Mass-mortality and its environmental and evolutionary consequences, Science, 216, 249-256, 1982.

Kastner, M., F. Asaro, H. V. Michel, W. Alvarez and L. W. Alvarez, The precursor of the Cretaceous-Tertiary boundary clays at Stevns Klint, Denmark, and DSDP Hole 465A, Science, 226, 137-143, 1984.

Kauffman, E.G., The ecology and biogeography of the Cretaceous-Tertiary extinction event, in:Christensen, W. K. and T. Birkelund, eds., Cretaceous-Tertiary Boundary Events, v.2, Copenhagen, Univ.Copenhagen, 29-37, 1979.

Kyte, F. T. and J. T. Wasson, The Cretaceous Boundary in GPC-3, an abyssal clay section, Geochim Cosmochim. Acta,in press.

Luterbacher, H. P. and I. Premoli-Silva, Note preliminaire sur une revision du profil du Gubbio, Italie, Rivista Ita. Paleontologia, 68, 253-288, 1962.

Lyell, C., Principles of Geology, London: Muray, v. 3, 398 pp., 1833.

Macellari, C. E. and W. J. Zinsmeister, Macropaleontology and sedimentology of the Cretaceous/Tertiary boundary in Antarctica (abstract), Gwatt Conf. on Rare Event, May, 1985, Proceeding Volume, Zurich ETH Geol. Institut.

Officer,C.B. and C.L.Drake, The Cretaceous-Tertiary transition,Science, 219, 1383-1390, 1983.

Perch-Nielsen,K.,J.A.McKenzie and Q.He, Biostratigraphy and the "catastrophic" extinction of calcareous nannoplankton at the Cretacoues/Tertiary boundary, in: Silver, L. T. and P. H. Schultz, Geological Implications of Impacts of Large Asteroids and Comets on the Earth, Geol. Soc. America, Spec. Paper 190, 353-372, 1982.

Russell,D.A.,The enigma of the extinction of dinosaurs, Ann. Rev. Earth & Planet. Sci., 7, 163-192, 1979.

Shimmelmann,A. and M.J. DeNiro, Ekemental and stable isotope variations of organic matter from a terrestrial sequence containing the Cretaceous/Tertiary boundary at York Canyon, New Mexico, Earth and Planet. Sci. Letters, 69, 392-398, 1984.

Smit,J. and F.T.Kyte, Siderophile-rich magnetic spheroids from the Cretaceous-Tertiary in Umbria, Italy, Nature, 330,.403-405,1984.

Surlyk,F. and M.B.Johanssn, End-Cretaceous brachiopod extinction in the chalk of Denmark, Science, 223, 1174-1177, 1984.

Thierstein, H.R., Late Cretaceous nannoplankton and the change at the Cretaceous-Tertiary boundary, Soc. Econ. Mineralogists & Paleontologists, Spec. Publ. 32, 355-394, 1981.

Tschudy, R. H., C. L. Pillmore, C. J. Orth, J. S. Gilmore, J. D. Knight, Disruption of the terrestrial plant ecosytem at the Cretaceous-Tertiary boundary, Western Interior, Science, 225, 1030-1032, 1984.

Wolbach, W. S., R. S. Lewis, and E. Anders, Cretaceous extinctions: Evidence for wildfire and search for meteoric material, Science, 230, 167-170, 1985.

Worsley, T. R., Cretaceous-Tertiary boundary event in the ocean, Soc. Econ. Mineralogists & Paleontologists, Spec. Publ. 20, 94-125, 1974.

PALEOCENE - EOCENE PALEOCENOGRAPHY

Hedi Oberhänsli and Kenneth J. Hsü

Geological Institute, Swiss Federal Institute of Technology, Zurich

Abstract. Distribution patterns of carbonates, siliceous and phosphatic sediments, fluctuations of the CCD stable isotopic patterns, as well as the oceanic microplankton and terrestrial fauna records hold the key to major paleoceanographic problems. These patterns allow a better understanding of changes in global temperature and oceanic circulation. These paleoceanographic tools led to the following reconstruction: The Paleocene and Early Eocene climate was charcterized by generally warm and stable conditions. A steplike climatic deterioration started by the early Middle Eocene and continued through the Eocene /Oligocene boundary. The climatic changes and changes in the current pattern are triggered, to a certain extent, by the reorganization of the ocean/continent configuration such as the separation between Greenland and Scandinavia (beginning C 24), Australia and Antarctica and the beginning closure of the western Tethys. The driving force controlling oceanic circulation was most probably dominantly dominated by halokinetic processes before it was changed to the predominately thermohaline processes during the Late Eocene.
The Late Paleocene positive $\delta^{13}C$ event, observed in surface and bottom water environments, coincides with a most significant benthic faunal turnover. The benthic faunal system quickly recovered some 2 Ma later.

Introduction

The Paleogene was a time of change, from the non-glaciated to a glaciated world. The climate was highly equitable during the Late Cretaceous. Stepwise cooling, with minor fluctuations, began in Middle Eocene, continued till Early Oligocene, and the glacial mode dominated the Neogene. The exact timing and the initial extent of the onset of polar glaciation are still subjects of much speculation. There are indications that Antarctic ice cap may have already existed in Late Eocene or Early Oligocene (Le Masurier, 1972; Matthews and Poore, 1980; Webb et al., 1984). But other interpretations placed the event to the Middle Miocene some 14-15 Ma (Shackleton and Kennett, 1975; Savin et al., 1981).

Paleogeographic and paleoceanographic reconstructions portray an early Paleogene world not much different from that of Cretaceous (e.g. Haq, 1981). The circum-equatorial surface currents efficiently separated the gyre circulation systems, which had been established in the Northern and Southern Hemispheres. Density increase due to evaporation may have played an important role in a world without polar ice caps. The staedy progress of plate displacement was, however, to change the geography and current circulation on the surface of the globe significantly before the end of the Paleogene. The equatorial circulation of the ancient Tethys became increasingly restricted, before the deep connection between the Indo-Pacific and Atlantic was finally severed in Early Miocene, when Africa collided with Eurasia in the Eastern Tethys (Adams,1973; Adams et al., 1983; Hsü et al, 1978). Already by the Late Eocene the Tethyan Ocean became successively more restricted north and east of the Indian Plate and in the Western Tethys the passage way to the Atlantic Ocean became narrower and shoaler (Biju-Duval et al., 1977). The establishment of an inland sea or the Proto-Mediterranean may have given rise to the first Atlantic shallow-intermediate wartermass west of Gibraltar by the Late Oligocene (Biolzi, 1983).

The separation between Greenland and Scandinavia became wider and deeper when the first ocean crust was formed there toward the beginning of Eocene (Anomaly 24 time, Talwani and Udintsev, 1976). A Norwegian Sea overflow representing an ancestral North Atlantic Deep Watermass (NADW), started in Late Eocene and provided high-salinity flow at mid-depths in the Central and South Atlantic (Miller and Curry, 1982; Johnson, 1985).

In the Southern Hemisphere, Australia moved steadily away from Antarctica during the early Paleogene. Circum-Antarctic currents may have originated in Paleocene (Barker, et al., 1977). With the drastic cooling culmiminating in early Oligocene, when extensive sea ice should have been formed around the Antarctic. This led inturn to the production of cold and saline waters which descended to drive a global bottom circulation like the present-day Antarctic Bottom Watermass (AABW). The circulation pattern changed again, probably in middle or late Oligocene time,when the Drake Passage was opened to permit deep circum-polar circumlation (Kennett et al., 1975).

Changes in the circulation pattern are manifested by biogeographic and sediment-distribution patterns, and find also an expression in sedimentation rates. In oceanic environments sedimentation during the Early and Middle Eocene was generally characterized by high carbonate accumulation rates (Worsley and Davies, 1981). Biosiliceous sediments, which are mainly encountered in the equatorial zone of the Atlantic, and Pacific Oceans, are common in deposits ranging from Late Paleocene to Middle Eocene age (Pisciotto, 1981). The sedimentation of those calcareous and siliceous plankton is a sensitive indicator of changes in ocean chemistry and ocean dynamics.

Reviews of the Cenozoic paleoceanography are numerous (e.g. Berggren and Hollister, 1977; Berger, 1979; Arthur, 1979; Van Andel, 1979; Schnitker, 1980; Berger et al., 1981; Haq, 1981). Regional syntheses have been published by Van Andel et al. (1975); Kennett (1977, 1978), Mc Coy and Zimmermann (1977); Kidd and Davies (1978); Mc Gowran (1978) and Hsü et al. (1984a). This paper reviews the aspects of the paleoceanographic and climatic history during the Paleocene and Eocene time interval. Special emphasis is placed by us to a reconstruction of major events as revealed by the records of isotope stratigraphy, of biotic evolution, and of sediment-distribution patterns.

The Sedimentary Record

Carbonate Sedimentation Rate and CCD

Rates of carbonate sediments in the oceans are mainly controlled by biological productivity and by dissolution. The productivity depends very much upon nutrient supply, which is related to weathering and erosion. Those in turn are influenced by climate and topography. Dissolution rates of bottom sediments in the oceans are related to the chemistry of circulating currents. A common pattern of increasing calcite-dissolution with depth has been discerned, because of the presence of more corrosive water at greater depth. A second dissolution maximum at the oxygen-minimum depth has been noted, probably because of the enhanced carbon-dioxide content there (Moore et al., 1984; Hsü, 1985).

Calculated average carbonate deposition rates have been compiled by Davies and Worsley (1981) for the last 60 Ma. The three oceans, Pacific, Atlantic and Indian, show the same general pattern of fluctuating rates, but the magnitude of the extremes is different. A close examination of Figure 1 reveals, however, some subtle difference which suggests ocean/ocean fractionation. The accumulation rates in the Pacific and Atlantic Oceans remained, for example, low during the Late Paleocene/Early Eocene, but increased sharply during the Middle Eocene (56-45 Ma; approx. Zones NP 11-16). The maximum rates were reached just before and after the earliest Middle Eocene (48 Ma; Zone NP 15). The Indian Ocean data show, however, a sudden increase during the Late Paleocene already. The rates in all oceans dropped sharply soon after the maxima, and minima were registered during Late Eocene and Early Oligocene before the Late Oligocene rise.

A more direct approach to estimate the changing productivity and dissolution is to determine the fluctuation of calcite compensation depth (CCD). Using various indicators of dissolution, the changing CCD of the South Atlantic since the Early Eocene has been deciphered and shown by Figure 2 (Van Andel, et al., 1977; Hsü, et al., 1984a). Both accumulation rate and CCD are related to the supply and dissolution of calcareous plankton, and the rate should be higher when supply is more plentiful, or when dissolution is reduced, namely at times of a deep CCD. Such a correlation has been observed in the Neogene record (e.g.,Hsü, et al., 1984a). The Paleogene record, on the other hand, seems to contradict this prediction: The CCD was shallow, for example, in the Atlantic at times of high accumulation rates in the Early and Middle Eocene, and it became much deeper in Late Eocene and Early Oligocene when the accumulation rates were minimal (Van Andel et al., 1975; Hsü et al., 1984a). It does not make sense that the average rate should be high when productivity is less and/or dissolution more. The key to the paradox lies perhaps in the meaning of the average: One should ask what samples have been included to calculate the average values?

One should recognize that the CCD is a direct measure of the compensation of dissolution by production, while the average sedimentation rate is influenced by many other factors as well. There is the possibility of sampling bias, if the average includes too much more samples from the sites where the sedimentation rate is high.

Probably the apparently high Eocene accumulation rates are misleading because little dissolved sequences are over-represented in the computation of the average by Davies and Worsley. Also we must not forget that accumulation rate is not only influenced by chemical dissolution, but can also be reduced by mechanical erosion. One can postulate that the lower than expected Late Eocene and Early Oligocene accumulation rates resulted from the presence of obscure erosional unconformities in sedimentary sequences of those ages. The balance of production and dissolution should have yielded a high sedimentation rate during those times when the CCD was deep, if there had been no mechanical erosion, but the net accumulation rate is low because of the removal by current erosion. These considerations led us to the conclusion that we should rely more on the history of CCD fluctuation than the "average" accumulation rates as a monitor of productivity and dissolution of calcareous plankton.

The CCD record suggests that the production of calcareous plankton in the oceans increased from Middle Eocene onward till Early Oligocene (Hsü, et al., 1984a). The increase is related to available nutrient supply. One could postulate a greater total supply in Oligocene because sealevel drop had increased land area exposed to erosion (Fisher and Arthur, 1977; Berger,1979), or that more supply became available to oceanic calcareous plankton when the nutrient precipitation on continental shelf and/or nutrient demand from siliceous plankton were reduced (Hsü, et al., 1984a).

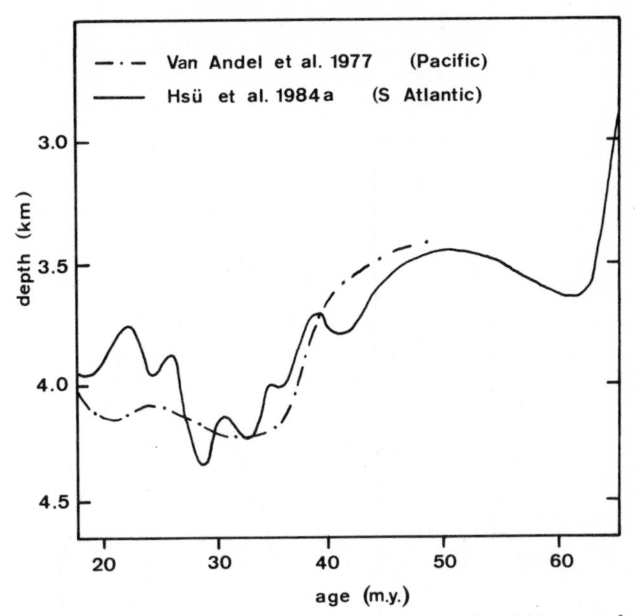

Fig. 2. CCD fluctuations in the Pacific and Southern Atlantic Oceans during Paleogene (after Van Andel et al., 1977; Hsü et al., 1984a).

Silica Sedimentation

Siliceous sediments accumulate preferentially in equatorial oceans, in waters at high latitudes, and in zones of coastal upwellings (e.g. Riedel, 1971). The accumulation rate of opal in sediments gives a measure of biogenic activities in surface-water environments (Leinen, 1979), but the rate is also governed by silica-dissolution.

The early Tertiary siliceous sediments were laid down mainly in an equatorial ocean -a continuous high-fertility belt - which extended across the Indian, Pacific and Atlantic Oceans.

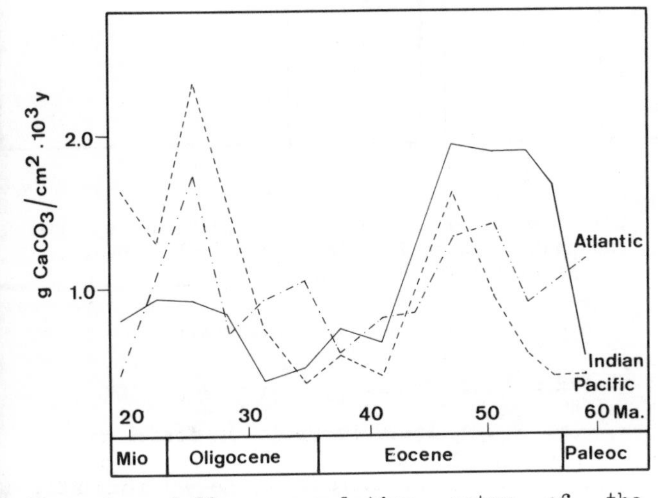

Fig. 1. $CaCO_3$ accumulation rates of the Atlantic, Pacific and Indian Oceans during Paleogene time (after Davies and Worsley, 1981).

Local occurrences of siliceous sediments have been reported from the eastern North Atlantic and from the Argentine Basin (Leinen, 1979; Riech and von Rad, 1979). Siliceous oozes of Early Paleocene to Middle Eocene age are present at DSDP Sites 167 and 283, located in the Central and the Southwest Pacific respectively; those of Late Paleocene to Early Eocene age at DSDP Site 245 in Southern Indian Ocean, and those of Early Eocene age at DSDP Site 405 in North Atlantic (Pisciotto, 1981). The available data (Table 1) are sufficient to show the greatly variable productivity during the Paleogene: Accumulation rate is as low as 3 g/cm^2/ Ma at Site 167, and as high as 630-945 g/cm^2/Ma at Site 405.

Phosphate Sedimentation

Phosphates deposits occur preferentially along coasts with significant upwelling activity. The amount of sedimentary phosphate deposited nearshore during various geologic epochs are vastly different (Table 1). During the Cenozoic, phosphate sedimentatation reached a maximum in the Late Paleocene and Early Eocene, became reduced in the Oligocene before reaching another maximum in the Middle Miocene (Cook and McElhinny, 1979; Arthur and Jenkyns,1981).

A naive supposition that the relative abundance of phosphate occurrence reflects terrestrial phosphorous flux would predict more phosphate deposition at times of low sea-level stand, when more land was exposed to erosion. In fact, the contrary is true, the Eocene and Miocene epochs of maximal depositional rates

TABLE 1. Compilation of (1) accumulation rates of C_{org} (McArthur and Jenkyns, 1981), (2) phosphatic deposits (Cook and McElhinny, 1979), (3) carbonate accumulation rate (Davies and Worsley, 1981), (4) siliceous deposits (Pisciotto, 1981) and major paleoceanographic and tectonic events during Paleocene and Eocene time.

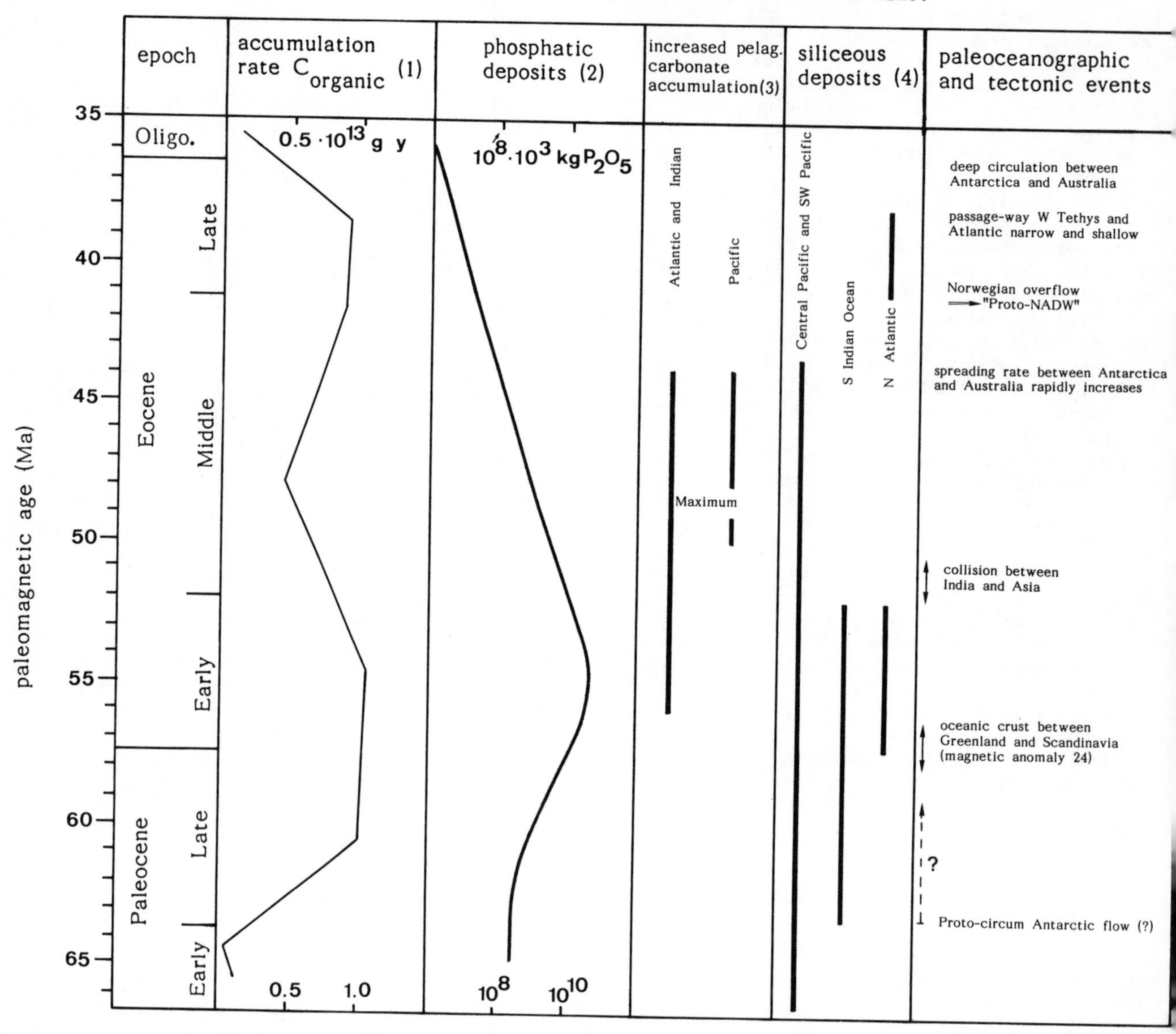

were times of transgression, not regression (Arthur and Jenkyns, 1981). The configuration of land and sea, the pattern of oceanic circulations, and other factor determine the degree of phosphorous fractionation. When continental shelves are drowned and when coastal upwelling is unusually active, the conditions are optimum for phosphate sedimentation. Phosphate deposition in nearshore upwelling zones results, however, in a reduced supply of this nutrient to the open oceans. Since phosphorous is a limiting nutrient controlling plankton-fertility, decreased supply leads to reduced productivity. It is, therefore, not surprising that the Eocene and Miocene, epochs of optimum phosphate deposition nearshore, also register the highest CCD levels of the Cenozoic Era.

Stable Isotope Stratigraphy

Oxygen Isotope Record

The oxygen isotopic ratio of fossil shells provides an excellent record of relative temperature change in ancient marine environments, because the $\delta^{18}O$ value of a calcareous fossil, when formed under conditions of isotopic equilibrium, is a function of the temperature.

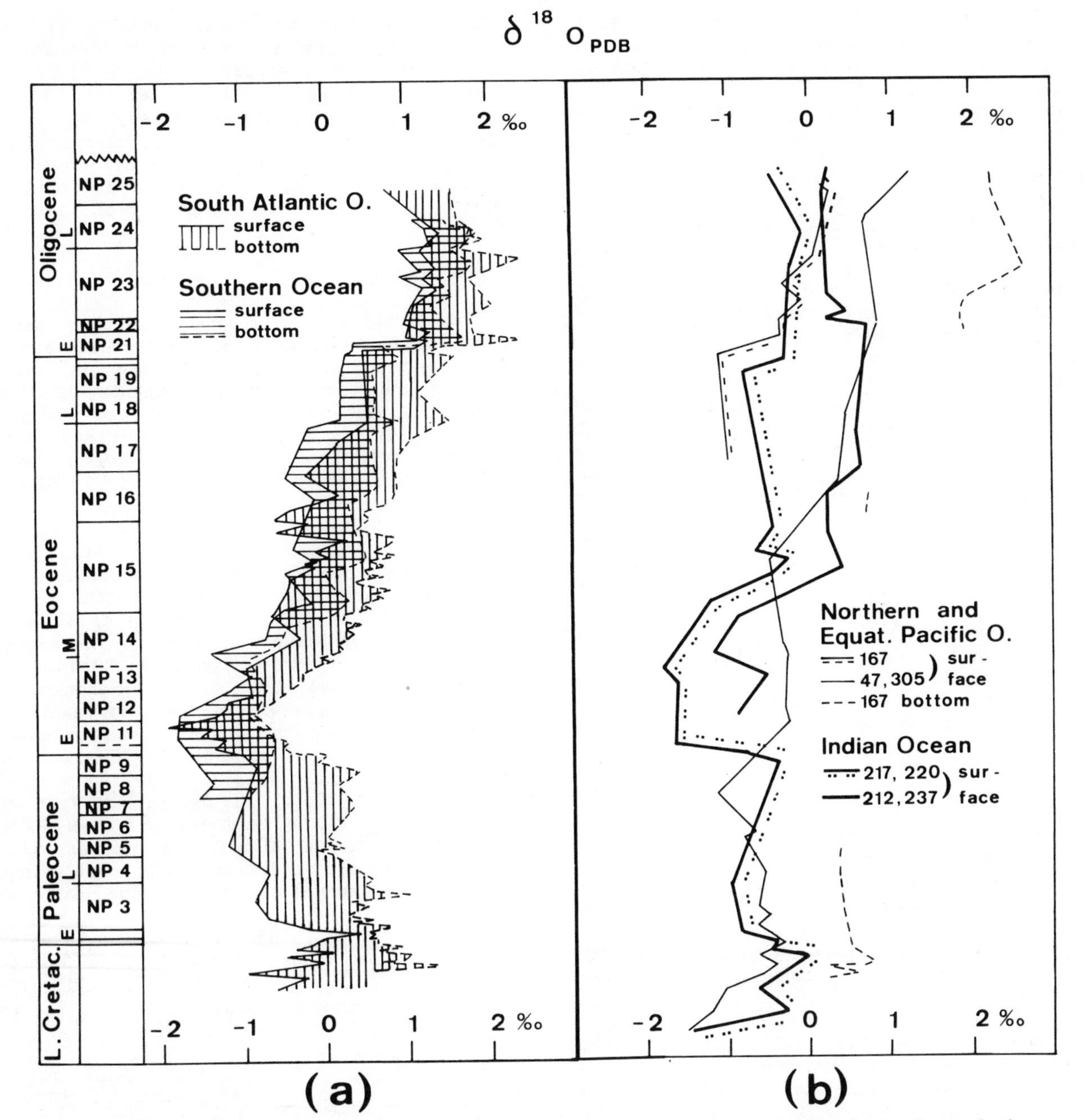

Fig. 3.a) Oxygen isotope pattern based on monospecific foraminiferal samples of the Southern Ocean and South Atlantic surface and bottom water environments (Shackleton and Kennett, 1975; Williams et al., 1983; Oberhänsli et al., 1984; Poore and Matthews, 1984; Shackleton et al., 1984; Oberhänsli and Toumarkine, 1985) and b) oxygen isotope pattern based on monospecific planktic foraminiferal samples and bulk samples from the Northern and Equatorial Pacific and Indian Oceans (Douglas and Savin, 1973, 1975; Oberhänsli, 1985).

But the oxygen isotope value also reflects the isotopic composition of the sea water in which the fossil organism grew. Variations in isotopic composition could thus also be interpreted in terms of salinity changes of the oceans, e.g. induced by the growth or waning of polar ice-caps or continental ice-sheets.

The Paleogene oxygen-isotope data from the Southern Ocean (DSDP Site 277; Shackleton and Kennett, 1975), the Northern and Equatorial Pacific (DSDP Sites 47, 167, 171 and 305; Douglas and Savin, 1973;1975), the South Atlantic (DSDP Sites 516F, 522- 524, 525, 527-529; Williams et al., 1983; Oberhänsli et al., 1984;

Poore and Matthews, 1984; Shackleton et al., 1984; Oberhänsli and Toumarkine, 1985) and the Indian Ocean (Oberhänsli, 1985) show comparable trends (Figure 3a ,b). The patterns of the Atlantic and the Southern Oceans are partiularly clear, and the trends have been registered by fossil organisms living in both the surface and bottom environments. The Paleocene was an epoch of general decrease of oxygen isotope values, signifying a warming trend. However, the oxygen-isotope data from the North and South Atlantic and from the North Pacific Oceans all indicate a relatively cool earliest Paleocene (Douglas and Savin, 1973; Vergnaud-Grazzini et al., 1978; Boersma and Premoli-Silva, 1983; Williams et al., 1983). After a peak value was reached during the earliest Eocene (Zone NP11), the $\delta^{18}O$ values began to increase and increased sharply by about 1.8 ‰ during the Early and early Middle Eocene. The increase became less pronounced during the late Middle and Late Eocene. Since the polar ice volume and its change are believed to have been insignificant during the Paleocene and Eocene, the oxygen-isotope signals are probably an expression of ocean cooling, especially at high and middle latitudes. The next major step-like increase of $\delta^{18}O$ values took place at the beginning of the Oligocene (NP 21). Whether the change indicates a drastic decrease of ocean temperature, or an increase of polar ice volume is a controversial issue (Shackleton and Kennett, 1975; Matthews and Poore, 1980). The results of South Atlantic drilling seemed to favor the latter alternative (Poore and Matthews, 1984; Hsü et al., 1984a).

The Pacific and the Indian Ocean records are on the whole similar: Low isotopic values (warm ocean temperatures) for Paleocene and Early Eocene, a sharp increase of values during the Middle Eocene, resulting in an early Oligocene minimum. A close look at those records show, however, considerable differences registered by nannoplankton samples collected from sites at different latitudes. The record of lower latitude sites (217, 220 and 167; Figure 3b) definitly indicates warmer temperature (or less likely lower salinity) than that of the mid latitude sites (212, 237, 305 and 47; Figure 3b). Although both the low and mid latitude oxygen-18 records show the same general trend as that of the Atlantic and the Southern Oceans (Figure 3a), the low-latitude record of the Indian Ocean suggests a somewhat delayed climatic deterioration taking place during the Middle Eocene (NP14/15). The mid-latitude record of the Pacific shows an earlier (NP8/11) and a later (NP15/16) cooling.

The difference in the $\delta^{18}O$ values between surface- and bottom-water inhabitants gives a measure of the vertical temperature gradient. The difference ranges between 1‰ to 1.5 ‰ during much of the Paleocene at mid-latitude Atlantic and Pacific sites. Assuming that the $\delta^{18}O$ values are a function of temperature only, we could conclude that the Paleocene Ocean had a modest vertical temperature gradient, with the difference between the surface and bottom (at 1500 to 3000 m paleodepth) in the range of about 4 to 7°C. The difference for the Eocene sites of the Atlantic and Southern Oceans was even smaller, probably of the order or 3 or 4 °C only.

Carbon Isotope Record

The carbon-isotope record of benthic fossils gives some indication of land/ocean biomass fractionation and the difference in values between the planktonic and benthic skeletons could serve as a monitor of organic productivity (e.g. Deuser and Hunt, 1969; Berger, 1971; Kroopnick et al., 1977; Shackleton, 1977; Bender and Keigwin, 1979; Vincent et al., 1980; Broecker and Peng, 1983; Hsü and McKenzie, 1985). The possibility that the values are influenced by vital effect is, however, a not completely resolved problem.

The carbon-isotope composition of the dissolved CO_2 in the oceans today is shown by Figure 4 (Kroopnick et al., 1977). The surface water is depleted in carbon-12 because of the utilization of the light carbon by plankton. A slight C-12 maximum within the oxyygen-minimum zone is noted and the increase can be attributed to addition of biogenic CO_2 derived from oxidation of organic matter.

The $\delta^{13}C$ plots (Figure 5a, b) from the South Atlantic, the Southern and the Indian Oceans show several significant changes within the Paleocene-Eocene interval. The most evident feature, documented by a positive carbon-13 peak in surface and bottom skeletons is noted during the latest Paleocene in the South Atlantic (maximum enrichment Zones NP8 - NP9, or top P4 - P5 respectively; Figure 5a). Calibration with the magnetostratigraphy indicates that the

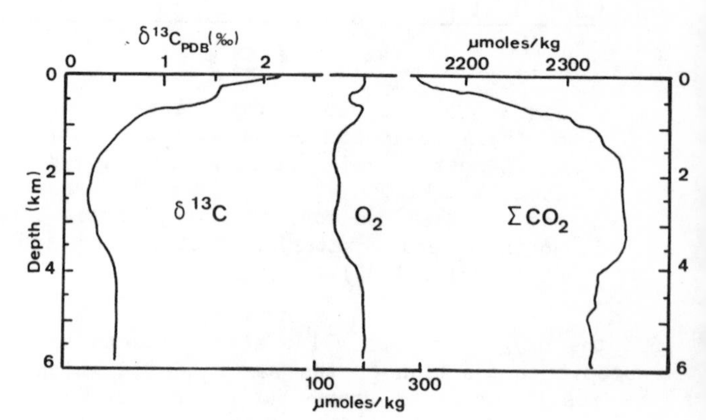

Fig. 4. Carbon isotopic pattern of the dissolved CO_2 in surface and deep ocean waters today (after Kroopnick et al., 1977)

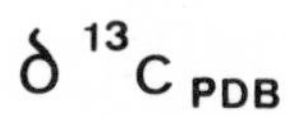

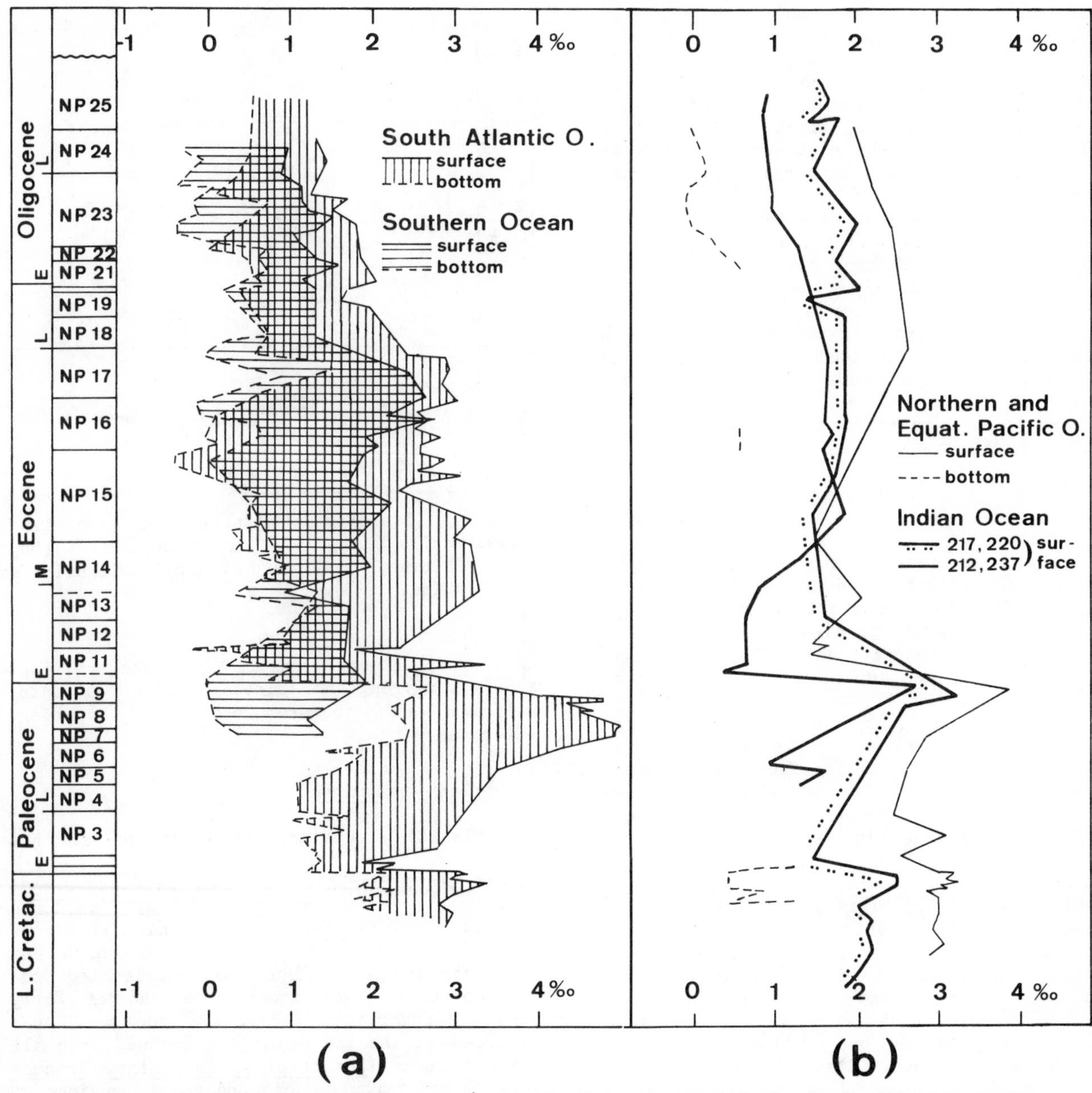

Fig. 5. Carbon isotopic pattern of a) the Southern Ocean and the South Atlantic Ocean surface and bottom water environments (Shackleton and Kennett, 1975; Williams et al., 1983; Oberhänsli et al., 1984; Poore and Matthews, 1984; Shackleton et al., 1984; Oberhänsli and Toumarkine, 1985) and b) the Northern and Equatorial Pacific and Indian Oceans (Douglas and Savin, 1973, 1975; Oberhänsli, 1985).

positive carbon isotope "spike" occurs in the interval from late C-25-R to early C-24-R (Oberhänsli and Toumarkine, 1985), with a magnetostratigraphic age of about 61 to 58 Ma (Hsü et al., 1984b; Berggren et al., in press). This C-13 enrichment is evident in both the planktonic and benthic deposits of all the oceans. A comparision of $\delta^{13}C$ values obtained from benthic foraminifers with those of surface-dwelling morozovellids indicates that at times when $\delta^{13}C$ values were most positive the surface-to-deep water gradient was at its maximum (Shackleton and Hall, 1984; Schackleton, et al., 1984). This facts suggest that the $\delta^{13}C$

maximum of the Late Paleocene was related to a certain extent to enhanced ocean productivity which removed light carbon through photosynthesis from the oceanic system. However, the positive anomaly of benthic fossils indicate that the whole ocean was depleted in carbon 12. The anomaly is not, as in the case of Cretaceous anoxic event, associated with widespread occurrences of black shales, and we cannot explain the C-13 enrichment through an assumption of preferential burial of C-12-rich organic sediments. The alternative is to invoke abnormal ocean/land fractionation as the cause of the anomaly. Such a fractionation implies an unusual abundance of land plants. More remarkable than the anomaly peak is the sudden collapse in $\delta^{13}C$ values at the end of the Paleocene. Was there a rapid decrease of the biomass on land? If so, what was the cause? Was there a catastrophic destruction of temperate forests on the land mass in high latitudes in response to a climatic change? Or was there an unusual marine transgression which drowned the tropical forests on coastal lowlands? We have not enough geological data to answer those questions until e. g. the subglacial Antarctic is explored.

Analyses of planktic and benthic skeletons of the South Atlantic and the Southern Oceans provide information on vertical carbon-isotope gradient. Very large difference, up to 3‰ or more, between the surface and bottom values has been registered by samples from the interval spanning Zones NP 12 to NP 16 (Figure 5a). One may invoke the simple explanation that the large vertical gradient reflects high biological productivity in surface water environments. Yet the CCD fluctuation and the phoshate data, as we have discussed, do not suggest unusually high plankton fertility during the Eocene. Furthermore, the vertical gradient should be greater at high-latitude (Southern Ocean) sites of high fertility than at mid-latitude (South Atlantic) gyre sites, if productivity alone had been responsible for the gradient. An alternative is to relate the gradient to the efficacy of ocean-mixing. A sluggish Eocene ocean with a turnover rate twice as long as the one at the present may have become more stratified, allowing isotope fractionation to become twice as advanced. This may be a better explanation of the large vertical C-isotope gradient of the Eocene (especially that of the mid-latitudes), as compared to the smaller gradient during the Oligocene and Neogene, when vigorous circulations of proto-AABW and proto-NADW contributed to accelerate the turnover time of ocean waters (Hsü et al., 1978; 1984a; Johnson, 1985).

The Paleontologic Record

Benthic and planktic faunal and floral assemblages are environmental indicators, and

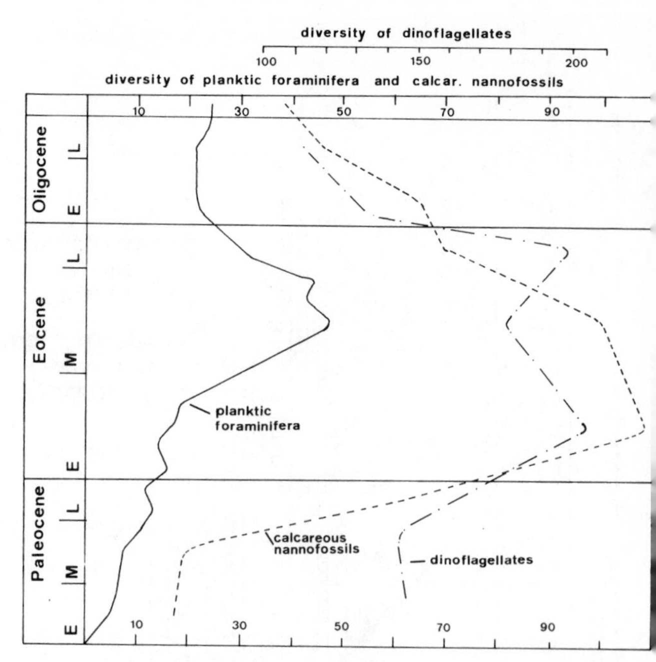

Fig. 6. Diversity patterns of calcareous nannofossils (Haq, 1973), dinoflagellates (Bujak and Williams, 1979) and planktic foraminifers (Toumarkine, 1983).

their population dynamics reflects nutrient supply, oxygen and carbon-dioxide contents, salinity and temperature of the ambient watermass.

Microplankton Record

Planktic foraminiferal evolution did not undergo dramatic crisis during the Paleocene-Eocene time. Planktic foraminiferal assemblages show mostly gradual changes. Following an accelerated recovery after the terminal Cretaceous crisis, new species continued to evolve and diversity increase (Figure 6). Accelerated speciation took place toward the end of Early Eocene (Toumarkine, 1983). But the trend was reversed at the end of Middle Eocene, when all spinose forms (*Morozovella, Acarinina, Truncorotaloides*) virtually disappeared in regions of lower latitudes (Toumarkine and Luterbacher, 1985). The patterns of nannoplankton and dinoflagellate evolution show the same genral trend, although, as shown by Figure 6, the accelerated increase of diversity of both took place somewhat earlier and the rapid decline of dinoflagellate later (Haq, 1973; Bujak and Williams, 1979). Diversity change has been commonly interepreted in terms of temperature variations. A comparison of figures 3 and 6 indicates that the diversity maximum (Early /Middle Eocene) came somewhat later than the postulated temperature-maximum (latest Paleo-

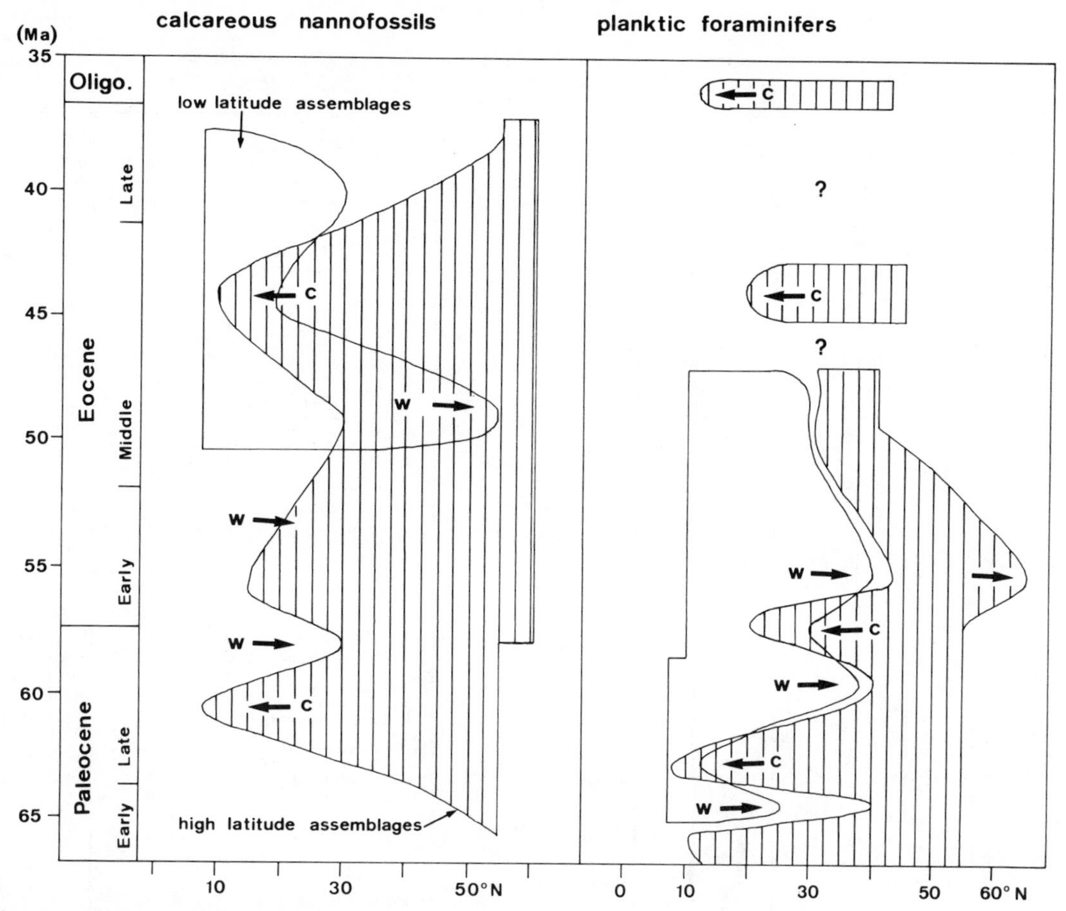

Fig. 7. Latitudinal migration patterns of calcareous nannofossils and planktic foraminifers from the Northern Atlantic during the Paleocene through Eocene (w= warming trend, c= cooling trend; after Haq et al. 1977).

cene/Early Eocene), suggesting a delayed faunal response. One must, however, remember that diversity change is probably closer related to fluctuating trophic resources (Valentine,1971). Diversity changes might indicate changes of the extent of faunal provinces (Schopf, 1984).

Faunal and floral provinces have been recognized by micropaleontologists, and latitudinal shift of faunal and floral distribution has been interpreted as signs of climatic change (e.g. Haq et al., 1977) and the interpretation by Haq et al. (1977) is shown by Figure 7 and Table 2. Sancetta (1979) recognized a tropical and a temperate province in the Southern Hemisphere during the Middle Paleocene (NP4 to NP5) and three (tropical, transitional and temperate) in the southern Indo-Pacific realm during the Middle and Late Eocene (NP15 to NP17). Provinciality increased during the late and Early Oligocene (NP22/23), when five faunal and floral provinces were differentiated. Whereas both the fossil and isotope data indicate a Late Paleocene/Early Eocene warming phase, the Early Eocene/Middle Eocene warming phase suggested by the fossil migration patterns is not registered by the isotope record (Figures 3). We do not have an adequate answer for this discrepancy in interpretations, except noting that the shift of provincial boundaries could be influenced by complex factors connected with changing oceanic circulations (Schopf, 1984) .

Sancetta (1979) placed emphasis on the increase of provinciality: The change from two to five reflects a more differentiated latitudinal temperature gradient in response to climatic cooling, and a most significant step took place only during the Late Eocene and Early Oligocene.

Benthic Fauna Record

Deep-sea benthic foraminifera evolved in a distinctly different way from that of near-surface dwelling calcareous microplankton (Table 2). The turnover of planktic species reached catastrophic proportions across the Cretaceous/Tertiary boundary, and was very ra-

TABLE 2. Major stable isotopic, biologic and biogeographic events as well as the climatic evolution deduced from terrestrial vegetation (1) Wolfe, 1978).

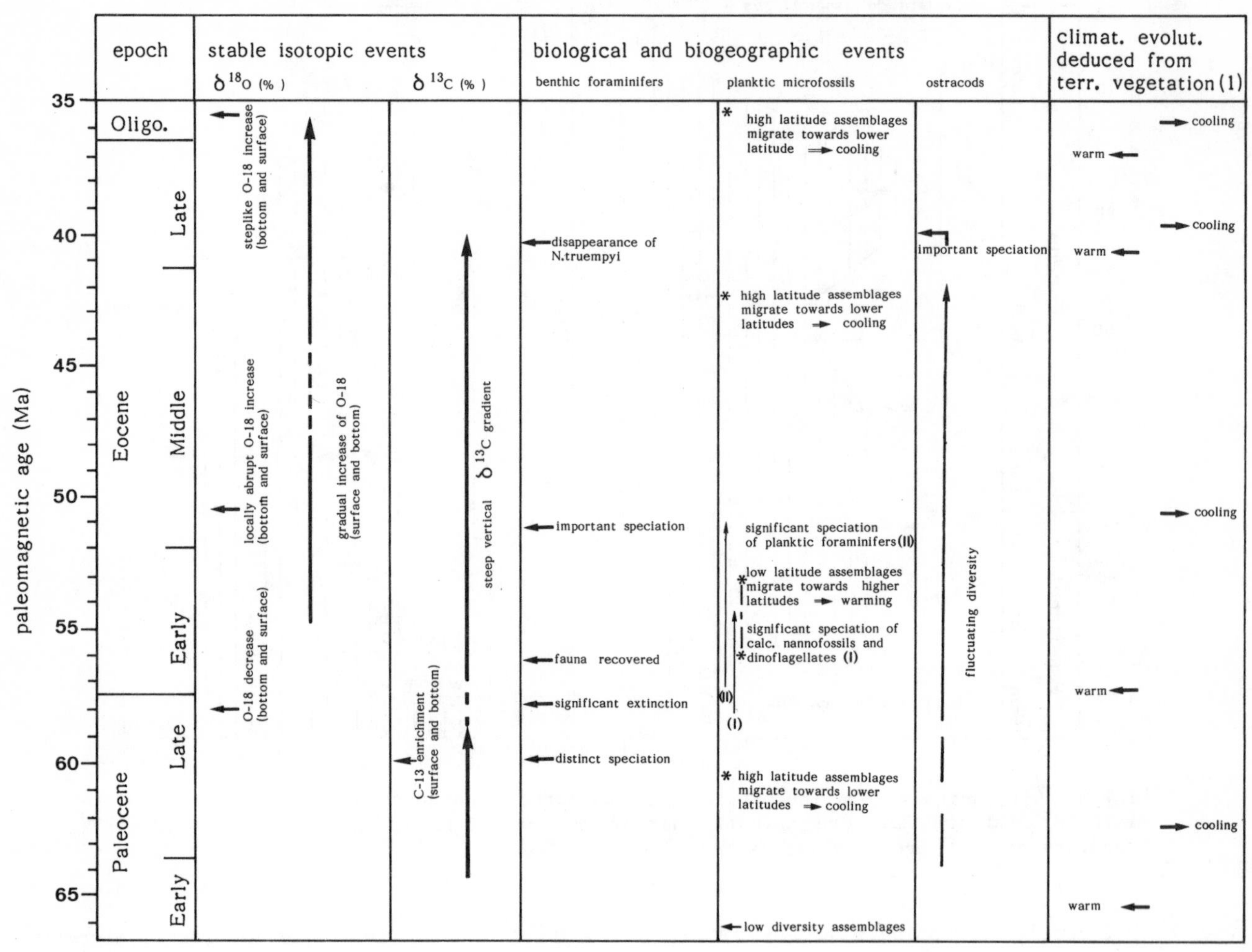

pid in late Middle Eocene and at the end of Eocene. In contrast, benthic assemblages in the deep sea remained little affected by these evolutionary events. The Paleocene benthic foraminiferal asemblages are considered relic from the Late Cretaceous; about 80 % of the benthic species survived the terminal Cretaceous crisis and continued into the Paleocene (Beckmann, 1960; Webb, 1973). The Paleocene was a period of remarkable faunal stability with appearently few appearances of new benthic species and even fewer extinctions (Tjalsma and Lohmann, 1982). The stasis persisted the Late Paleocene when several new species appeared in Zones P4 to P5. A most significant faunal turnover occurred during the latest Paleocene, at the beginning of Zone P6a (Beckmann, 1960; Tjalsma, 1977; Schnitker, 1979). This benthic faunal turnover has been regisitered in the records of the Equatorial and South Atlantic, of the Pacific, of the Tethys and of the Indian Ocean (von Hillebrand, 1962; Rona, 1973; Vincent et al., 1974; Sigal, 1974; Braga et al., 1975; Douglas, 1979). The damage seems to have been particularly severe to the shallower-water benthic faunas (Tjalsma and Lohmann, 1982). The benthic faunal system quickly recovered, however, some 2 Ma later (Zone P6b), when the fastest speciation rate for the Paleocene/Eocene interval was recorded (Tjalsma and Lohmann, 1982). A new assemblage, well described from the South Atlantic, appeared just after the first appearance of Morozovella subbotinae, which marked the beginning of Zone P6b (Tjalsma, 1977). The fast faunal turnover was accompanied by an usually rapid migration of depth-habitats, resulting in large changes of bathymetric ranges of various taxonomic groups (Douglas, 1979; Tjalsma and Lohmann, 1982). Another phase of accelerated speciation took place somewhat later, during the late Early Eocene to the early Middle

Eocene (Zones P9 - P10), when a considerable number of new taxa first appeared. After that, according to Tjalsma and Lohmann (1982), the benthic foraminiferal assemblages remained stable, with relatively few new appearances and few extinctions during the Middle/Late Eocene time except for the remarkable disappearance of the deep-water index species Nuttallides truempyi at the end of the Middle Eocene. Douglas and Woodruff (1981) found evidence that the benthic foraminiferal assemblages evolved rapidly in waters shallower than 2000 meters depth during the MIddle Eocene (Zones P 13/14), but more slowly in deeper waters. Parkers et al. (1984) studying the benthic foraminiferal assemblages of the South Atlantic found also a fast turnover rate in the middle Middle Eocene time (P12, NP16), and another in latest Eocene. Those changes involve a small increase of local extinctions, a large reduction of local originations, a dramatic shift in species relative abundances, and an abrupt increase in species equitability. They concluded that the psychrospheric fauna was established in latest Eocene.

Studies of ostracodes have led to a similar conclusion. The benthic ostracode populations were high during the Cretaceous, but fluctuated radically during the Paleocene. The Eocene population was on the whole low. Starting with an impoverished fauna in the Early Eocene, the population increased considerably till the Middle Eocene, only to drop suddenly again in the Late Eocene, at about 40 Ma. The population became again high during the Oligocene. The faunal diversity fluctuated from Cretaceous till the late Middle Eocene, when the total number of genera and species increased considerably, especially in the South Atlantic and the Indo-Pacific Oceans.

A major faunal turnover , involving a change from thermospheric to psychrospheric fauna, took place during the Late Eocene (Benson et al., 1985). The extent and the rate of the change are, however, not the same in all the oceans, nor was the event synchronous everywhere. This replacement of the warm-loving bottom-dwelling ostracodes by the cold-loving confirmed the foraminiferal evidence for an intrusion of cold bottom waters to the world oceans during the Late Eocene (Benson et al., 1985).

Terrestrial Plants

Fossil plants and their distribution are useful paleoclimatic indicators (Table 2). Fossil floras of North America, especially their physiognomic aspects, have indicated warmer intervals during Early Paleocene, latest Paleocene to Early Eocene, late Middle Eocene and latest Eocene and cooler intervals during the early Late Paleocene, the Early/Middle Eocene transition and the early Late Eocene (Axelrod and Bailey, 1969; Wolfe, 1978; Wolf and Poore, 1982). The fluctuation resulted in an overall deterioration of climate, yet the Paleocene/Eocene mean annual temperatures were nevertheless higher than those of today in North America, while the mean annual ranges of temperature variation were less. A similar pattern of climatic changes is known from studies of plant remains found in Western Europe (Dorf, 1964; Schwarzbach, 1968) and southern Australia (Kemp, 1978). Climatic cooling had reached significant proportion prior to the beginning of the Oligocene. The land record is thus in general accord with the marine data.

Paleoclimate, Paleotectonic Events and Paleocirculation

The ocean climate in Early Paleocene was relatively cool. A warming trend, with local fluctuations, during the Paleocene led to the Cenozoic climatic optimum at the beginning of the Eocene (Zone NP11). The broadening of the low-latitude planktic-fossil province suggests that latitudinal temperature gradient was not pronounced at this time. The Eocene tropical/subtropical belt had a width twice that of the present (Wolfe, 1978), and tropical rainforests were widespread on the continents of both the Northern and Southern Hemispheres (Wolfe, 1978; Kemp, 1981).

The cooling trend started at the earliest Middle Eocene. This climatic deterioration is well documented by the isotope records, especially those from the Atlantic, Indian and Southern Oceans. The surface temperatures dropped during the Middle Eocene at mid- and high latitudes about just as much as the bottom temperatures did. A climatic deterioration at this time is also evident in the record of fossil plants on land (Wolfe, 1978).

The Middle Eocene cooling did not lead to a significant reduction of latitudinal gradient, which had a value about half of that prevailing today (Shackleton and Boersma, 1981). The vertical temperature gradient was also not very pronounced.

The ocean climate continued to deteriorate during the late Middle and Late Eocene. Cooling of the surface water, however modest, is shown by the isotope record, while the bottom environments were less affected, the temperature was in the range of 1 or 2 °C in the Southern Ocean realm. Then came the sudden drop of temperature by several degrees at the beginning of the Oligocene (Corliss, this volume).

The ocean /continent configuration during the Early Paleogene (Table 1) was mainly conrolled by the movement of the Indian and Australian plates and to a lesser extent by that of the North American, European and African plates. Seafloor spreading of the Norwegian/Greenland Sea region began in latest Paleocene or earliest Eocene (Anomaly 25 or 24; Talwani and Eldholm, 1977). The Early/Middle

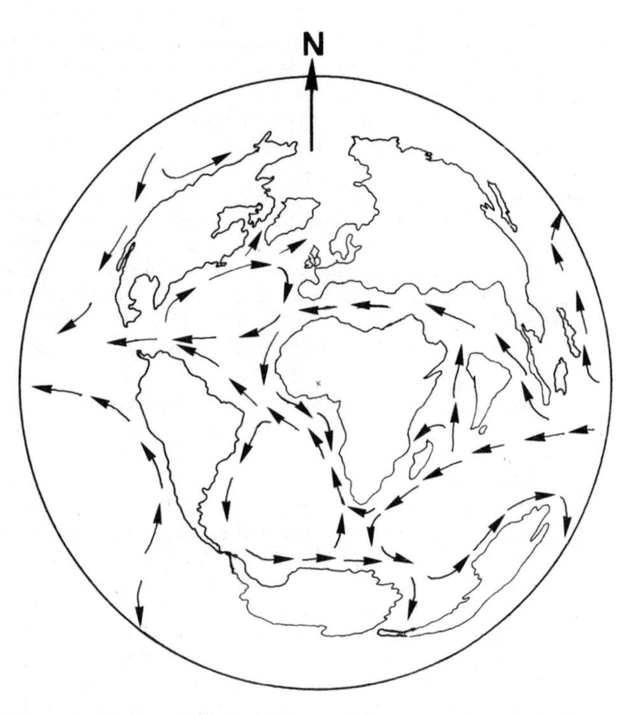

Fig. 8. Paleocirculation model reconstructed for the Late Paleocene to Early Eocene (after Haq, 1981).

Eocene plate movement in the Indo-Pacific realm led to 1) Collision of India with Asia (Anomaly 22, approximately 53-50 Ma; Curray and Moore, 1974; McGowran, 1978), and 2) separation of Australia from Antarctica during the late Middle Eocene. Progressive spreading resulted in the subsidence of the Tasman Rise to allow deep circulation by the Late Eocene. Meanwhile in the North Atlantic the Iceland-Faroe Sill sank below the sea-level during the Late Eocene (Talwani, et al., 1976). The Svalbard-Greenland Strait started to open after the beginning of the Oligocene (post-Anomaly 13 time, Schrader, et al., 1976), permitting a water exchange between the Arctic and the North Atlantic. Collision of the African and European Plates in the Alpine realm in Late Eocene severely restricted deep water connections between the Tethys and the Atlantic (Biju-Duval et al., 1977).

Plate tectonic events and the accompanying climatic changes led to reorganization of oceanic circulations. Halokinetic circulations predominated during the Cretaceous, Paleocene and Early Eocene, when deep circulation was induced by the descent of saline, but warm, bottom waters (Berger, 1970; 1979; Brass et al., 1982; Johnson, 1985). A thermohaline circulation was probably not fully established until the Early Oligocene (Johnson,1985). Deep thermohaline convection during the Paleogene was probably salinity-limited rather than temperature-limited; it seems that the essential condition governing the initial formation of the AABW was related to the injection of a high-salinity at intermediate depth (i.e. proto-NADW; Miller and Curry, 1982).

A most prominent feature of the earlier Paleogene circulation was the Tethys current, effecting a circum-Equarorial transport of surface watermass (Figure 8). Anti-clockwise and clockwise gyres are inferred to have existed in the Southern and Northern Hemisphere respectively (Haq, 1981). By the end of the Eocene, however, the circum-Equatorial current was weakened due to the collision of Africa and Eurasia. In the North Atlantic, the ancestral Gulf Stream made its appearance during the Eocene, driving surface currents toward the Labrador and Norwegian Seas, as the southen Labrador Passage and the Norwegian/Greenland Sea were being widened by seafloor spreading (Pinet and Popenoe, 1985).

In the Southern Hemisphere, circum-Antarctic surface currents was established before the Late Eocene (Kennett, et al., 1975; Harwood, et al, 1983; Webb et al., 1984; Wise et al., 1985). But the deep circumpolar current was only established when the deep Drake passage was opened, probably in the Late Oligocene (Kennett et al., 1975), although one school of thought suggested that a circumpolar flow at depth greater than 1000 m came into existence as early as the Paleocene (Barker et al., 1977).

The psychrospheric circulation may have had its start in Late Eocene, and should have been fully established after the severe climatic deterioration at the beginning of the Oligocene. Nutrient-rich waters begun to flow northward for the first time from the newly established bottom water sources (Benson, 1975; Shackleton and Kennett, 1975) where "Proto-AABW"formed. The AABW circulation through the basins of the southwestern Atlantic was certainly in full force during the Early Oligocene (Anomaly 12 time, approximately 32 Ma; Johnson, 1985) an event documented by an unconformity in the Vema Channel and the Brazil Basin (Horizon A of Ewing et al., 1975).

We have found indications that upwelling activities enhancing biologic productivity increased locally in the South Atlantic during the Late Paleocene, resulting in the deposition of silica-rich oozes near the Falkland Plateau (Gombos, 1977). Rich radiolarian and diatom floras are also known the Paleocene of the Cape Basin (Gombos, 1984), of the southeastern Indian Ocean (Murkhina, 1976) and of Central Pacific (Leinen, 1979). In the Northern Atlantic upwelling was apparently not active until the Early Eocene (Riech and von Rad, 1979).

The major speciation events in the time interval Late Paleocene/Middle Eocene (Table 2) should have been related to in someway to the paleoceanographic events just described. The relatively faster evolutionary rates in shallower depths suggests that rapid changes of the circulation patterns of the intermediate watermasses (Douglas and Woodruff, 1981). The complete replacement of the thermospheric faunas by the pschrospheric fauna indicated the dramatic conversion from halokinetic to thermohaline circulations during the Late Eocene/Early Oligocene crisis.

References

Adams, C.G., Some Tertiary foraminifera, In: Hallam, A. (ed) Atlas of palaeobiogeography, Elsevier, Amsterdam, 453-468, 1973.

Adams, C.G., Gentry, A.W. and Whybrow, P.J., Dating of the terminal Tethyan event, Utrecht Micropaleont. Bull., 30, 273-298, 1983.

Arthur, M.A., Paleoceanographic events - recognition, resolution, and reconsideration, Rev. Geophys. Space Phys., 17, 1474-1494, 1979.

Arthur, M.A. and Jenkyns, H.C., Phosphorites and paleoceanography, Oceanol. Acta, No.Sp., 83-96, 1981.

Axelrod, D.H. and Bailey, H.P., Paleotemperature analysis of Tertiary floras. Palaeogeogr. Palaeoclimat. Palaeoecol., 6, 163-195, 1969.

Barker, P., Dalziel, W.D., Dinkelman, M.G., Elliot, D.H., Gombos, A.M., Lonardi, A., Pflafker, G., Tarney, J., Tompson, R.W., Tjalsma, R.G., Von der Borch, C.C., Wise, S.W., Harris, W. and Sliter, W.V., Evolution of the southwestern Atlantic Ocean basin, Init. Reports. DSDP, 36, 993-1014, 1977.

Beckmann, J.-P., Distribution of benthonic foraminifera at the Cretaceous-Tertiary boundary of Trinidad (West-Indies), 21st Int. Geol. Congr. Proc., Sec. 4, 57-69, 1960.

Bender, M.L. and Keigwin, L.D. jr., Speculations about the upper Miocene change in abyssal Pacific dissolved bicarbonate d13 C, Earth Planet. Sci. Lett., 45, 383-393, 1979.

Benson, R.H., The origine of the psychrosphere as recorded in changes of deep-sea ostracode assemblages, Lethaia, 6, 69-83, 1975.

Benson, R.H., Chapman, R.E. and Deck, L.T., Evidence from the ostracode of major events in the South Atlantic and world-wide over the past 80 million years, In: Hsü, K.J. and Weissert, H. (eds) South Atlantic paleoceanography, Cambridge Univ. Press, 325-350, 1985.

Berger, W.H., Biogenous deep-sea sediments: fractionation by deep-sea circulation, Geol. Soc. America Bull., 81, 1385-1402, 1970.

Berger, W.H., Sedimentation of planktonic foraminifera, Marine Geology, 11, 325-358, 1971.

Berger, W.H., Biogenous deep-sea sediments: production, preservation and interpretation, In: Riley, J.P. and Chester, R. (eds) Treatise on chemical oceanography, Academic Press, London, 265-388, 1976.

Berger, W.H., Impact of Deep-Sea Drilling in Paleoceanography, In: Talwani, M. Hay.W. and Ryan, W.B.F. (eds) Results in the Atlantic Ocean: Continental margins and paleoenvironment. American Geophys. Union, Washington D.C. Maurice Ewing Series 3, 297-314, 1979.

Berger, W.H. and Vincent, E., Chemostratigraphy and biostratigraphic correlation: exercises in systematic stratigraphy, Oceanol. Acta, Sp.No., 115-127, 1981.

Berger, W.H., Vincent, E. and Thierstein, H.R., The deep-sea record: Major steps in Cenozoic ocean evolution, Soc.Econ.Pal.Min., Spec.-Publ., 32, 489-504, 1981.

Berggren, W.A. and Hollister, C.D., Plate tectonics and paleocirculation-commotion in the ocean, Tectonophys., 38, 11-48, 1977.

Berggren, W.A., Kent, D.V. and Flynn, J.J., Paleogene geochronology and chronostratigraphy, In: Snelling, N.J. (ed) Geochronology and the geological record, Geol. Soc. London Spec. Paper, in press.

Biju-Duval, B., Delcourt, J. and LePichon, X., From the Tethys Ocean to the Mediterranean Sea: A plate tectonic model of the evolution of the western alpine system, In: Biju-Duval, B. and Montadert, L. (eds) Internat. Sympos. Struct. History Mediter. Basins. Split (Jugoslavia), 25-29th Oct. 1976, 143-164, 1977.

Biolzi, M., Stable isotopic study of Oligocene-Miocene sediments from DSDP Site 354, Equatorial Atlantic, Marine Micropaleontol., 8, 121-139, 1983.

Boersma, A. and Premoli Silva, I., Paleocene planktonic foraminiferal biostratigraphy and the paleoceanography of the Atlantic Ocean, Micropaleontol. 29/4, 355-381, 1983.

Braga, G., de Biase, R., Grünig, A. and Proto Decima, F., Foraminiferi bentonici del Paleocene ed Eocene della Sezione di Possagno, Schweiz. Paläontol. Abh., 97, 85-96, 1975.

Brass, G.W., Southam, J.R. and Petersen, W.H., Warm saline bottom water in ancient oceans, Nature, 296, 620-623, 1982.

Broecker, W.S., Chemical oceanography, Harcourt Brace Jovanovitch, New York, 1-214, 1974.

Broecker, W.S. and Peng, T.-H., Tracers in the sea, Lamont-Doherty Geol. Observat. Publ., Columbia University, 1-690, 1982.

Buchardt, B., Oxygen isotope paleotemperatures from the Tertiary period in the North Sea, Nature, 275, 121-123, 1978.

Bukry, D., Paleogene paleoceanography of the Arctic Ocean is constrained by the middle and late Eocene age of USGS Core Fl.-422: Evidence from silicoflagellates, Geology 12/4, 199-201, 1984.

Bujak, J.P. and Williams, G.L., Dinoflagellate diversity through time, Marine Micropaleontol., 4, 1-12, 1979.

Cook, P.J. and McElhinny, M.W., A re-evaluation of the spatial and temporal distribution of sedimentary phosphate deposits in the light of

plate tectonics, Econom. Geol., 74, 315-330, 1979.
Curray, J.R. and Moore, D.G., Sedimentary and tectonic processes in the Bay of Bengal deep-sea fan and geosyncline, In: Burk, C.A. and Drake, C.L. (eds) The geology of the continental margins, Springer Verlag, New York, 617-627, 1974.
Deuser, W.G. and Hunt, J.M., Stable isotope ratios of dissolved inorganic carbon in the Atlantic, Deep Sea Res., 16/2, 221-225, 1969.
Dorf, E., The use of fossil plants in paleoclimatic interpretations, In: Nairn, A.E.M. (ed.) Problems in paleoclimatology, Interscience Publ. New York, 13-30, 1964.
Douglas, R.G., Benthic foraminiferal ecology and paleoecology: A review of concepts and methods, Soc.Econ.Pal.Min. Short Course, 6, Houston, Texas, 21-53, 1979.
Douglas, R.G. and Savin, S.M., Oxygen and carbon isotope analyses of Cretaceous and Tertiary foraminifera from the Central North Pacific, Init. Repts. DSDP, 17, 591-605, 1973.
Douglas, R.G. and Savin, S.M., Oxygen and carbon isotope analyses of Tertiary and Cretaceous microfossils from Shatsky Rise and other sites in the North Pacific Ocean, Init. Repts. DSDP, 32, 509-520, 1975.
Douglas, R.G. and Woodruff, F., Benthic foraminifera in the deep sea, In: Emiliani, C. (ed) The Secular variations in the pelagic realm, Soc. Econ. Pal. Min., Spec. Publ. 25, 19-50, 1977.
Fisher, A.G. and Arthur, M.A., Secular variations in the pelagic realm, SEPM, Spec. Publ. 25, 19-50, 1977.
Gombos, A.M., Paleogene and Neogene diatoms from the Falkland Plateau and Malvinas outer basin: Leg36 Deep Sea Drilling Project, Init. Repts. DSDP, 36, 575-602, 1977.
Gombos, A.M., Late Paleocene diatoms in the Cape Basin, Init. Repts. DSDP, 73, 495-511, 1984.
Haq, B.U., Transgressions, climatic change and diversity of calcareous nannoplankton, Marine Geology, 15, 15-20, 1973.
Haq, B.U.,Paleogene paleoceanography:Early Cenozoic oceans revisted, Oceanol.Acta, Sp.No.,71-82, 1981.
Haq, B.U., Premoli-Silva, I. and Lohman, G.P., Calcareous plankton paleobiogeographic evidence for major climatic fluctuations in the Early Cenozoic Atlantic Ocean, J. Geophys. Res., 82, 3861-3876, 1977.
Hillebrandt, A. von, Das Paleozän und seine Foraminiferenfauna im Becken von Reichenhall und Salzburg, Bayr. Akad. Wiss. Math.-Naturwiss. Kl. Abh., N.Ser. 108, 1-182, 1962.
Hsü, K.J., 1985.
Hsü, K.J., Montadert, L., Bernoulli, D., Cita, M.B., Erison, A., Garrison, R.E., Kidd, R.B., Melières, F., Müller, C. and Wright, R., History of the Mediterranean salinity crisis, Init. Repts. DSDP, 42/1, 1053-1078, 1978.
Hsü, K.J., McKenzie, J., Oberhänsli, H. and Wright, R.C., South Atlantic Cenozoic paleoceanography, Init. Repts. DSDP, 73, 771-785, 1984a.
Hsü, K.J., Percival jr., S.F., Wright, R.C., Petersen, N.P., Numerical ages of magnetostratigraphically calibrated biostratigraphic zones, Init. Repts. DSDP, 73, 623-635, 1984b.
Hsü, K.J. and McKenzie, J.A., The "Strangelove" Ocean in the earliest Tertiary. In: Sundquist, E.T. and Broecker, W.S. (eds) The carbon cycle and atmospheric CO_2: Natural variations Archean to Present, Geophys.Monogr., 32, 487-492, 1985.
Johnson, D.A., Abyssal teleconnections II. Initiation of Antarctic bottom water flow in the Southwestern Atlantic, In: Hsü, K.J. and Weissert, H. (eds) South Atlantic paleoceanography, Cambridge Univ. Press, 243-281, 1985.
Kemp, E.M., Tertiary palaeogeography and the evolution of Australian climate, In: Keast, A. (ed) Ecological biogeography of Australia, Dr. W. Junk Publishers, The Hague, 33-49, 1981.
Kemp, E.M., Tertiary climatic evolution and vegetation history in the southeast Indian Ocean region, Palaeogeogr. Palaeoclimat. Palaeoecol., 24, 169-208, 1978.
Kennett, J.P., Cenozoic evolution of Antarctic glaciation, the Circum-Antarctic Ocean, and their impact on global paleoceanography, J. Geophys. Res., 82/27, 3843-3860, 1977.
Kennett, J.P., The development of planktonic biogeography in the southern ocean during the Cenozoic, Marine Micropaleontol., 3, 301-345, 1978.
Kennett, J.P., Houtz, R.E., Andrew, P.B., Edwards, A.R., Gostin, V.A., Hajos, M., Hampton, M., Jenkins, D.G., Margolis, S.V., Ovenshine, A.F. and Perch-Nielsen, K., Cenozoic paleoceanography in the southwest Pacific Ocean, Antarctic glaciation and the developement of the Circum-Antarctic current, Init. Repts. DSDP, 29, 1155-1169, 1975.
Kidd, R.B. and Davies, T.A., Indian Ocean sediment distribution since late Jurassic, Marine Geology, 26, 49-70, 1978.
Kroopnick, P.M., Margolis, S.V. and Wong, C.S., d13 C variations in marine carbonate sediments as indicators of the CO2 balance between the atmosphere and oceans, In: Andersen, N.R. and Malahoff, A. (eds) The fate of fossil fuel CO2 in the oceans, Marine Sci., 6, Plenum Press, New York, 295-321, 1977.
Leinen, M., Biogenic silica accumulation in the central Equatorial Pacific and its implications for Cenozoic paleoceanography, Geol. Soc. America Bull., 90, 1310-1376, 1979.
Matthews, R.K. and Poore, R.Z., Tertiary d18 O record and glacioeustatic sea level fluctuations, Geology, 8, 501-504, 1980.
McCoy, F.W. and Zimmerman, H.B., A history of sediment lithofacies in the South Atlantic, Init. Repts. DSDP, 39, 1047-1079, 1977.

McGowran, B., Stratigraphic record of Early Tertiary oceanic and continental events in the Indian Ocean region, Marine Geolgy, 26, 1-39, 1978.

Miller, K.G. and Curry, W.B., Eocene to Oligocene benthic foraminiferal isotopic record in the Bay of Biscay, Nature, 296, 347-350, 1982.

Moore, T.C.jr., Rabinowitz, P.D., Borella, P.E., Shackleton, N.J. and Boersma, A., History of the Walvis Ridge, Init.Rept.DSDP., 74, 873-894, 1984.

Murkhina, V.P., The Paleocene diatom ooze in the eastern part of the Indian Ocean, Okeanologiya, 14, 852-858, 1976.

Oberhänsli, H., Latest Cretaceous-Early Neogene oxygen and carbon isotopic record at DSDP sites in the Indian Ocean, Marine Micropaleontol., Spec. issue on stable isotopes,1985.

Oberhänsli, H., McKenzie, J. Toumarkine, M. and Weissert, H., A paleoclimatic and paleoceanographic record of the Paleogene in the Central South Atlantic. (Leg 73, Sites 522, 523, 524), Init. Repts. DSDP, 73, 737-747, 1984.

Oberhänsli, H. and Toumarkine, M., The Paleogene oxygen and carbon isotope history of DSDP Sites 522, 523 and 524 from the central South Atlantic Ocean, In: Hsü, K.J. and Weissert, H. (eds.) South Atlantic paleoceanography, Cambridge Univ. Press, 125-147, 1985.

Pinet, P.R. and Popenoe, P., A scenario of Mesozoic-Cenozoic ocean circulation over Blake Plateau and its environs, Geol. Soc. America Bull., 96, 618-626, 1985.

Pisciotto, K.A., Distribution, thermal histories, isotopic compositions, and reflection characteristics of siliceous rocks recovered by the Deep Sea Drilling Project, Soc. Econ.Pal.Min., Spec. Publ. 32, 129-147, 1981.

Poore, R.Z. and Matthews, R.K., Late Eocene-Oligocene oxygen and carbon isotope record from the South Atlantic Ocean DSDP Site 522, Init. Repts. DSDP, 73, 725-735, 1984.

Riech, U. and von Rad, U., Silica diagenesis in the Atlantic Ocean: diagentic potential and transformations, In: Talwani, M., Hay, W. and Ryan, W.B.F. (eds) Implications of deep drilling results in the Atlantic Ocean, Amer. Geophys. Union, M.Ewing Series, 2, 315-341, 1979.

Riedel, W.R., The occurrence of pre-Quarternary radiolaria in deep-sea sediments, In: Funnel, B.M. and Riedel, W.R. (eds) Micropaleontology of oceans, Cambridge Univ. Press., 567-594, 1971.

Rona, P.A., Worldwide unconformities in marine sediments related to eustatic changes of sea level, Nature; Phys. Sci., 244/132, 25-26, 1973.

Saltzman, E.S. and Barron, E.J., Deep-circulation in the Late Cretaceous: Oxygen isotope paleotemperatures from Inoceramus remains in DSDP cores, Palaeogeogr. Palaeoclimat. Palaeoecol., 40, 167-181, 1982.

Sancetta, C., Paleogene Pacific microfossils and paleoceanography, Marine Micropaleontol., 4, 363-398, 1979.

Savin, S.M., Douglas, R.G., Keller, G., Killingley, J.S., Shaughnessy, L., Sommer, M.A., Vincent, E. and Woodruff, F., Miocene benthic foraminiferal isotopic records: a synthesis, Marine Micropaleontol., 6, 423-450, 1981.

Schnitker, D., Cenozoic deep-water benthic foraminifera, Bay of Biscay, Init. Repts. DSDP, 48, 377-413, 1979.

Schnitker, D., Global paleoceanography and its deep water linkage to the Antarctic glaciation, Earth Sci. Rev., 16, 1-20, 1980.

Schopf, T.J.M., Climate is only half the story in the evolution of organisms through time, In: Brenchley, P. (ed) Fossils and climate, John Wiley & Sons, New York, 279-289, 1984.

Schrader, H.-J., Bjorklund, K., Manum, S., Martini, E. and van Hinte, J., Cenozoic biostratigraphy, physical stratigraphy and paleoceanography in the Norwegian-Greenland Sea, DSDP Leg 38 Paleoceanographical synthesis, Init. Repts. DSDP, 38, 1197-1211, 1976.

Schwarzbach, M., Tertiary temperature curves in New Zealand and Europe, Tuatara, 16, 1968.

Shackleton, N.J., Carbon-13 in Uvigerina: Tropical rainforest history and the Equatorial Pacific carbonate dissolution cycles, In: Anderson, N.R. and Malahoff, A. (eds) The fate of fossil fuel CO2 in the ocean, Marine Sci., 6, Plenum Press, New York, 401-427, 1977.

Shackleton, N.J. and Boersma, A., The climate of the Eocene Ocean, J. Geol. Soc. London, 138, 153-157, 1981.

Shackleton, N.J. and Hall, M.A., Carbon isotope data from Leg 74 sediments, Init. Repts. DSDP, 74, 613-617, 1984.

Shackleton, N.J. and Kennett, J.P., Paleotemperature history of the Cenozoic and the initiation of Antarctic glaciation: Oxygen and carbon isotope analyses in DSDP Sites 277, 279, and 281, Init. Repts. DSDP, 29, 743-755, 1975.

Shackleton, N.J., Hall, M.A. and Boersma, A., Oxygen and carbon isotope data from Leg 74 foraminifers, Init. Repts. DSDP, 74, 599-611, 1984.

Sigal, J., Comments on Leg 27 sites in relation to the Cretaceous and Paleogene stratigraphy in the eastern and southeastern African Coast and Madagascar regional setting, Init. Repts. DSDP, 27, 687-723, 1974.

Talwani, M. and Udintsev, G., Tectonic synthesis, Init. Repts. DSDP, 38, 1213-1242, 1976.

Talwani, M.J. and Eldholm, O., Evolution of the Norwegian-Greenland Sea, Geol. Soc. America Bull.,88, 969-999, 1977.

Tjalsma, R.C., Cenozoic foraminifera from the South Atlantic DSDP Leg 36, Init. Repts. DSDP, 36, 493-517, 1977.

Tjalsma, R.C. and Lohman,G.P., Paleocene-Eocene bathyal and abyssal benthic foraminifera from the Atlantic Ocean, Micropaleotology, Spec. Publ. 4, 1-90, 1982.

Toumarkine, M., Les foraminifères planctoniques de l' Eocène moyen et supérieur des régions tropicales à tempérées chaudes, Thèse d' Etat., Mém. Sci. Terre, 83-05, Univ. P. & M. Curie, Paris, 1-219, 1983.

Toumarkine, M. and Luterbacher, H. P., Paleocene and Eocene planktic foraminifera, In: Bolli, H.M., Saunders, J. and Perch-Nielsen, K. (eds) Plankton stratigraphy, Cambridge Univ. Press, 87-154, 1985.

Valentine, J.W., Plate tectonics and shallow marine diversity and endemism, an actualistic model, System. Zool., 20, 253-264, 1971.

Van Andel, T.H., An eclectic overview of plate tectonics, paleogeography and paleoceanography, In: Gray, J. and Boucot, A.J. (eds) Historical biogeography, platetectonics and changing environments, Oregon State Univ. Press., 9-25, 1979.

Van Andel, T.H., Heath, G.R. and Moore jr., T.L., Cenozoic tectonics, sedimentation and paleoceanography of the central Equatorial Pacific, Geol. Soc. America Bull., Mem.,143, 1-134, 1975.

Van Andel, T.H., Thiede, J., Sclater, J.G. and Hay, W.W., Depositional history of the South Atlantic Ocean during the last 125 million years, J. Geology, 85, 651-698, 1977.

Vergnaud Grazzini, C., Pierre, C. and Letolle, R., Paleoenvironment of the Northeast Atlantic during the Cenozoic: oxygen and carbon isotope analyses at DSDP Sites 398, 400 and 401, Oceanol. Acta, 1/3, 381-390, 1978.

Vincent, E., Gibson, J.M. and Brun, L., Paleocene and Early Eocene microfacies, benthonic foraminifera and paleobathymetry of Deep Sea Drilling Project Sites 236 and 237, western Indian Ocean, Init. Repts. DSDP, 24, 859-885, 1974.

Vincent, E., Killingley, J.S. and Berger, W.H., The magnetic epoch-6 carbon isotope shift: a change in the oceans 13 C/ 12 C ratio 6.2 million years ago, Marine Micropaleontol.,5, 185-203, 1980.

Webb, P.N., Paleocene foraminifera from DSDP Site 283, South Tasman Basin, Init. Repts. DSDP, 29, 833-844, 1973.

Webb, P.N., Harwood, D.M., McKelvey, B.C., Mercer, J.H. and Stott, L.D., Cenozoic marine sedimentation and ice-volume variation on the East Antarctic craton, Geology, 12, 287-291, 1984.

Williams, D.F., Healy-Williams, N., Thunell, R.C. and Leventer, A., Detailed stable isotope and carbonate records from the upper Maestrichtian-Lower Paleocene section of Hole 516F (Leg 72) including the Cretaceous /Tertiary boundary, Init. Repts. DSDP, 72, 921-930, 1983.

Wise, S.W., Gombos, A.W. and Muza, J.P., Cenozoic evolution of polar water masses, Southwest Atlantic Ocean, In: Hsü, K.J. and Weissert, H. (eds) South Atlantic paleooceanography, Cambridge Univ. Press, 283-324, 1985.

Wolfe, J.A., A paleobotanical interpretation of Tertiary climates in the northern hemisphere, American J. Sci. 66, 694-703, 1978.

Wolfe, J.A. and Poore R.Z., Tertiary marine and nonmarine climatic trends, In: Crowell, J.C. and Berger, W.H. (eds) Climate in earth history, National Academy of Sciences, Studies in Geophysics, 154-158, 1982.

Worsley, T. and Davies, T., Paleoenvironmental implications of oceanic carbonate sedimentation rates, Soc. Econ. Pal. Min., Spec. Publ. 32, 169-179, 1981.

EOCENE-OLIGOCENE PALEOCEANOGRAPHY

Bruce H. Corliss and Lloyd D. Keigwin, Jr.

Department of Geology, Duke University, Durham, NC 27708 and
Woods Hole Oceanographic Institution, Woods Hole, MA 02543

Abstract. Recent studies of Eocene-Oligocene paleoceanography based on the analysis of deep-sea microfossils are reviewed. Distinct changes in foraminiferal stable isotope compositions and biotic patterns occurred in the ocean near the Eocene/Oligocene boundary, but the data do not suggest a catastrophic extinction event took place at this time. The geological events can best be explained in terms of changing paleoceanographic and paleoclimatic conditions and do not support a bolide impact event having a major influence on the earth's biosphere near the boundary. The lack of a mass extinction of marine microfossils at the Eocene/Oligocene boundary argues against the existence of 26 m.y. cyclicity of extinctions in the Cenozoic.

Introduction

In 1833 Lyell subdivided the Tertiary period into four subdivisions, Older Pliocene, Newer Pliocene, Miocene, and Eocene, based on the proportion of living to extinct mollusc species which occur in geological formations in England and continental Europe [Lyell, 1833]. This subdivision was amended by Beyrich [1854] who erected the Oligocene, created from the upper part of Lyell's Eocene and lower part of the Miocene. Since that time, the Eocene/Oligocene boundary has been of continuing interest to geologists, in large part because of numerous associated geological events, including major changes in sedimentation patterns and chemistry of the oceans [Heath, 1969; van Andel et al., 1975; Moore et al., 1978], biogeographic distributions [Haq and Lohmann, 1976; Kennett, 1978; Sancetta, 1979], as well as a dramatic global climatic cooling [Margolis et al., 1975; Savin et al.,1975; Shackleton and Kennett, 1975; Kennett and Shackleton, 1976; Boersma and Shackleton, 1977; Buchardt, 1978; Vergnaud-Grazzini et al., 1979; Keigwin, 1980; Cavelier et al., 1981; Miller and Curry, 1982; Oberhänsli et al., 1984; Shackleton et al., 1984; Oberhänsli and Toumarkine, 1985]. In this paper, we review recent paleoceanographic research dealing with the Eocene/Oligocene boundary based on material from DSDP sites and land sequences, and present a synthesis of paleoceanographic events associated with the boundary.

Stable Isotopes

The first stable isotope analyses of Tertiary foraminifera were reported by Emiliani [1954]. At that time, he realized that increasing $\delta^{18}O$ through time in low latitudes must reflect cooling climate at high latitudes, although there was hardly enough data even several years later [Emiliani, 1961] to constitute a time series or to identify the time when greatest changes occurred. By the late 1960's there was sufficient stable isotope data from analyses of macrofossils [for example, Dorman, 1966; Devereux, 1967] to document a cooling trend within the Tertiary as well as identify the Eocene/Oligocene boundary as a time of abrupt cooling.

Long and sometimes continuous sequences of sediment from the world ocean were made available for paleoclimatic study by the Deep Sea Drilling Project beginning in the late 1960's. Two principal papers emerged several years later which established the trends of Cenozoic paleoceanographic change with samples spaced about every million years [Savin et al., 1975; Shackleton and Kennett, 1975]. For the first time it became apparent that the $\delta^{18}O$ increase near the Eocene/Oligocene boundary was one of several major steps in the evolution of today's climate. The exact timing of the Eocene/Oligocene oxygen isotope change, however, was in dispute. The record of Shackleton and Kennett [1975] from the southwest Pacific Ocean showed it to be an early Oligocene event, whereas the results of Savin et al. [1975], mostly from the central and north Pacific Ocean, suggested a late Eocene age. A late Eocene age was also suggested by the first isotope record from an Atlantic site [Boersma and Shackleton, 1977]. More detailed study of two DSDP sites spanning 70° of paleolatitude in the Pacific revealed that the $\delta^{18}O$ enrichment began at the Eocene/Oligocene boundary [Keigwin, 1980] and this pattern was subsequently verified for the Atlantic Ocean [Miller and Curry, 1982; Oberhänsli et al., 1984; Poore and Matthews, 1984; Snyder et al., 1984;

Oberhänsli and Toumarkine, 1985] and Indian Ocean [Oberhänsli, in prep; Keigwin and Corliss, 1986.]

Although the nature of the time series of Eocene/Oligocene $\delta^{18}O$ is reasonably well-established as seen in Figure 1, its interpretation is not. Prior to 1980, most workers assumed that the early Tertiary world lacked significant continental ice and thus $\delta^{18}O$ variations reflected primarily temperature changes. This view was challenged by Matthews and Poore [1980] who argued that the $\delta^{18}O$ record could be reinterpreted assuming significant and variable ice volume, and constant sea surface temperatures in the tropics. This interpretation is supported by three lines of evidence.

The first involves covariance in the $\delta^{18}O$ records of benthonic and planktonic foraminifera. Since increased continental ice volume rapidly enriches the ocean in $\delta^{18}O$ and since the ocean is rapidly mixed, its isotope effect should be evident in benthonic and planktonic foraminifera everywhere. The covariance of benthonic and planktonic foraminifera is shown in the recent studies of Poore and Matthews [1984] and Keigwin and Corliss [1986]. Other workers have measured large $\delta^{18}O$ values in earliest Oligocene benthonic foraminifera from which they infer there must have been continental ice [Miller and Fairbanks, 1983; Shackleton et al., 1984]. These workers reason that the enrichment in benthic $\delta^{18}O$ is so great that it would require deep waters in the ocean to have been close to freezing. This would mean surface waters in high latitudes were close to freezing, so polar temperatures were low enough to promote glaciation. A third line of evidence is that planktonic foraminiferal isotopic values indicate high-latitude surface temperatures to be too warm to be compatible with microfossil data if ice-free conditions existed [Poore and Matthews, 1984].

Recently, workers have become interested in isotope events later in Oligocene time, since the change at the base of that epoch is well-known. From an isotope viewpoint, the Oligocene is of interest for several reasons: there is more, better-preserved $CaCO_3$ than in Eocene sections; it is a time when there exists independent evidence for a sea level drop and cooler climates; and long, relatively continuous sequences are available for study, especially from the North Atlantic Ocean. These features have allowed generation of long time-series of data capable of resolving new paleoceanographic and paleoclimatic trends. From their study of equatorial Pacific DSDP Site 77B, Keigwin and Keller [1984] showed evidence of cooling and glaciation beginning about 32 m.y. ago and reaching maximum levels 29 m.y. ago. Similar results have been found in North Atlantic DSDP site 558 [Miller and Fairbanks, in press] documenting the global nature of the $\delta^{18}O$ enrichment. This observation is in agreement with seismic stratigraphic evidence for pronounced sea level lowering 29 m.y. ago [Vail et al., 1977] which probably resulted from glaciation. It also allows us to propose a solution to the problematical observation of a major 32 m.y. old unconformity (rather than 29 m.y.) of Olsson et al. [1980]: since the middle Oligocene glaciation began 32 m.y. ago and increased for 3 m.y., its effect should be seen at different times on different continental margins, depending on local sedimentary conditions.

At about the same time, Miller and Fairbanks [1983] and Keigwin and Keller [1984] published the first evidence of coherent changes in $\delta^{13}C$ of Oligocene benthonic foraminifera. Miller and Fairbanks [1983] proposed that the increased $\delta^{13}C$ in North Atlantic benthonic foraminifera of earliest and latest Oligocene age reflected increased production of northern-source deep water which would not have accumulated large amounts of CO derived from the degradation of organic matter. On the other hand, Keigwin and Keller [1984] thought lowered $\delta^{13}C$ values in middle Oligocene time might have resulted from lowered sea level exposing organic matter trapped in estuaries and on continental shelves to erosion and oxidation. Such speculations about the meaning of $\delta^{13}C$ will undoubtedly continue for some time. The most recent data compilation [Miller and Fairbanks, in press] proposes the existence of $\delta^{13}C$ cycles with periods of 7 to 12 m.y. in Oligocene and Miocene foraminifera which may be due in part to cyclic changes in the ratio of organic carbon to carbonate carbon buried in marine sediments.

The Deep-Sea Environment

Initial studies of Eocene/Oligocene deep-sea benthonic foraminifera, reviewed by Tjalsma and Lohmann [1983], were largely taxonomic in nature, and based on outcrops from land sequences. These initial studies were followed by the analysis of Cenozoic sequences from DSDP Sites 116 and 117 [Berggren and Aubert, 1976], Sites 167 and 171 [Douglas, 1973], Site 357 [Boersma, 1977] and Sites 400 and 401 [Schnitker, 1979]. In these studies, a faunal turnover was noted near the Eocene/Oligocene boundary, which was suggested to be related to an inferred bottom-water cooling based on the enrichment in $\delta^{18}O$ observed in benthonic foraminifera. The nature of the faunal turnover was considered by Corliss [1979] in a detailed study of late Eocene-early Oligocene material from Site 277. A gradual turnover in benthonic foraminifera was noted, rather than a catastrophic change, and it was suggested that bottom-water circulation changes occurred as a series of events during middle Eocene-early Oligocene time. A review of previously published benthonic foraminiferal biostratigraphic data corroborated the observation of a gradual faunal turnover [Corliss, 1981].

The most comprehensive study of Eocene benthonic foraminifera was presented by Tjalsma and Lohmann [1983] in a study of Paleocene-Eocene material from Atlantic DSDP sites. A major faunal turnover was documented at the Paleocene/Eocene

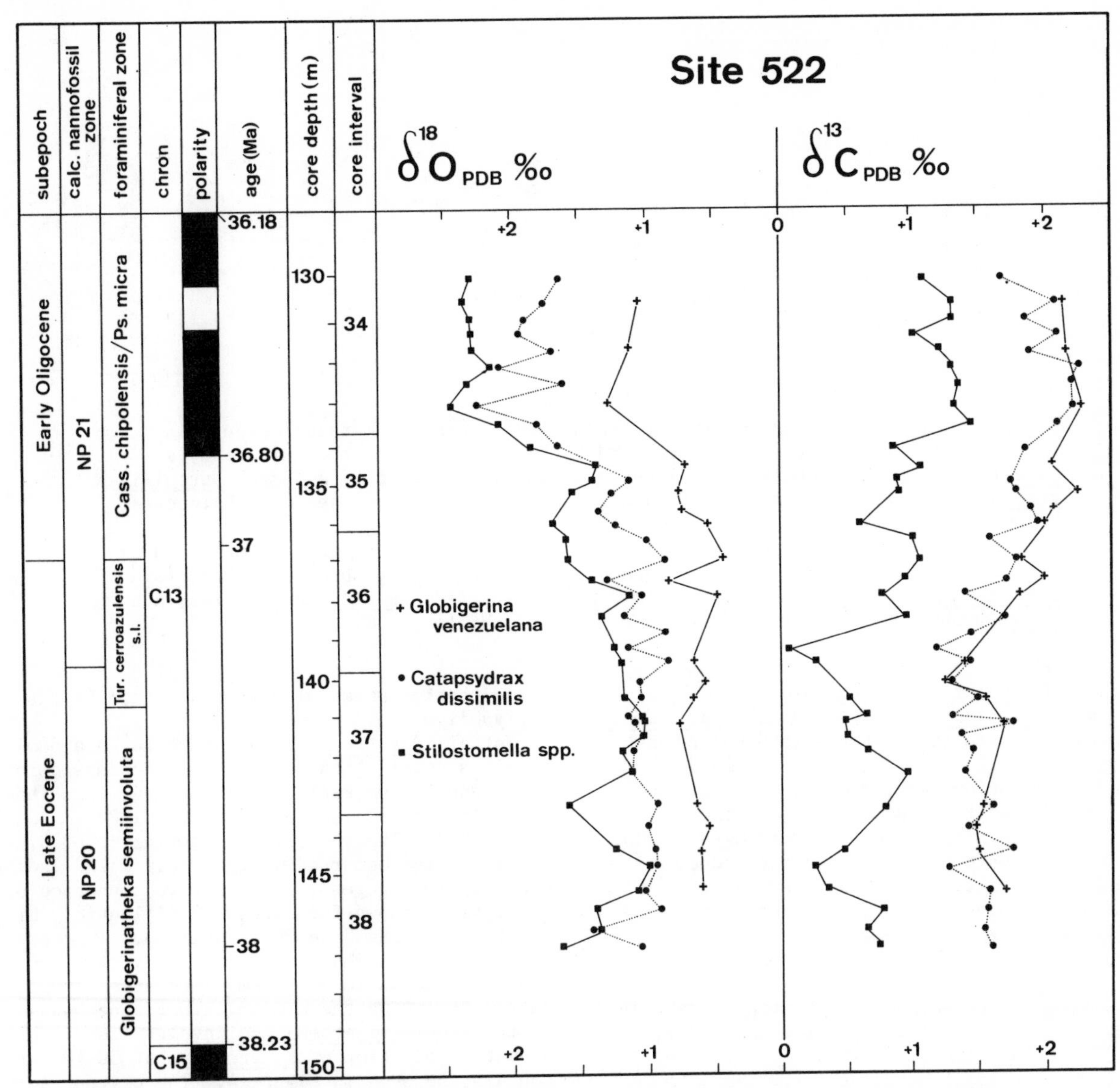

Figure 1. Stable isotope data from DSDP Site 522 in the southeast Atlantic Ocean (Oberhänsli and Toumarkine, 1985). An enrichment in benthonic and planktonic $\delta^{18}O$ occurs across the Eocene/Oligocene boundary and is similar to that found at other DSDP sites in all of the ocean basins.

boundary when at least 19 species became extinct. A similar faunal turnover at this time had been noted earlier from the Lizard Springs Formation in Trinidad [Beckman, 1960], land sections in Austria [von Hillebrandt, 1962] and Italy [Braga et al., 1975], and Sites 400 and 401 in the eastern North Atlantic [Schnitker, 1979]. A second faunal event of lesser magnitude occurred in the late middle Eocene with the decrease in abundance of *Nuttallides truempyi*.

The Paleocene fauna was composed of a *Gavelinella beccariiformis* assemblage and a deeper *Nuttallides truempyi* assemblage (Figure 2). The *G. beccariiformis* assemblage became progressively restricted to shallower depths and was replaced by the *N. truempyi* assemblage. This depth migration culminated in the extinction of numerous taxa from the *G. beccariiformis* assemblage at the end of the Paleocene.

In the Eocene, numerous taxa appeared, including costate forms of the genus *Uvigerina*, and plano-convex *Cibicidoides* became an important component. The *Nuttallides* assemblage migrated into deeper water, followed by the disappearance of *N. truempyi* in the late Eocene. An assemblage dominated by "*Cibicidoides ungerianus*",

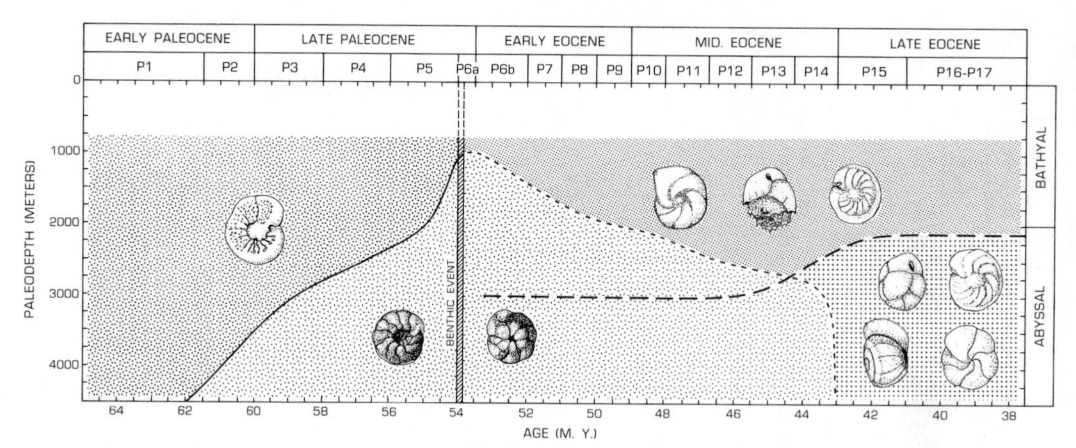

Figure 2. Schematic presentation of benthonic foraminiferal faunal events in Paleocene-Eocene time (after Tjalsma and Lohmann, 1983). In the Paleocene, a Gavelinella beccarriformis assemblage is gradually replaced by a deeper Nuttallides truempyi assemblage. The extinction of 19 species took place near the Paleocene/Eocene boundary and is the most significant event in the Paleogene record. In the Eocene, the N. truempyi assemblage is gradually restricted to deeper water and is replaced by wide ranging and stratigraphically long ranging species Globoccassidulina subglobosa, Oridorsalis species, Gyroidinoides species, Lenticulina species, and Cibicidoides species (Tjalsma and Lohmann, 1983; Miller, 1983).

Oridorsalis umbonatus, Globocassidulina subglobosa, Gyroidinoides species and Lenticulina species became important in the late Eocene and continued into the early Oligocene. Based on the faunal distribution data, Tjalsma and Lohmann [1983] suggested that the lack of distinct benthonic foraminiferal depth assemblages in the Paleocene may have reflected a homogeneous, unstably stratified deep ocean, whereas in the Eocene a heterogeneous, stable water mass stratification was present.

Benthonic foraminiferal data from North Atlantic DSDP sites were combined with lithologic, stable isotope and seismic stratigraphic data in a series of papers by Miller [Miller and Curry, 1982; Miller and Tucholke, 1983; Miller, 1983; Snyder et al., 1984; Miller et al., 1985] to determine deep-water circulation conditions during the late Eocene-early Oligocene. Analysis of Sites 119, 400A and 401 revealed a major change between early middle Eocene and earliest Oligocene with a number of Eocene taxa (N. truempyi, Clinapertina spp., Abyssamina spp.) becoming extinct, and an increase in abundance in bathymetrically wide-ranging and stratigraphically long-ranging species G. subglobosa, Oridorsalis spp., Gyroidinoides spp., and "C. ungerianus". At Sites 548 and 549 N. truempyi was replaced by a buliminid assemblage, which in turn was replaced by the long-ranging early Oligocene assemblage. The faunal, isotopic and sedimentological data were interpreted to reflect old, relatively warm, corrosive and sluggish Eocene bottom water in the North Atlantic, which was replaced by younger, colder, less corrosive, more vigorously circulating bottom water of northern origin by the early Oligocene.

Nuttallides truempyi was shown to be a dominant Eocene species which dramatically decreased in abundance in the earliest late Eocene throughout the Atlantic [Tjalsma and Lohmann, 1983; Miller et al., 1985]. The reduction in dominance of N. truempyi also occurred in the Pacific and Indian Oceans. In Figure 3, the relative abundance of N. truempyi is shown for Sites 363 (Atlantic), Site 219 (Indian) and Site 292 (Pacific). In these sites with detailed sample coverage, the precipitous reduction in N. truempyi occurred immediately following the middle/late Eocene boundary in the Globigerinatheka semiinvoluta zone. This dramatic reduction of N. truempyi at 38-40 m.y. can also be seen in Figure 4. The oxygen isotope data show an enrichment of about $1^{o}/oo$ near the boundary, and a comparison of faunal and isotopic data show that the faunal change coincides with the isotopic enrichment. Based on this correlation, we suggest that the reduction in dominance of N. truempyi in all of the ocean basins was in response to a bottom-water cooling event of $2\text{-}4^{o}C$ near the middle/late Eocene boundary.

Tjalsma [1983] studied Site 516 which had an estimated paleodepth of 500 to 1500m and was shallower than sites considered earlier by Tjalsma and Lohmann [1983]. The late Eocene assemblage was dominated by Cibicidoides, Lenticulina and buliminids. High rates of new species appearances were observed during late Eocene and Oligocene time in contrast to low rates observed in the deeper abyssal faunas [Tjalsma and Lohmann, 1983]. At the Eocene/Oligocene boundary, no generic frequency

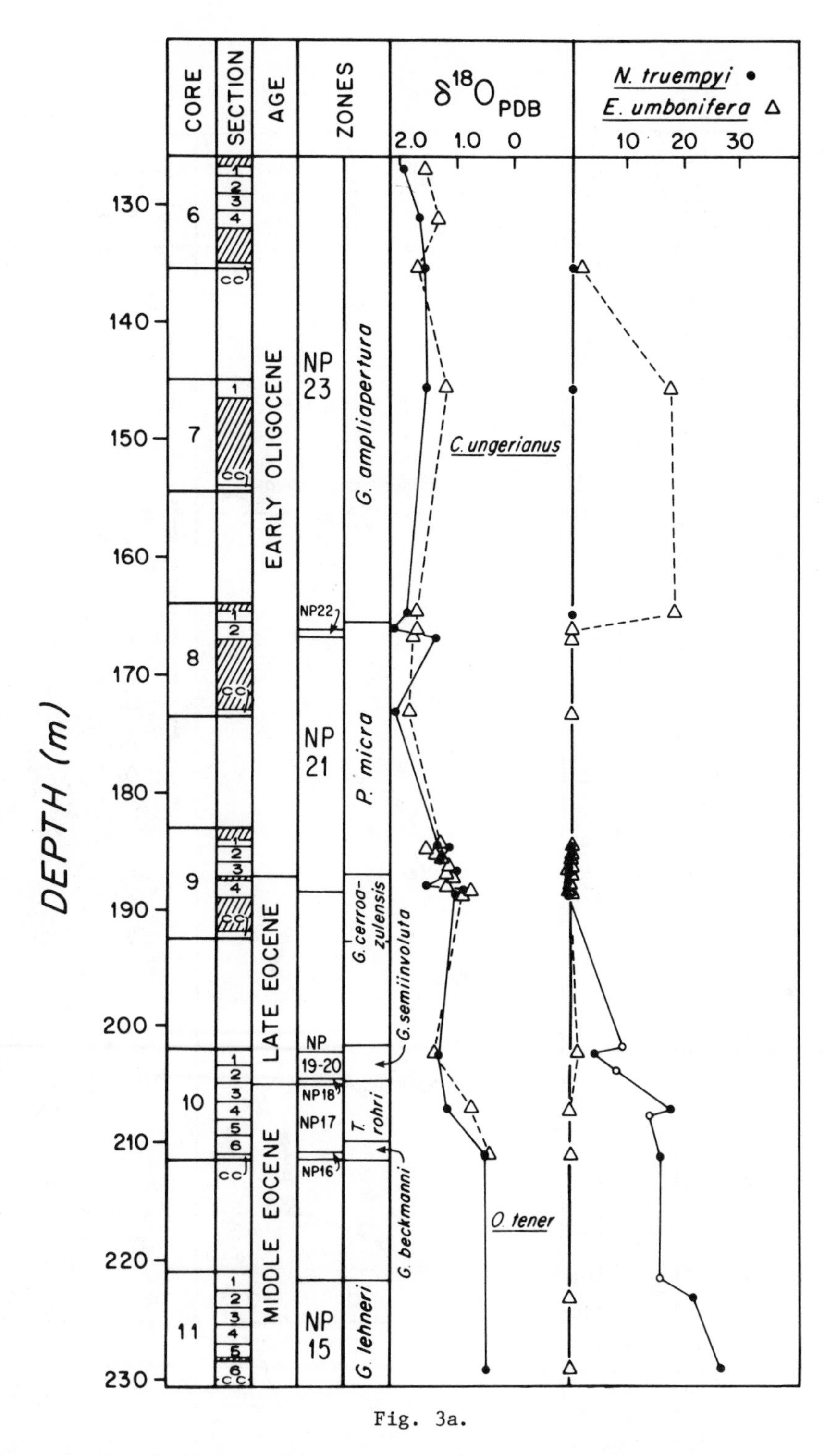

Fig. 3a.

Figure 3. The distribution of Nuttallides truempyi and oxygen isotopic data in a) Site 363, b) Site 219, and c) Site 292. The precipitous reduction of N. truempyi occurs in the G. semiinvoluta zone following the middle/late Eocene boundary, and coincides with a benthonic foraminiferal isotopic enrichment of about 1°/oo and an inferred bottom water cooling of 2-4°C.

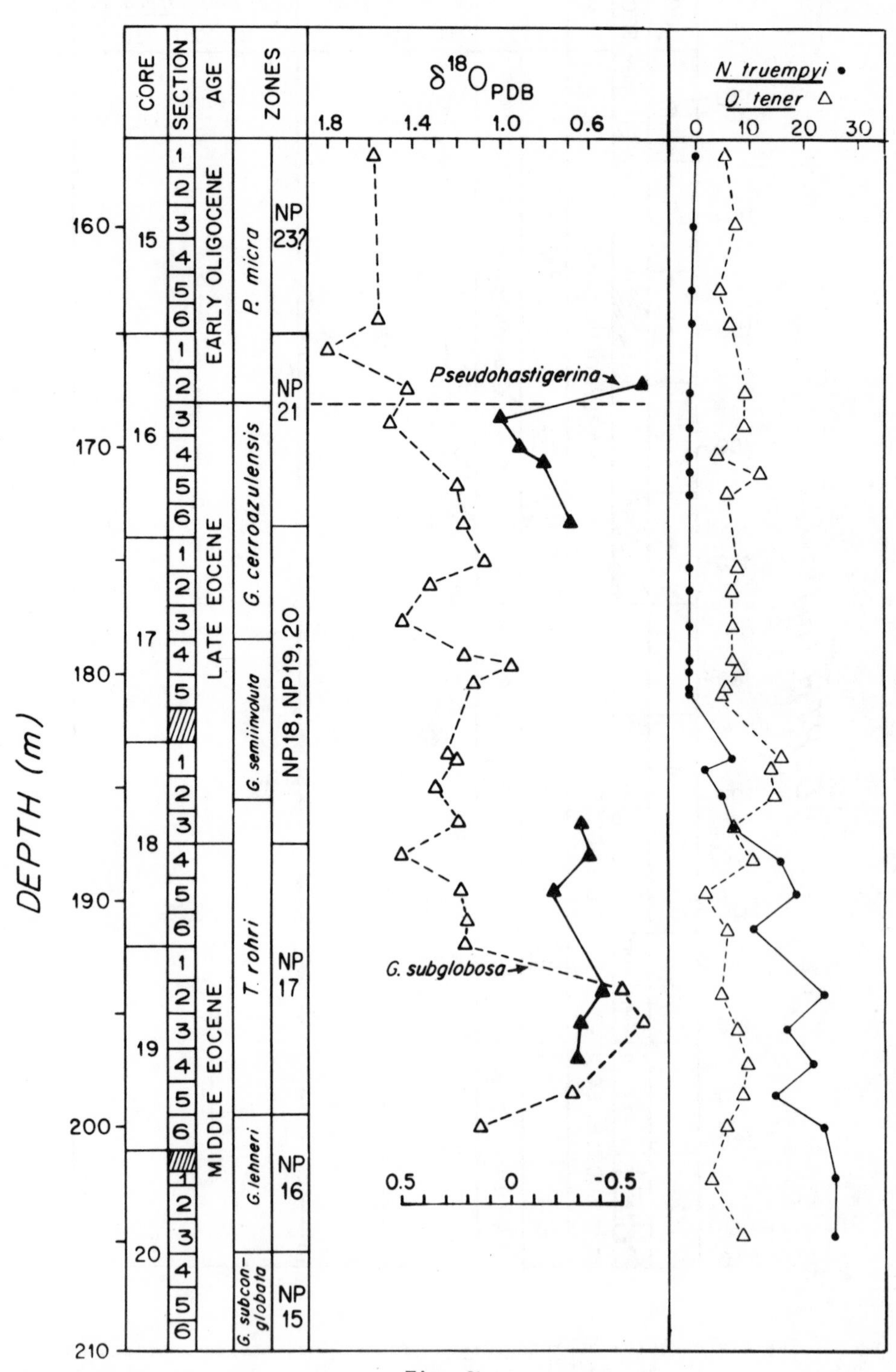
CORE
SECTION
AGE
ZONES
δ18O PDB
1.8
1.4
1.0
0.6
N. truempyi •
O. tener △
0
10
20
30
DEPTH (m)
160
170
180
190
200
210
15
16
17
18
19
20
EARLY OLIGOCENE
LATE EOCENE
MIDDLE EOCENE
P. micra
G. cerroazulensis
G. semiinvoluta
T. rohri
G. lehneri
G. subcon- globata
NP 23?
NP 21
NP18, NP19, 20
NP 17
NP 16
NP 15
Pseudohastigerina
G. subglobosa
0.5
0
-0.5

Fig. 3b.

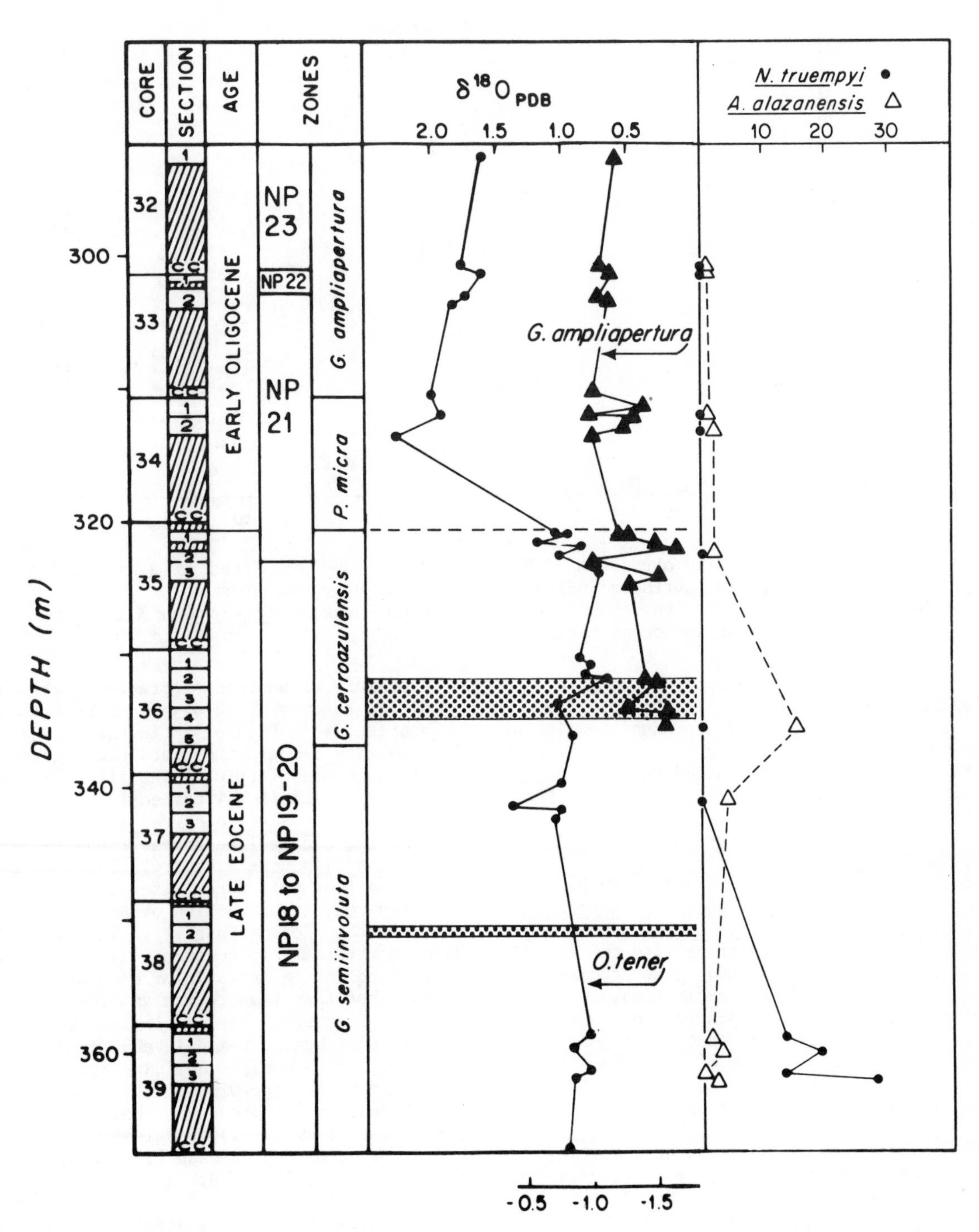
CORE
SECTION
AGE
ZONES
$\delta^{18}O_{PDB}$
2.0
1.5
1.0
0.5
N. truempyi
A. alazanensis
10
20
30
EARLY OLIGOCENE
LATE EOCENE
NP 23
NP22
NP 21
NP18 to NP19-20
G. ampliapertura
P. micra
G. cerroazulensis
G. semiinvoluta
O. tener
DEPTH (m)
300
320
340
360
-0.5
-1.0
-1.5

Fig. 3c.

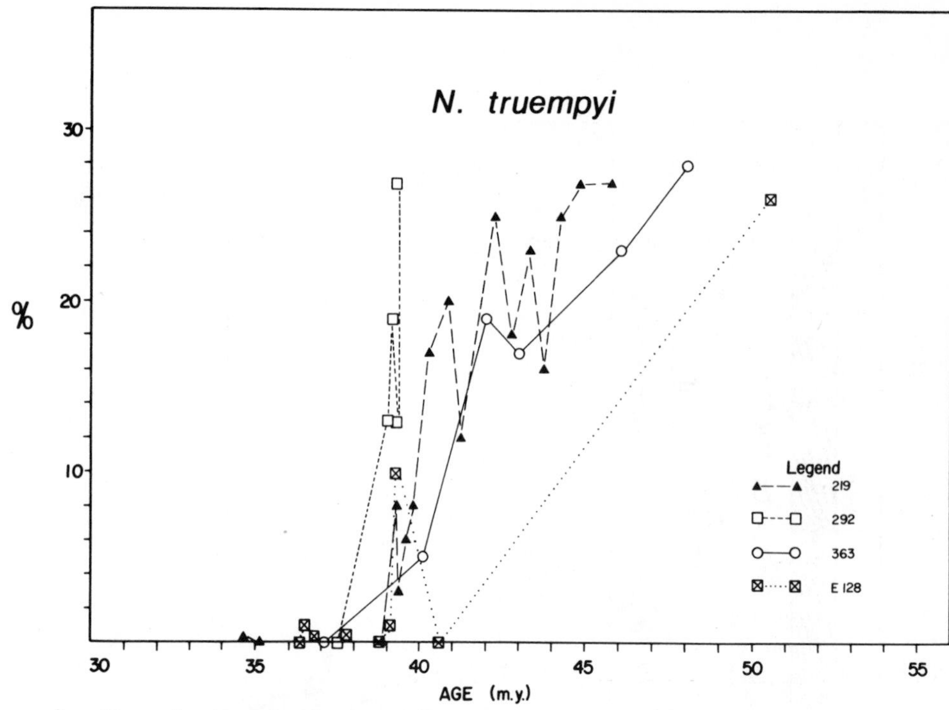

Figure 4. The relative abundance of N. truempyi vs. age from Sites 363 (southeast Atlantic Ocean), 219 (northern Indian Ocean), 292 (western equatorial Pacific) and Eureka 128 (Gulf of Mexico) showing that the reduction of N. truempyi is isochronous and occurs in all of the ocean basins.

change or catastrophic extinction was noted, although there was an increase in the relative abundance of Cibicidoides and a gradual decline of buliminids.

Clark and Wright [1984] studied the benthonic foraminifera from Sites 522, 523, and 524 in the southeast Atlantic and found similar faunal patterns as in other Atlantic studies. In the late Eocene, Nuttallides crassaformis-truempyi decreased in abundance, with an increase in Nuttallides umbonifera, G. subglobosa, O. umbonatus, Gyroidinoides spp. and "C. ungerianus" occurring in the late Eocene and continuing into the early Oligocene. Parker et al. [1984] also studied these sites for origination, extinction, species diversity, equitability, and carried out a factor analysis of species abundance data. At the boundary, there was an almost complete cessation of originations, a small increase in extinctions, a change in abundance patterns and an increase in species equitability. The changes were suggested to reflect an initiation of a northern source of deep water based on the occurrence of taxa in both the southeast Atlantic and Bay of Biscay. Although a similar faunal assemblage in both areas would indicate similar paleoenvironmental conditions (i.e. water masses), it is unclear why the data would indicate a source and direction of the deep water, and this conclusion by Parker et al. [1984] of a northern source of deep water must be viewed as speculative.

A summary of benthonic foraminiferal faunal events is shown with an oxygen isotope curve schematically in Figure 5. In the Paleocene, bottom-water temperatures were considered relatively warm with little variability. Based on the benthonic foraminiferal oxygen isotope data, bottom-water coolings occurred in the early Eocene, the middle/late Eocene boundary, and at the Eocene/Oligocene boundary. The most dramatic Paleogene benthonic foraminiferal faunal event was the simultaneous extinction of numerous taxa near the Paleocene/Eocene boundary. The oxygen isotope data show relatively constant values at this time, indicating that a variable or combination of variables other than temperature was responsible for the faunal crisis. The middle/late Eocene cooling coincides with a decrease in abundance of N. truempyi, which was the dominant species prior to this. At the Eocene/Oligocene boundary, no major catastrophic event occurs as shown by the studies of Corliss [1979, 1981], Douglas and Woodruff [1981], Tjalsma and Lohmann [1983], Miller [1983], Tjalsma [1983], Synder et al. [1984], and Corliss et al. [1984].

We suggest that this sequence of events was a result of the elimination during the early Paleogene of species with narrow environmental tolerances. The cooling at the Eocene/Oligocene boundary did not have a major effect on the faunas because the sensitive species had been removed during previous paleoenvironmental events, leaving

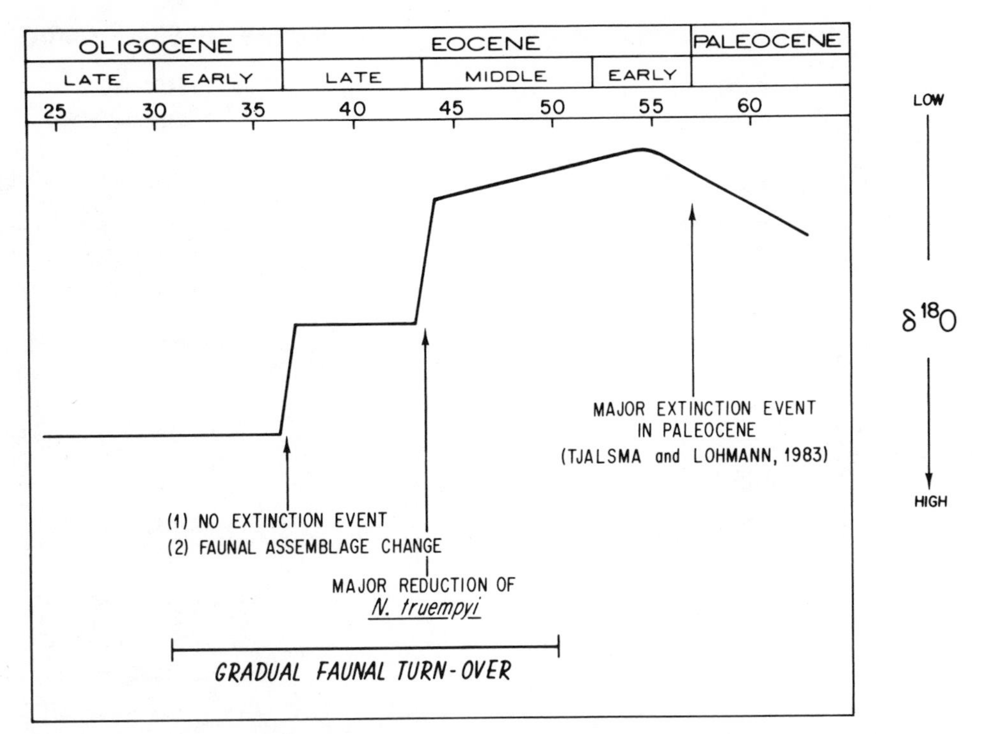

Figure 5. A schematic diagram of Paleogene benthonic foraminiferal faunal events and the oxygen isotopic record. The lack of an extinction event at the Eocene/Oligocene boundary is suggested to be a result of the previous elimination of environmentally-sensitive species, with the remaining environmentally-tolerant species able to withstand a 2-3°C bottom water temperature decrease at the Eocene/Oligocene boundary.

species with wide environmental tolerances able to withstand a 2-3°C bottom water temperature decrease. The faunal record at the Eocene/Oligocene boundary should not be viewed by itself, but rather as part of a sequence of environmental events which had progressively less influence on the benthonic foraminiferal faunas.

A study of Cenozoic deep-sea ostracodes was recently completed by Benson et al. [1984], based on the analysis of DSDP material. In the Paleogene major faunal turnovers occurred at the Cretaceous/Tertiary boundary (mostly extinctions) and originations at 40 to 36 m.y. Generic diversity decreased at 66 to 60 m.y. and 36 to 35 m.y. and increased from 44 to 36 m.y. An analysis of the faunal similarity, based on a Rho-groups analysis, showed that before 40 m.y. faunal similarities were low, reflecting provinciality of the faunas. In the interval younger than 40 m.y., faunal similarity increased, reflecting stratified, anisotropic deep-sea environments with increased interchange across geographic regions. The most significant event was at the Cretaceous/Tertiary boundary, followed by the event at about 40 m.y. The 40 m.y. event began first in deep water and migrated into shallow water, and was marked by a sharp increase in mean abundance and diversity. The 40 m.y. event was suggested to mark the origin of the psychrosphere and the modern two-layer ocean [Benson, 1975], and was the transition from a warm, sluggish, saline-driven water mass system to a cold, strongly stratified, psychrospheric water-mass system with higher velocities and nutrient content. A decrease in diversity and abundance values at 36-35 m.y. indicated a cessation of this water mass mixing. The terminal Eocene event was viewed as a shallow water stage of the 40 m.y. event, and not as a separate, catastrophic event in itself.

The Planktonic Realm

Early Cenozoic biogeographic patterns of calcareous nannoplankton and planktonic foraminifera [Haq and Lohmann, 1977; Haq et al., 1979] from Atlantic Ocean material showed fluctuations in Eocene/Oligocene times. These changing distributional patterns were interpreted to reflect latitudinal migrations of the planktonic fauna and flora in response to changing climatic conditions. Cooling intervals were inferred during the late Eocene and earliest Oligocene as shown in Figure 6. The early Oligocene cooling was marked by the equatorward migrations of a relatively high-latitude foraminiferal globigerinid assemblage and by high-latitude nannofloral assemblages

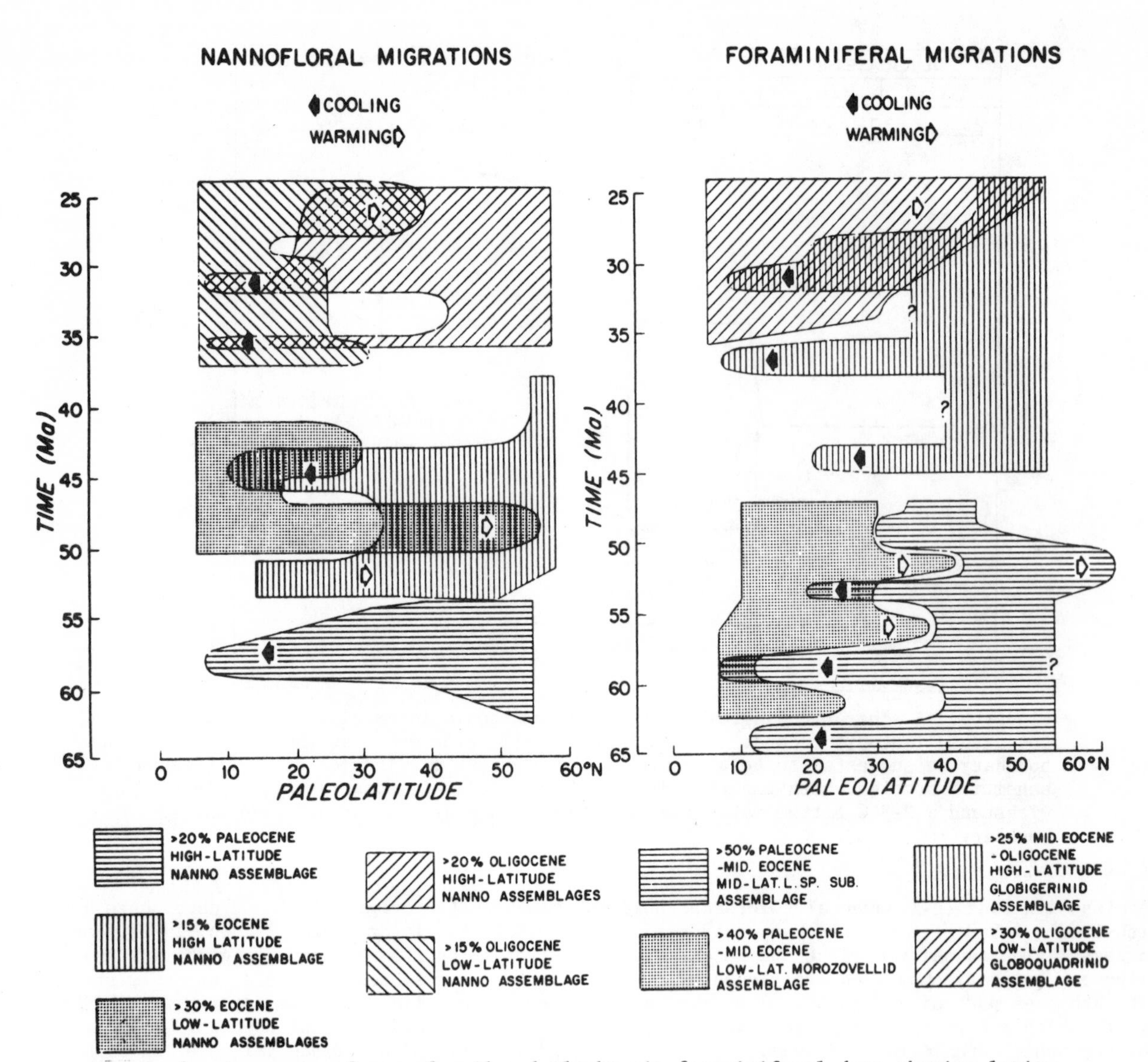

Figure 6. A summary of nannofossil and planktonic foraminiferal data showing latitudinal assemblage migrations in the early Cenozoic (after Haq et al., 1979). The arrows indicate times of major latitudinal shifts of assemblage patterns, which were interpreted to reflect paleoclimatic fluctuations.

(reticulofenestrid) and Coccolithus pelagicus.

Biogeographic patterns of Pacific planktonic microfossils were presented by Sancetta [1979] for six Paleogene time-slices, including the late Eocene and the early Oligocene intervals (Figure 7). The late Eocene time-slice had 4 biotic provinces, whereas 6 provinces existed in the early Oligocene including an Antarctic province and a Subantarctic Province. The Antarctic province was composed of radiolaria and diatoms with the genera Prunopyle, Antarctissa, Eucyrtidium, Denticula and Stephanopyxis being important. The Subantarctic province had a mixture of calcareous and siliceous genera, with Subbotina angiporoides, Chiloguembelina species, chiasmoliths, reticulofenestrids, Zygrhablithus bijugates, and siliceous genera of the Antarctic province. Sancetta [1979] suggested that the presence of these two well-developed provinces reflected the development of the Antarctic convergence at this time.

Detailed time-series studies of middle Eocene to early Oligocene sequences from the Atlantic, Indian and Pacific Oceans have recently been carried out for the planktonic foraminifera by Keller [1983a,b] and for the calcareous nannoplankton by Aubry [1983]. Analysis of the planktonic foraminifera in Site 363 (Atlantic), Sites 77B, 277, 292 (Pacific) and Sites 219, 253 (Indian) showed major faunal changes interpreted

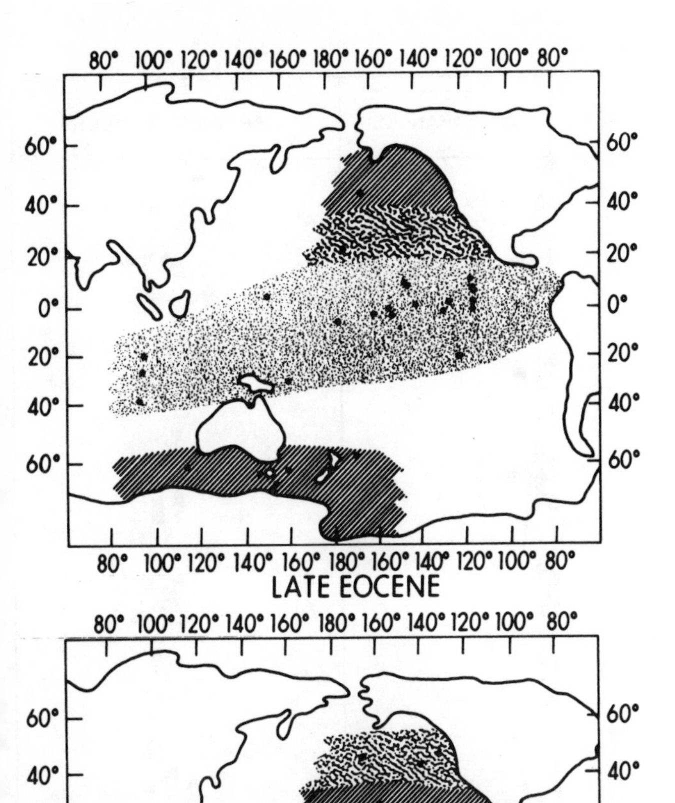

Figure 7. Pacific Ocean planktonic assemblage distributions for the late Eocene and the early Oligocene (after Sancetta, 1979). Late Eocene plankton provinces: stippled = Tropical Province; irregular lines = Transitional Province; diagonal lines = Temperate Province. Early Oligocene plankton provinces: stippled = Tropical Province; diagonal lines = Subtropical Province; irregular lines = Transitional Province; vertical lines = Subantarctic Province; horizontal lines = Antarctic Province. In the early Oligocene the Antarctic and Subantarctic provinces appeared for the first time, reflecting the development of the Antarctic Convergence at this time.

to be cooling episodes in the middle Eocene (44-43 m.y.), middle/late Eocene boundary (41-40 m.y.), late Eocene (39-38 m.y.), the Eocene/Oligocene boundary (37-36 m.y.) and the late Oligocene (31-29 m.y.). The Eocene/Oligocene boundary was marked by increasing abundances of cool-water species and not major species extinction events as found during the other cooling episodes. The Eocene/Oligocene cooling was suggested to be part of a cooling trend that began during the middle Eocene.

Aubry [1983] studied Paleogene calcareous nannoplankton diversity and found a 70% reduction in diversity over a 7 m.y. interval with approximately 50% of this reduction occurring between the late middle Eocene and the early late Eocene. A reduction in diversity of 20% occurred across the Eocene/Oligocene boundary. Quantitative analysis of Sites 292, 277 (Pacific), 219 (Indian) and Site 363 (Atlantic) revealed a sequential, step-like change in the floras during the late Eocene to early Oligocene similar to that found among the planktonic foraminifera.

An analysis of planktonic diatoms was carried out by Fenner [1985] from 17 DSDP sites. First and last occurrences were found throughout the late Eocene-early Oligocene, but no mass extinction event was found within the late Eocene-early Oligocene interval. A floral assemblage change does occur, with relatively large and robust, low to mid-latitude species of the genus <u>Cestodiscus</u> increasing from 5-20% in the late Eocene to 60-80% in the early Oligocene [Fenner, 1985]. The strong assemblage change occurred in equatorial sites as well as high latitudes, indicating that paleoceanographic changes had a major impact on low-latitude surface waters.

Extra-terrestrial Event(s) at the Eocene-Oligocene Boundary

The discovery of an iridium anomaly at the Cretaceous/Tertiary boundary in European land sections led to the suggestion that the massive extinction event observed at the end of the Cretaceous was a result of an extra-terrestrial impact event [Alvarez et al., 1980]. Iridium is found in greater abundance in extra-terrestrial material than in the earth's mantle or crust, and a large iridium peak at the Cretaceous/Tertiary boundary in land sections was inferred to reflect a bolide impact event [Alvarez et al., 1980]. This bolide event at the Cretaceous/Tertiary boundary was suggested to have led to massive extinctions by either screening in-coming radiation, which disrupted food chains [Alvarez et al., 1980], or by causing a global temperature change [Emiliani et al., 1981], or by poisoning of many taxa by the release of cyanide upon impact [Hsu, 1980]. The simultaneous deposition of the North American tektite field and the extinction of five radiolarian species near the Eocene/Oligocene boundary [Glass and Zwart, 1979] led Ganapathy [1982] and Alvarez et al. [1982] to determine iridium abundances in deep-sea sediment sequences to evaluate whether or not an extra-terrestrial event occurred at the boundary. Analysis of late

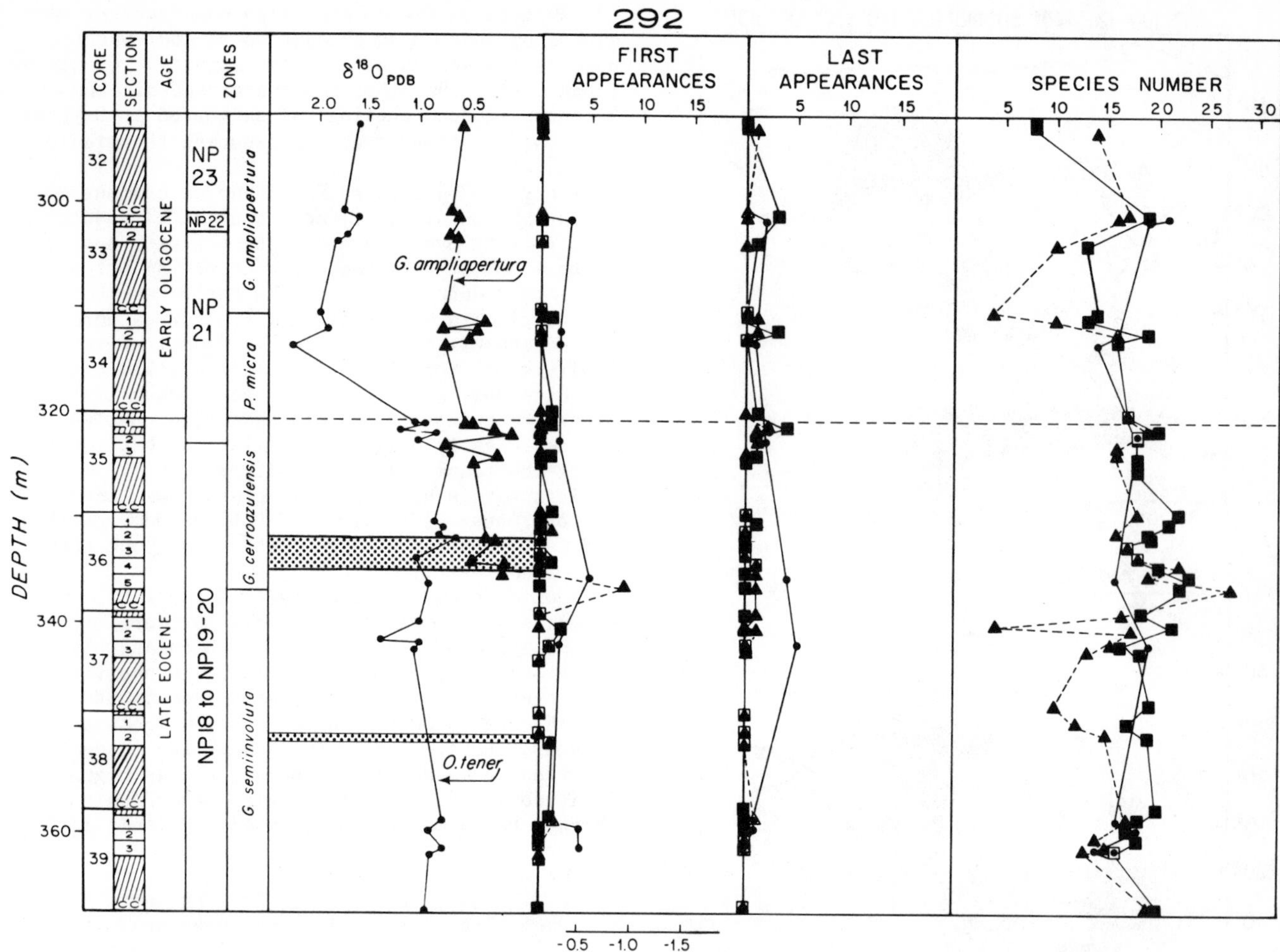

Figure 8. Summary of isotopic, floral, and fauna data from Site 292 (Corliss et al., 1984), with the location of microtektites (stippled) also shown (Glass and Crosbie, 1982; Keller et al., 1983). The first and last appearance data and the species number per sample are shown for benthonic foraminifera (circles), planktonic foraminifera (triangles), and calcareous nannoplankton (squares). The dashed line indicates the location of the Eocene/Oligocene boundary at 36.6 million years ago. The microtektites are found preceding the Eocene/Oligocene cooling, and no major biotic event is associated with the microtektites.

Eocene sediments from piston core RC9-58 [Ganapathy, 1982] and DSDP Site 149 [Alvarez et al., 1982] from the Caribbean Sea showed that peaks in iridium coincided with peak abundances of microtektites, and it was suggested that a bolide event was responsible for massive marine and terrestrial extinctions near the Eocene/Oligocene boundary.

Recent evidence presented by Keller et al. [1983] indicates that more than one tektite layer exists within the late Eocene-early Oligocene interval. Eight deep-sea sedimentary sequences were analyzed and multiple tektite layers were found in five of the eight sites, suggesting that the North American tektite field does not reflect a unique Paleogene event. One of the cores with multiple tektite occurrences was Site 292 which has two tektite layers in the late Eocene (Figure 8). One occurrence at 351 m is in the G. semiinvoluta zone [Keller et al., 1983] and a second between 332 and 335 m [Glass and Crosbie, 1982] is within the G. cerroazulensis zone. The microtektites are found preceeding the Eocene/Oligocene boundary cooling, and no major biotic changes are associated with the tektites [Keller et al., 1983; Corliss et al., 1984]. The interpretation of the microtektite data by Keller et al. [1983] has prompted discussion on whether or not multiple extra-terrestrial events occurred during Eocene-Oligocene time [Glass, 1984; Keller et al., 1984]. This question will not be resolved until quantitative microtektite data from DSDP sequences and land sections and geochemical analyses of the microtektites are presented.

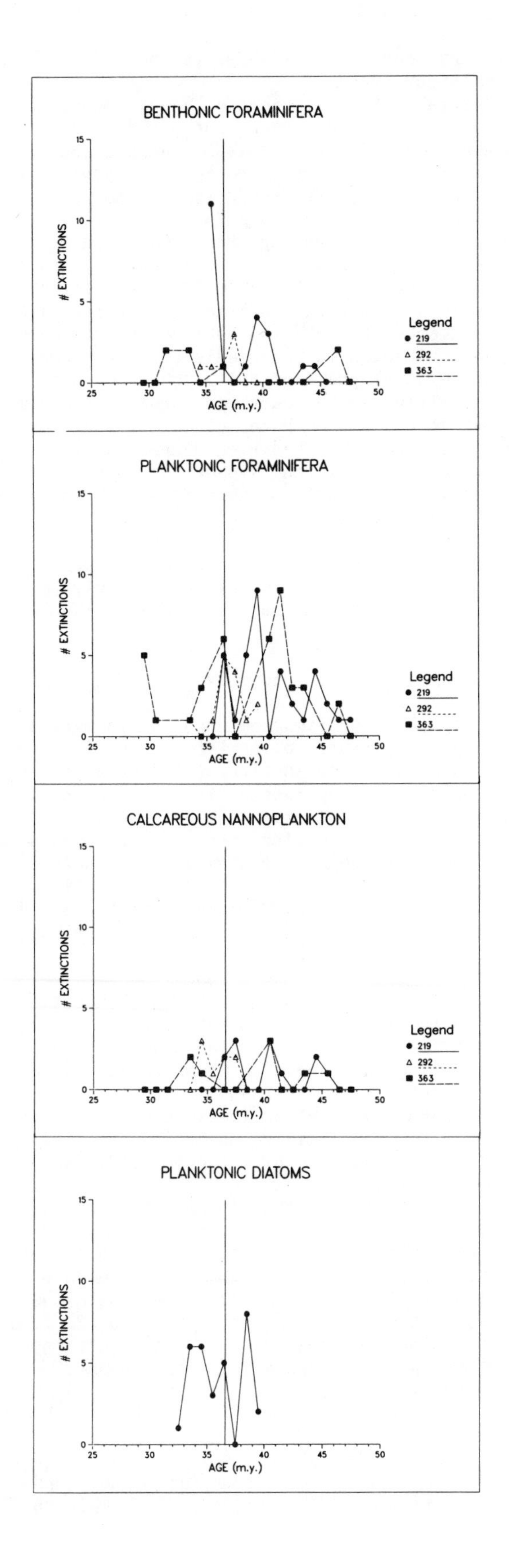

The Lack of a Catastrophic Extinction Event at the Eocene/Oligocene Boundary

The iridium anomaly found near the Eocene/Oligocene boundary was suggested to be associated with marine and terrestrial mass extinctions [Alvarez et al., 1982]. A recent study by Corliss et al. [1984] examined the nature of faunal and floral changes at the Eocene/Oligocene boundary. Detailed records of middle Eocene to early Oligocene planktonic and benthonic foraminifera, calcareous nannoplankton and planktonic diatoms were determined in DSDP sites from the Atlantic, Indian and Pacific Oceans. The number of extinctions for each fossil group was calculated and is shown in Figure 9.

The data show no dramatic change in the number of extinctions of benthonic and planktonic taxa across the Eocene/Oligocene boundary when compared with data from the middle Eocene to early Oligocene interval. The magnitude of the changes at the boundary is similar to those found before and after the boundary, reflecting a sequential faunal and floral turnover and not a catastrophic event. These findings are similar to those of Snyder et al. [1984] based on the analysis of Site 549 from the North Atlantic, Parker et al. [1984] based on Sites 522, 523 and 524, and the conclusions of McKenna [1983]. This sequential biotic change is seen in the land record as well [Wolfe, 1978; Collinson et al., 1981].

The inferred bolide event at the Eocene/Oligocene boundary was suggested to have occurred at the time of a terrestrial mammalian extinction event, the "Grande Coupure". The work of Hartenberger [1973], Vige and Vianey-Liaud [1979], and Prothero [1985] demonstrate, however, that the turnover spanned the late Eocene and early Oligocene time and was not catastrophic.

A quantitative analysis of the diversity of the Phanerozoic fossil record at the family level by Raup and Sepkoski [1984] revealed 12 extinction events with a 26 m.y. cyclicity in extinctions. In the Cenozoic the times identified as mass extinctions were at the Cretaceous/Tertiary boundary, the Eocene/Oligocene boundary and within the mid-Miocene. The high number of extinctions at the Cretaceous/Tertiary boundary is well documented, but the work summarized in this paper clearly shows that no mass extinction event of planktonic and deep-sea benthonic microfossils occurred at the Eocene/Oligocene boundary.

Figure 9. The number of extinctions of benthonic and planktonic foraminifera, calcareous nannoplankton and planktonic diatoms plotted against time (Corliss et al., 1984). The data show no dramatic change in the number of extinctions when compared with data from the middle Eocene to early Oligocene interval. The number of extinctions of benthonic foraminifera in the early Oligocene in Site 219 is probably an overestimate due to the sporadic occurrence of rare species.

Furthermore, no data have been presented to support the idea of a mass extinction event in the Miocene. In the Cenozoic record, two of the three mass extinction events noted by Raup and Sepkoski [1984] do not exist in the marine microfossil record. These identified intervals are times of biotic turnover, however, suggesting that the nature of these biotic events has been misinterpreted. Furthermore, the methodology used in determining extinction patterns may influence the results as discussed by McKenna [1983]. It is premature to conclude that the data presented by Raup and Sepkoski [1984] represent mass extinction events until detailed studies at the species level across the critical intervals are evaluated to determine the nature of the biotic events.

Sedimentation and Deep-Water Circulation Changes in the Oceans

One of the most striking changes in sedimentation at the Eocene/Oligocene boundary was the depth change of the carbonate compensation depth (CCD) in the oceans as documented by the analysis of DSDP sequences [Heath, 1969; Berger, 1973; van Andel et al., 1975; van Andel, 1975 among others]. The depth of the CCD increased in all of the oceans across the Eocene/Oligocene boundary, and these dramatic changes were suggested to reflect more vigorous oceanic circulation and higher biological productivity.

The occurrence of hiatuses at the Eocene/Oligocene boundary was noted in early shallow-water studies and Rona [1973] demonstrated that the occurrence of hiatuses at this time was widespread in the deep-sea as well. More detailed regional studies [Davies et al., 1975; Kennett et al., 1972] corroborated these earlier observations, and it has been suggested that these hiatuses resulted from an increase in the vigor of deep-water circulaton due to the incipient production of Antarctic Bottom Water. Moore et al. [1978] presented a comprehensive analysis of hiatuses in DSDP sites and showed that the occurrence of Eocene hiatuses increased from a minimum within the early Eocene to a maximum in the late Eocene after which time the occurrence of hiatuses gradually decreased. Geographic differences in the timing of hiatuses were noted and were in response to differing regional circulation conditions [Moore et al., 1978; Tucholke and Embley, 1984].

Miller and Tucholke [1983] studied lithologic, sedimentological, stable isotopic and micropaleontological data from North Atlantic DSDP sites, and suggested a sequence of deep water circulation events in the North Atlantic. These include: a) an increase in bottom water current strength during the latest Eocene to early Oligocene, b) followed by a decrease during the Oligocene. The increase in current strength at the boundary was suggested to result from formation of deep water with northern origins and coincided in timing with increased circulation of southern-source bottom-water [Kennett and Shackleton, 1976].

Initial interpretations of deep-sea ostracode [Benson, 1975] and oxygen isotopic data [Shackleton and Kennett, 1975; Kennett and Shackleton, 1976] suggested that the psychrosphere developed near the Eocene/Oligocene boundary, reflecting a decrease in Antarctic surface water temperatures to freezing, the production of the first extensive sea ice near Antarctica, and the inception of production of cold, vigorous bottom water. This view was modified by Corliss [1981] who suggested that Eocene bottom-water circulation began to develop in the middle to late Eocene, with the development of the psychrosphere at the Eocene/Oligocene boundary representing one component of the bottom water circulation history. This interpretation has been supported by recent isotopic studies which have demonstrated that the Eocene isotopic enrichments initially observed in the Southern Ocean were global and not regional phenonema. The isotopic and faunal studies showing Eocene changes reflect deep-water circulation events beginning in the early Eocene and continuing throughout the Cenozoic [Hsü et al., 1984].

Summary

1. The Eocene/Oligocene boundary is marked by a benthonic foraminiferal enrichment in $\delta^{18}O$ reflecting some combination of deep-sea cooling and build-up of continental ice on Antarctica. Planktonic foraminiferal $\delta^{18}O$ data show a latitudinal pattern with greatest enrichments in the high latitudes reflecting greater surface water cooling than in low latitudes. Biotic changes include biogeographical migrations and extinctions of planktonic and benthonic microfossils due to the change in surface and deep thermal structure of the oceans.

2. The biotic data do not support a simultaneous and massive extinction event at the boundary or at any time during middle Eocene to middle Oligocene time. The observed biotic changes are suggested to be in response to changing paleoceanographic and paleoclimatic conditions, probably due to changes in ocean-continent configurations as discussed by Berggren [1981] and Kennett [1982].

3. The lack of a catastrophic extinction event argues against the existence in the Cenozoic of 26 m.y. extinction cycles as recently proposed by Raup and Sepkoski [1984]. Detailed studies of critical intervals are needed to evaluate the nature of biotic events in the Phanerozoic fossil record, since the suggested biotic events may represent gradual turnovers, rather than catastrophic extinctions.

4. A benthonic foraminiferal $\delta^{18}O$ enrichment at the middle/late Eocene boundary reflects a deep-water cooling of the same or larger magnitude (2-4°C) as that at the Eocene/Oligocene boundary. The timing of the cooling coincides with the

dramatic decrease in abundance of the benthonic foraminifer N. truempyi and the most significant Cenozoic change in the composition of deep-sea ostracodes. These biotic changes are suggested to be due to changes in the deep-sea environment reflected by the isotopic enrichment.

5. It is suggested that the cooling at the Eocene/Oligocene boundary did not have a major effect on the benthonic foraminifera because the sensitive species had been eliminated during previous paleoenvironmental events, leaving species with wide environmental tolerances able to withstand a 2-3° bottom water temperature decrease.

Acknowledgements. We thank M.-P. Aubry, W.A. Berggren, G. Keller, D. Schnitker, and R.C. Tjalsma for discussions on the research, W.A. Berggren, B. Glass, G. Keller, K.G. Miller, and H. Oberhänsli for reviewing the manuscript, M.E.H. Jeglinski for technical assistance, G. Pierce for typing the manuscript, and K.J. Hsü for inviting us to contribute this paper. We also thank the National Science Foundation for Deep Sea Drilling Project samples. This research was supported by NSF grant OCE8008879 and partially funded by a grant from a consortium of oil companies (Atlantic-Richfield Co., British Petroleum Corp., Chevron U.S.A. Inc., Cities Service, Elf-Aquitaine, Exxon Production Research Co., Gulf Oil Co., Mobil Oil Corp., Phillips Petroleum, Shell Oil Co. (International), Shell Oil Co. (U.S.A.), Texaco, Inc., Union Oil Company of California).

References

Alvarez, L. W., W. Alvarez, F. Asaro, and H. V. Michel, Extraterrestrial cause for the Cretaceous-Tertiary extinction, Science, 208, 1095-1108, 1980.

Alvarez, W., F. Asaro, H. V. Michel, and L. W. Alvarez, Iridium anomaly approximately synchronous with terminal Eocene extinctions, Science, 216, 886-888, 1982.

Aubry, M.-P., Late Eocene to early Oligocene calcareous nannoplankton biostratigraphy and biogeography, Amer. Assoc. Pet. Geol. Bull., 67, 415, 1983.

Beckman, J. P., Distribution of benthonic foraminifera at the Cretaceous-Tertiary boundary of Trinidad (West Indies), Rep. Inter. Geol. Congress, 21, Norden, Copenhagen, Part 5, 57-69, 1960.

Benson, R. H., The origin of the psychrosphere as recorded in changes of deep-sea ostracode assemblages, Lethaia, 8, 69-83, 1975.

Benson, R. H., R. E. Chapman, and L. T. Deck, Paleoceanographic events and deep-sea ostracodes, Science, 224, 1334-1336, 1984.

Berger, W. H., Cenozoic sedimentation in the eastern tropical Pacific, Geol. Soc. Amer. Bull., 84, 1941-1954, 1973.

Berggren, W. A., Role of ocean gateways in climatic changes, Stockholm Contributions in Geology, 37(2), 9-20, 1981.

Berggren, W. A., and J. Aubert, Late Paleogene (Late Eocene and Oligocene) benthonic foraminiferal biostratigraphy and paleobathymetry of Rockall Bank and Hatton-Rockall Basin, Micropaleontology, 22(3), 307-326, 1976.

Beyrich, E., Uber die Verbreitung tertiarer Albagerungen in der Gegend von Dusseldorf. Z., dt. Geol. Ges., 7, 451-452, 1854.

Boersma, A., Eocene to early Miocene Benthic Foraminifera, DSDP Leg 39, South Atlantic, In: Supko, P. R., K. Perch-Nielsen, et al., Init. Repts. DSDP: 39, Washington (U.S. Government Printing Office), 643-656, 1977.

Boersma, A., and N. Shackleton, Tertiary oxygen and carbon isotope stratigraphy, site 357 (Mid latitude South Atlantic), In: Supko, P. R., K. Perch-Nielsen, et al., Init. Repts. DSDP, 39, Washington (U.S. Government Printing Office), 911-924, 1977.

Braga, G., R. De Biase, A. Grunig, and F. Proto Decima, Foraminiferi bentonici del Paleocene e dell' Eocene della Sezione di Possagno, In Bolli, H. M., Ed., Monografia micropaleontologica sul Paleocene e l'Eocene di Possagno, Prov. di Treviso, Italia, Schweizenische Pal., Abh.-Mem. Suisse Pal., 97, 85-111, 1975.

Buchardt, B., Oxygen isotope paleotemperatures from the Tertiary period in the North Sea area, Nature, 275, 121-123, 1978.

Cavelier, C., J. J. Chateauneuf, C. Pomerol, D. Rabussier, M. Renard, and C. Vergnaud-Grazzini, The geological events at the Eocene/Oligocene boundary, Palaeogeography, Palaeoclimatology,Palaeoecology,18,223-248.1981.

Clark, M. W., and R. C. Wright, Paleogene abyssal foraminifers from the Cape and Angola Basins, South Atlantic Ocean, In: Hsu, K. J., J. L. LaBrecque, et al., Init. Repts. DSDP, 73, Washington (U.S. Government Printing Office), 459-480, 1984.

Collinson, M. E., K. Fowler, and M. C. Boulter, Floristic changes indicate a cooling climate in the Eocene of southern England, Nature, 291, 315-317, 1981.

Corliss, B. H., Response of deep-sea benthonic foraminifera to development of the psychrosphere near the Eocene/Oligocene boundary, Nature, 282, 63-65, 1979.

Corliss, B. H., Deep-sea benthonic foraminiferal faunal turnover near the Eocene/Oligocene boundary, Mar. Micro., 6, 367-384, 1981.

Corliss, B. H., M.-P. Aubry, W. A. Berggren, J. M. Fenner, L. D. Keigwin, Jr., and G. Keller, The Eocene/Oligocene boundary event in the deep sea, Science, v 226, 806-810, 1984.

Davies, T. A., O. E. Weser, B. P. Luyendyk, and R. B. Kidd, Unconformities in the sediments of the Indian Ocean, Nature, 253, 15-19, 1975.

Devereux, I., Oxygen isotope paleotemperature measurements on New Zealand Tertiary fossils, N.Z. Jour. Sci., 10, 988-1011, 1967.

Dorman, F. H., Australian Tertiary paleotemperatures, Jour. Geol., 74, 49-61, 1966.
Douglas, R. G., Benthonic foraminiferal biostratigraphy in the central North Pacific, Leg 17: Deep Sea Drilling Project, In: Winterer, E. L., J. I. Ewing, et al., Initial Repts. DSDP, 17, Washington (U.S. Government Printing Office), 591-605, 1973.
Douglas, R. G., and F. Woodruff, Deep sea benthic foraminifera, In: The Oceanic Lithosphere, Ed., Emiliani, C., Wiley-Interscience, N.Y., The Sea, 7, 1233-1327, 1981.
Emiliani, C., Temperatures of Pacific bottom waters and polar superficial waters during the Tertiary, Science, 119, 853, 1954.
Emiliani, C., The temperature decrease of surface seawater in high latitudes and of abyssal-hadal water in open oceanic basins during the past 75 million years, Deep Sea Res., 8, 144-147, 1961.
Emiliani, C., E. B. Kraus, and E. M. Shoemaker, Sudden death at the end of the Mesozoic, Earth and Planetary Science Letters, 55, 317-334, 1981.
Fenner, J., Eocene-Oligocene planktic diatom stratigraphy in the low and high southern latitudes, Micropaleontology, 30(4), 319-342, 1984.
Ganapathy, R., Evidence for a major meteorite impact on the earth 34 million years ago: implications for Eocene extinctions, Science, 216, 885-886, 1982.
Glass, B. P., Multiple microtektite horizons in upper Eocene sediments?, Science, 224, 309, 1984.
Glass, B. P., and J. R. Crosbie, Age of Eocene/Oligocene boundary based on extrapolation from North American microtektite layer, Amer. Assoc. Pet. Geol. Bull., 66(4), 471-476, 1982.
Glass, B. P., and M. J. Zwart, North American microtektites in Deep Sea Drilling Project cores from the Caribbean Sea and Gulf of Mexico, Geol. Soc. Amer. Bull., Part 1, 90, 595-602, 1979.
Haq, B. U., and G. P. Lohmann, Early Cenozoic calcareous nannoplankton biogeography of the Atlantic Ocean, Mar. Micro., 1, 119-194, 1976.
Haq, B. U., I. Premoli-Silva, and G. P. Lohmann, Calcareous plankton biogeographic evidence for major climatic fluctuations in the early Cenozoic Atlantic Ocean, Jour. Geophys. Res., 82, 3861-3876, 1977.
Hartenberger, J.-L., Les rongeurs de l'Eocene d'Europe. Leur evolution dans leur cadre biogeographique., Bull. Mus. Nat. Hist. Nat., 132, 49-70, 1973.
Heath, G. R., Carbonate sedimentation in the abyssal equatorial Pacific during the past 50 million years, Geol. Soc. Amer. Bull., 80, 689-694, 1969.
Hsü, K. J., Terrestrial catastrophe caused by cometary impact at the end of Cretaceous, Nature, 285, 201-203, 1980.
Hsü, K. J., J. A. McKenzie, and H. Oberhansli, South Atlantic Cenozoic paleoceanography, In: Hsü, K. J., J. L. LaBrecque, et al., Init. Repts. DSDP: 73, Washington (U.S. Government Printing Office), 771-785, 1984.
Keigwin, L. D., Jr., Paleoceanographic change in the Pacific at the Eocene-Oligocene boundary, Nature, 287, 722-725, 1980.
Keigwin, L. D., and B. H. Corliss, Stable isotopes in late middle Eocene to Oligocene foraminifera, Geol. Soc. Amer. Bull., 97(3), 335-345, 1986.
Keigwin, L., and G. Keller, Middle Oligocene cooling from equatorial Pacific DSDP Site 77B, Geology, 12, 16-19, 1984.
Keller, G., Biochronology and paleoclimatic implications of middle Eocene to Oligocene planktic foraminiferal faunas, Mar. Micro., 7, 463-486, 1983a.
Keller, G., Paleoclimatic analysis of middle Eocene through Oligocene planktonic foraminiferal faunas, Palaeogeography, Palaeoclimatology, Palaeoecology, 43, 73-94, 1983b.
Keller, G., S. D'Hondt, and T. L. Vallier, Multiple microtektite horizons in upper Eocene marine sediments: no evidence for mass extinctions, Science, 221, 150-152, 1983.
Keller, G., S. D'Hondt, and T. L. Vallier, Multiple microtektite horizons in upper Eocene marine sediments?--Reply, Science, 224, 309-310, 1984.
Kennett, J. P., The development of planktonic biogeography in the Southern Ocean during the Cenozoic, Mar. Micro., 3, 301-345, 1978.
Kennett, J. P., Marine Geology, Englewood Cliffs, N. J.: Prentice-Hall, Inc., 813 pp. 1982.
Kennett, J. P., and N. J. Shackleton, Oxygen isotope evidence for the development of the psychrosphere 38 Myr ago, Nature, 260, 513-515, 1976.
Kennett, J. P., et al., Australian-Antarctic continental drift, palaeocirculation changes and Oligocene deep-sea erosion, Nature Phys. Science, 239, 51-55, 1972.
Lyell, C., Principles of Geology, John Murray, London, 398 pp., 1833.
Margolis, S. V., P. M. Kroopnick, D. E. Goodney, W. C. Dudley, and M. E. Mahoney, Oxygen and carbon isotopes from calcareous nannofossils as paleoceanographic indicators, Science, 189, 555-557, 1975.
Matthews, R. K., and R. Z. Poore, Tertiary $\delta^{18}O$ record and glacio-eustatic sea-level fluctuations, Geology, 8, 501-504, 1980.
McKenna, M. C., Holarctic landmass rearrangement, cosmic events, and Cenozoic terrestrial organisms, Ann. Missouri Bot. Gard. 70, 459-489, 1983.
Miller, K. G., Eocene-Oligocene paleoceanography of the deep Bay of Biscay: benthic foraminiferal evidence, Mar. Micro., 7, 403-440, 1983.

Miller, K. G., and R. G. Fairbanks, Oligocene to Miocene global carbon isotope cycles and abyssal circulation changes, In: The carbon cycle and atmospheric CO_2: natural variations Archean to present. (Sundquist, E. J., and W. S. Broecker, eds. Geophysical Monograph 32, American Geophysical Union, Washington, D.C. 469-486.

Miller, K. G., and W. Curry, Eocene to Oligocene benthic foraminiferal isotopic records in the Bay of Biscay, Nature, 296, 347-352, 1982.

Miller, K. G., and R. G. Fairbanks, Evidence for Oligocene-Middle Miocene abyssal circulation changes in the western North Atlantic, Nature, 306, 250-253, 1983.

Miller, K. G., and B. E. Tucholke, Development of Cenozoic abyssal circulation south of the Greenland-Scotland Ridge, In: Structure and development of the Greenland-Scotland Ridge (Eds. Bott, M., S. Saxou, M. Talwani and J. Thiede), Plenum Publishing Corp., 549-589, 1983.

Miller, K.G., W. B. Curry, and D. R. Osterman, Late Paleogene (Eocene to Oligocene) benthic foraminiferal paleoceanography of the Goban Spur region, DSDP Leg 80, In: de Graciansky, P. C., C. W. Poag, et al., Init. Repts. DSDP, 80: Washington (U.S. Government Printing Office), 505-538, 1985.

Moore, T. C., Jr., T. H. van Andel, C. Sancetta, and N. Pisias, Cenozoic hiatuses in pelagic sediments, Micropaleontology, 24(2), 113-138, 1978.

Oberhänsli, H., and M. Toumarkine, The Paleogene oxygen and carbon isotope history of Sites 522, 523, and 524 from the central South Atlantic, In: Hsü, K. J., and H. Weissert, eds., Symposium on the South Atlantic Paleoceanography, Cambridge University Press, in press, 1985.

Oberhänsli, H., J. McKenzie and M. Toumarkine, A paleoclimatic and paleoceanographic record of the Paleogene in the central South Atlantic (Leg 73, Sites 522, 523, and 524), In: Hsü, K. J., J. L. LaBrecque, et al., Init. Repts. DSDP, 73: Washington (U.S. Government Printing Office), 737-747, 1984.

Olsson, R. K., K. G. Miller, and T. E. Ungrady, Late Oligocene transgression of middle Atlantic coastal Plain, Geology, 8, 549-554, 1980.

Parker, W. C., M. W. Clark, R. C., Wright, and R. K. Clark, Population dynamics, Paleogene abyssal benthic foraminifers, eastern South Atlantic, In: Hsu, K. J., J. L. La Brecque, et al., Init. Repts. DSDP, 73: Washington (U.S. Government Printing Office), 481-486, 1984.

Poore, R. Z., and R. K. Matthews, Oxygen isotope ranking of late Eocene and Oligocene planktonic foraminifers: implications for Oligocene sea-surface temperatures and global ice-volume, Mar. Micro., 9, 111-134, 1984.

Prothero, D. R., North American mammalian diversity and Eocene-Oligocene climate, Paleobiology, in press, 1985.

Raup, D. M., and J. J. Sepkoski, Jr., Periodicity of extinctions in the geologic past, Proc. Natl. Acad. Sci., 81, 801-805, 1984.

Rona, P. A., Worldwide unconformities in marine sediments related to eustatic changes of sea level, Nature Physical Science, 244, 25-26, 1973.

Sancetta, C., Paleogene Pacific microfossils and paleoceanography, Mar. Micro., 4, 363-398, 1979. Savin, S. M., R. G. Douglas, and F. G. Stehli, Tertiary marine paleotemperatures, Geol. Soc. Amer. Bull., 86, 1499-1510, 1975.

Schnitker, D., Cenozoic deep water benthic foraminifers, Bay of Biscay, In: Montadert, L., D. G. Roberts, et al., Init. Repts. DSDP, 48: Washington (U.S. Government Printing Office), 377-413, 1979.

Shackleton, N. J., and J. P., Kennett, Paleotemperature history of the Cenozoic and the initiation of Antarctic glaciation: oxygen and carbon isotope analyses in DSDP Sites 277, 279, and 281, In: Kennett, J. P., R. E. Houtz, et al., Init. Repts. DSDP, 29: Washington (U.S. Government Printing Office), 743-755, 1975.

Shackleton, N. J., M. A. Hall, and A. Boersma, Oxygen and carbon isotope data from leg 74 foraminifers, In: Moore, T. C., Jr., P. D. Rabinowitz, Init. Repts. DSDP, 74, Washington (U.S. Government Printing Office), 599-612, 1984.

Snyder, S. W., C. Muller, and K. G. Miller, Eocene-Oligocene boundary: biostratigraphic recognition and gradual paleoceanograhic change at DSDP Site 549, Geol. 12, 112-115, 1984.

Tjalsma, R. C., Eocene to Miocene benthic foraminifers from Deep Sea Drilling Project Site 516, Rio Grande Rise, South Atlantic, In: Barker, P. F., R. L. Carlson, D. A. Johnson, et al., Init. Repts. DSDP, 72: Washington (U.S. Government Printing Office), 731-755, 1983.

Tjalsma, R. C., and G. P. Lohmann, Paleocene-Eocene bathyal and abyssal benthic foraminifera from the Atlantic Ocean, Micropaleontology, Sp. Publ. 4, 1-90, 1983.

Tucholke, B. E., and R. W. Embley, Cenozoic regional erosion of the abyssal seafloor off South Africa, In: J. S. Schlee (Ed.), Interregional unconformities and hydrocarbon accumulation, Amer. Assoc. Pet. Geol. Memoir, 36, 145-164, 1984.

van Andel, T. H., Mesozoic/Cenozoic calcite compensation depth and the global distribution of calcareous sediments, Earth and Planet. Sci. Letts., 26, 187-195, 1975.

van Andel, T. H., G. R. Heath, and T. C. Moore, Jr., Cenozoic history and paleoceanography of the central equatorial Pacific, Geol. Soc. Amer. Mem., 143, 134 pp., 1975.

Vail, P. R., R. M. Mitchum, Jr., R. G. Todd, J. M. Widmier, S. Thompson, III, J. B. Sangree, J. N. Bubb, and W. G. Hatlelid, Seismic stratigraphy and global changes of sea level, In: Seismic Stratigraphy--Applications to Hydrocarbon Exploration, 49-212, Ed. C. E.

Payton, Amer. Assoc. Petr. Geol., Tulsa, Oklahoma, 1977.
Vergnaud-Grazzini, C., C. Muller, C. Pierre, R. Letolle, and J. P. Peypouquet, Stable isotopes and Tertiary paleontological paleoceanography in the northeast Atlantic, In: Montadert, L., D. G. Roberts, et al., Init. Repts. DSDP, 48: Washington (U.S. Government Printing Office), 475-491, 1979.
Vige, B., and M. Vianey-Liaud, Impropriete de la grande coupure de Stehlin comme support d'une Umite Eocene - Oligocene., Newsl. Stratigr., 8, 79-82, 1979.
von Hillebrandt, A., Das Paleozan und seine Foraminiferen-Fauna im Becken von Reichenhall and Salzburg, Bayer. Akad. Wiss. Math.-Naturw. Kl., Abh., n. ser., 108, 1-182, 1962.
Wolfe, J. A., A paleobotanical interpretation of Tertiary climates in the Northern Hemisphere, Amer. Sci., 66, 694-703, 1978.

MIOCENE PALEOCEANOGRAPHY AND PLANKTON EVOLUTION

James P. Kennett

Graduate School of Oceanography, University of Rhode Island, Kingston, R. I. 02881

Extended Abstract. During the Miocene, a number of major developments occurred in the evolution of the ocean and of its planktonic biota which are summarized in Table 1. Numerous workers of CENOP (Cenozoic Paleoceanography Program) and others have contributed towards better understanding of the Miocene Ocean; that which existed between the still unfamiliar Oligocene Ocean and the more familiar latest Cenozoic.

The development of a Circumantarctic current near the end of the Oligocene (22 to 25 Ma) and related continued restriction of low latitude oceanic circulation set the stage for Miocene oceanographic and biotic evolution. The Circumantarctic circulation system developed as southern land masses moved away, creating unrestricted latitudinal flow. Changing boundary conditions in this region included the opening of the Tasmanian Seaway as Australia moved northwards, the opening of the Drake Passage and the development of the Kerguelen Plateau. The initial formation, and later more complete development, of the Circumantarctic Current thermally isolated Antarctica by decoupling the warmer subtropical gyres from colder high latitude waters. The climatic history of the Southern Ocean is, in general, one of decreasing temperatures, including the development of increased Antarctic glaciation and later ice sheet formation, a climatic regime which itself had a profound effect on global environmental evolution, and planktonic biogeography.

Latitudinal Temperature Gradients

The Miocene Ocean experienced a major steepening of thermal gradients in surface waters as the Circumantarctic Current developed and intensified. Steepening of latitudinal isotopic temperature gradients in surface waters started to become conspicuous near the beginning of the Oligocene replacing broader temperature gradients of the Eocene (Murphy and Kennett, 1986). Subantarctic waters cooled relative to those at temperate latitudes. During the Oligocene, this surface water differentiation continued. Increasing isotopic offset between latitudinally distributed southwest Pacific sites is linked to the establishment and strengthening of the Circumantarctic Current previously considered to have developed during the Middle to Late Oligocene. The intensification of this current system progressively decoupled the warm subtropical gyres from Antarctic circulation, in turn leading to increased Antarctic glaciation.

The pole to equator surface-water temperature gradient continued to increase during the Miocene. In the equatorial Pacific, planktonic $\delta^{18}O$ values decreased during the Miocene, whereas in the higher southern latitudes, planktonic $\delta^{18}O$ values become more positive in response to cool-surface waters. $\Delta\delta^{18}O_{PH-PL}$ (high latitude-low latitude) planktonic values are a measure of the latitudinal temperature gradient. An increase during the Miocene indicates that the latitudinal temperature gradient increased by about $6^{o}C$ to a value of $12^{o}C$ in the latest Miocene between equatorial and Subantarctic sites (Loutit et al., 1983). Part of this increase is obviously due to a cooling of high-latitude surface waters, but also to warming of low-latitude surface waters. The latitudinal temperature gradient in the late Miocene was about three-quarters of the present day gradient and it doubled during the Miocene from about $6^{o}C$ to about $12^{o}C$ (Loutit et al., 1983).

Middle Miocene Glaciation

The Middle Miocene (about 14 Ma) represents a crucial stage in the development of global paleoceanography, for at this time much of the Antarctic ice sheet formed. This event is marked by a sharp increase in the $\delta^{18}O$ values of calcareous plankton and of benthonic foraminifera (Figure 1). This increase certainly reflects, in part, a major period of ice-sheet growth, and a drop in surface temperatures at the Antarctic coast. The existence of a widespread ice sheet from the Middle Miocene is supported by the presence of common and persistent ice-rafted sediments around the Antarctic continent from that time. Cooler planktonic assemblages also became more important.

Part of the oxygen isotopic change that oc-

TABLE 1. Major Paleoceanographic, Paleoclimatic and Biotic Events of the Miocene

Feature	Changes
Temperature Gradient	Pole to equator increase
Circum-Antarctic Current	Expansion & intensification
East Antarctic ice sheet	Major expansion ~14 Ma
Indonesian Seaway	Closure in Middle Miocene
Pacific Equatorial Undercurrent	Development in Middle Miocene
Pacific Equatorial Countercurrent	Strengthening in Middle Miocene
Gyral Circulation	Intensification and "Spin-up" of boundary currents
Planktonic biogeography	Increased latitudinal provincialism
Plankton	Major evolutionary radiation in Early Miocene
Antarctic High Productivity Zone	Increased development through Neogene
Equatorial High Productivity Zone	Decrease in Late Neogene
Latest Miocene events	Closure of Mediterranean Basin
	Carbon shift
	Antarctic ice volume expansion
	Surface-water cooling
	Northward expansion of Antarctic waters

curred during the Middle Miocene also resulted from temperature change. This is indicated by large changes at that time of deep-sea benthonic foraminiferal assemblages. Coincident with the isotopic change was the rapid replacement or increased dominance of numerous species originating in the Oligocene or earlier with forms that dominate the late Cenozoic and modern deep-sea environment. $\Delta\delta^{18}O_{B-P}$ (Benthonic-Planktonic) is a measure of the thermal structure of the water column. This increased during the Miocene with the greatest increase during the Middle Miocene at about 14 Ma. By the latest Miocene, the isotopic gradient at Site 289 in the equatorial Pacific approached the present day isotopic gradient (about 4-5°/oo).

The cause of development of a major Antarctic ice sheet in the Middle Miocene remains unexplained. No permanent ice sheet is believed to have formed before this even though temperatures were cold enough since the earliest Oligocene. Nevertheless it now seems likely that significant, but temporary ice sheets occurred periodically on Antarctica during the Oligocene. The heaviest oxygen isotopic values in all records cluster in the Middle Oligocene (~30 Ma) with oscillating episodes of >0.5^{o}/oo enrichments occurring most prominently in the Subantarctic record of Site 277. This interval can be interpreted as recording temporary accumulations of Antarctic ice and the coldest oceanic isotopic values of the Paleogene. About this time, rapid

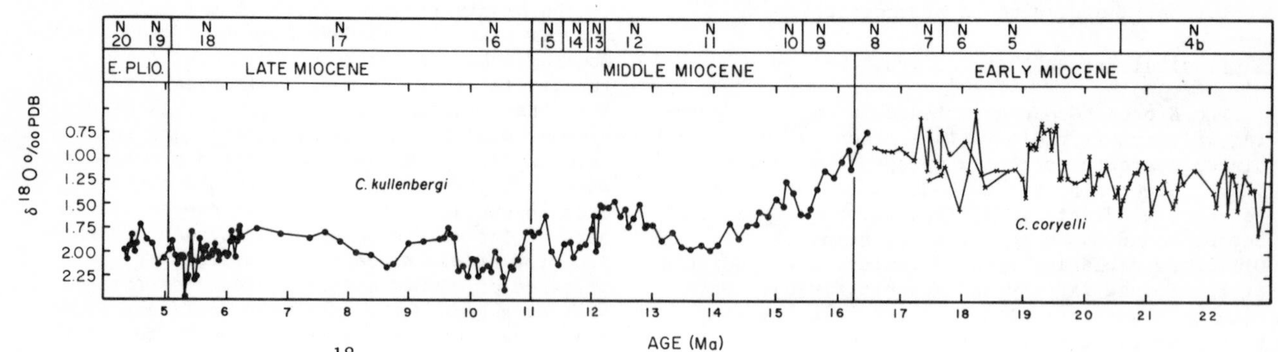

Fig. 1. Benthic $\delta^{18}O$ for Holes 588, 588A and 588C plotted against age at DSDP Site 588 and planktonic foraminiferal zones (from Kennett, 1986).

development occurred of latitudinal thermal differentiation of surface waters which actually warmed in the temperate areas.

A more extreme view, held by some workers, is that major Antarctic ice sheets have existed since the Eocene and perhaps during much of the Cretaceous, and that almost all ice volume was in place prior to the Middle Miocene. This interpretation is based on an assumption of constant tropical sea-surface temperatures during the late Phanerozoic. Strong evidence exists, however, to show that tropical sea-surface temperatures have both warmed and cooled during the Miocene. Faunal, floral and isotopic evidence suggest that Pacific surface waters in the tropical and especially the temperate regions of the North Pacific warmed during the Middle to Late Miocene, perhaps by as much as 4°C, close to present-day temperatures. If true, then tropical Pacific surface waters were probably close to 22-24°C in the Early Miocene (Loutit et al, 1983). Furthermore there is little independent sedimentological and paleobiological evidence in support of major, permanent ice sheets prior to the Middle Miocene.

Tropical Surface Water Changes

Biogeographic patterns of planktonic microfossils underwent distinct change during the Miocene and have assisted in evaluating paleocirculation patterns. Important differences are apparent between the Early and Late Miocene which resulted from changes in surface water circulation within the Pacific Ocean and between the tropical Pacific and Indian Oceans.

In the Early Miocene, tropical Pacific planktonic foraminifera are dominated by different taxa in the eastern and western areas, but by the Late Miocene the assemblages are similar across the entire tropical Pacific (Kennett et al., 1985). East to west faunal differences were probably due to differences in the surficial water-mass structure and temperature. It is likely that a deeper thermocline existed in the west favoring shallow water dwellers such as Globigerinoides and Globigerina angustiumbilicata, and a shallower thermocline in the east favoring slightly deeper-dwelling forms, especially Globorotalia siakensis and mayeri. During the Late Miocene a trans-equatorial assemblage developed, dominated by Globorotalia menardii-limbata and Globigerinoides groups. These faunal changes are interpreted to reflect both the development during the Middle Miocene, of the Equatorial Undercurrent system when the Indonesian Seaway effectively closed (as a result of Australian northward drift), and the general strengthening of the gyral circulation, and of the Equatorial Countercurrent that resulted from increased Antarctic glaciation and high-latitude cooling during the Middle Miocene.

The trans-equatorial planktonic foraminiferal distribution patterns typical of the Late Miocene did not persist to the present-day oceans when east-west differences are again evident. However, these differences are exhibited within forms that usually inhabit deeper waters. There is a successive changing dominance from west to east of Pulleniatina obliquiloculata to Globorotalia tumida to Neogloboquadrina dutertrei. The modern west to east differences in these deeper-dwelling forms reflect an intensification of the Equatorial Undercurrent system and its shallowing towards the east to depths well within the photic zone. Shallow-water forms, such as Globigerinoides, maintain trans-tropical distribution patterns in the modern ocean unlike the early Miocene which lacked an effective equatorial countercurrent system in the Pacific.

The distribution of faunas in the North Pacific indicates that the gyral circulation system was only weakly developed in the Early Miocene, but was strong by the Late Miocene. In the northwest Pacific, temperate faunas are displaced northward as the Kuroshio Current intensified in the Late Miocene. In the South Pacific, more distinct latitudinal faunal provinces appear during the Middle to Late Miocene along with a northward expansion of the polar-subpolar provinces and contraction of the tropical province. These faunal changes resulted from the continued areal expansion of the polar and subpolar water masses as Australia drifted northward from Antarctica and from the steepening of pole to equator thermal gradients related to increased Antarctic glaciation.

Late Miocene Events

Following the development of the major Antarctic ice sheet during the Middle Miocene, benthic $\delta^{18}O$ values exhibited distinct fluctuations, but the average values remained unchanged (Figure 1). The isotopic data (Kennett, 1986) show two distinct episodes of climatic cooling close to the Middle/Late Miocene boundary. The earliest of these events occurred between 12.5 and 11.5 Myr in the latest Middle Miocene. The second cooling event occurred from 11 to 9 Myr, and is marked by some of the highest $\delta^{18}O$ values of the entire Miocene. This was followed by relative warmth during the middle part of the late Miocene. The terminal Miocene and earliest Pliocene (6.2 to 4.5 Myr) was marked by relatively elevated $\delta^{18}O$ values indicating increased cooling and glaciation.

The discovery of a marked isochronous depletion of δC^{13} values (0.5-0.8°/oo) in marine carbonates of the Indo-Pacific region at about 6.2 Ma has stimulated speculation about the origin of this event and of its possible interrelations with the terminal Miocene event. Several workers have suggested that this decrease may reflect changes in circulation patterns and upwelling dynamics or an increase in the amount of particulate phosphorous reaching the deep ocean, while others have suggested that the shift reflected a sudden increase in the rate of supply of organic carbon from shallow water areas exposed by regression as well as changes in deep circulation patterns and ocean fertility. Kennett (1986) believed that this shift was caused by a glacioeustatic sea-level lowering that exposed continental margins via regression, and ultimately increased the flux of organic carbon to the deep-sea. The shift was reinforced also by an increase in oceanic fertility. An increase in $\delta^{13}C$ values during the early Pliocene (~5 to 4 Myr) resulted from marine transgression during a time of global warmth.

Towards the end of the Miocene and during the earliest Pliocene, a distinct and rapid northward movement (300 km) occurred in the siliceous biogenic sediment belt in both the southeast Pacific and Indian Ocean regions. This event is equated with a rapid northward movement of the Antarctic Convergence and Antarctic surface water mass and is considered to be related to a major expansion of the Antarctic ice cap at this time and a northward extension of the Ross Ice Shelf much beyond present-day limits.

An extraordinary number of important oceanographic changes are now known to have occurred during the latest Miocene in association with the carbon shift. These include a drop in sea level, cooling of surface waters, intensification of bottom-water circulation, increased biogenic productivity over large areas of the ocean (especially the Southern Ocean), coupled with a decrease in biogenic silica deposition in the eastern equatorial Pacific, the isolation and dessication of the Mediterranean basin and increased Antarctic glaciation. The intensification of oceanic circulation at this time is believed by some to have resulted from the development of the west Antarctic ice sheet and the formation of the first major production of true AABW which permanently altered global abyssal circulation.

Planktonic Foraminiferal Evolution

Paleoceanographic changes in the Neogene played a principal role in the evolution of the planktonic biota. The evolution of Neogene planktonic foraminifera was principally propelled by paleoceanographic changes and regulated by diversity-dependent mechanisms to approach the equilibrium state compatible with both abiotic and biotic environment (Wei and Kennett, 1986).

The Oligocene/Miocene boundary is not marked by any crisis in Oligocene planktonic foraminiferal or other microfossil assemblages. Instead it is marked by the beginnings of a major evolutionary radiation leading to Neogene assemblages. The evolution of Neogene planktonic foraminifera occurred in three stages: a diversification stage (22 to 16 Ma, Early Miocene): an equilibrium stage (16 to 5 Ma, Middle and Late Miocene); and a declining stage (5 Ma to Recent, Pliocene and Quaternary). Early Miocene diversification reflects an initial exponential stage of phylogenetic radiation into new niches as the Miocene Ocean conditions began to develop. Following this, the assemblages maintained approximate evolutionary equilibrium, although increased turnover rates reflect faunal reorganization during times of more intense paleoceanogrpahic change. A diversity decline in the Pliocene and Quaternary seems to have resulted from adverse effects imposed by high-frequency paleoceanographic oscillations related to Northern Hemisphere glaciations.

Acknowledgments. This research was supported by the U. S. National Science Foundation.

References

Kennett, J. P., Miocene-Early Pliocene Oxygen and Carbon Isotopic Stratigraphy in the Southwest Pacific: DSDP Leg 90, in Kennett, J. P., von der Borch, C., et al., eds., Initial Reports of the Deep Sea Drilling Project, Volume 90, Washington, D. C., (U. S. Government Printing Office), 1383-1411, 1986.

Kennett, J. P., G. Keller, and M. S. Srinivasan, Miocene Planktonic Foraminiferal Biogeography and Paleoceanographic Development of the Indo-Pacific Region, in Kennett, J. P., ed., Evolution of the Miocene Ocean and its biogeography, G.S.A. Memoir, 163, 197-236, 1985.

Loutit, T. S., J. P. Kennett, and S. M. Savin, Miocene Equatorial and Southwest Pacific Paleoceanography from Stable Isotope Evidence. Marine Micropaleontology, 8, 215-233, 1983.

Murphy, M. G., and J. P. Kennett, Development of Latitudinal Thermal Gradients during the Oligocene: Oxygen Isotopic Evidence from the South Pacific, in Kennett, J. P. and C. von der Borch et al., eds., Initial Reports of the Deep Sea Drilling Project, Volume 90, Washington, D. C., (U. S. Government Printing Office), 1347-1360, 1986.

Wei, K.-Y., and J. P. Kennett, Taxonomic Evolution of Neogene Planktonic Foraminifera and Paleoceanography, Paleoceanography, 1(1), 1986.

THE TERMINAL MIOCENE EVENT

Maria Bianca Cita

Department of Earth Sciences, University of Milan, 20133 Milan, Italy

Judith A. McKenzie

Department of Geology, University of Florida, Gainesville, FL 32611 USA

Abstract. The terminal Miocene paleoceanographic event is most certainly connected with the salinity crisis that affected the Mediterranean, although the cause/effect relationships are still debated. The hydrologic budget of the Mediterranean is, and presumably was, strongly negative. Therefore, when permanent connections with the Atlantic were interrupted due to relative motions between Africa and Europe and/or eustatic sea-level lowerings, the Mediterranean basins became evaporitic. The Mediterranean paleogeography underwent drastic changes; the basin margins were subject to subaerial erosion, and the main rivers in an attempt to adapt their thalweg to the rapid fall in base level incised deep and narrow canyons, which were subsequently filled by Plio-Pleistocene marine sediments. Calculations show that over one million cubic kilometers of halite and other evaporitic minerals were laid down in the Mediterranean, lowering the salinity of the oceans by 2 per mil.

Looking for oceanographic expressions of the Messinian Mediterranean salinity crisis in deep-sea sediments, paleoceanographers have found (a) an extended regression on continental margins, (b) a sudden global lightening of the isotopic composition of carbon in oceanic sediments and (c) oxygen-isotopic evidence for several periods of lowered sea level. The regression is well expressed and results in the interruption of carbonate buildups; however, it is not well dated in most cases. On the other hand, the "carbon shift" is well calibrated. Its numerical age is 5.9-6.1 m.y., occurring just after (above) the FAD of Amaurolithus primus in the lower part of paleomagnetic Epoch 6, (Haq et al., 1980). Its significance is still in part obscure but is possibly related to the partial restriction of Mediterranean waters during the non-evaporative Messinian. During Epoch 5 and the lowermost Gilbert Epoch (the evaporative Messinian), high frequency changes interpreted as glacial maxima and minima have been recorded in the oxygen-isotope record of pelagic sediments from both the North and South Atlantic and southwest Pacific Oceans (Cita and Ryan, 1979; McKenzie and Oberhansli, 1985; Hodell et al., 1986; and Keigwin et al., in press). The maxima may represent significant eustatic sea-level drops which could have restricted entirely or partially the inflow of sea water from the Atlantic to the Mediterranean. The flow of water from the Mediterranean to the Atlantic was undoubtedly terminated during the evaporative Messinian. An exact correlation of the various intra-Messinian events recorded in the Mediterranean and the open ocean remains only tentative awaiting better stratigraphic control from both realms.

Introduction

The new science of paleoceanography recognizes both long-term, transitional changes and short-term, rapid changes often manifested as catastrophic events. The terminal Miocene event belongs to the second category, being an event which lasted less than one million years (probably 0.5 million years) and is related to the salinity crisis that affected the Mediterranean during Messinian times. The influence of global climatic changes during the latest Miocene upon the Mediterranean salinity crisis versus a potential response of the world's oceans to the crisis have been the subject of numerous studies.

One activity of the Messinian Project of the International Geological Correlation Programme (IGCP No. 96 "Correlation of the Messinian in the Mediterranean and outside the Mediterranean") was to investigate the response of the world's oceans to the Mediterranean salinity crisis. R. C. Wright, a member of this Working Group, organized a symposium on the global recognition of Late Miocene (Messinian) events in conjunction with the 1978 GSA Annual Con-

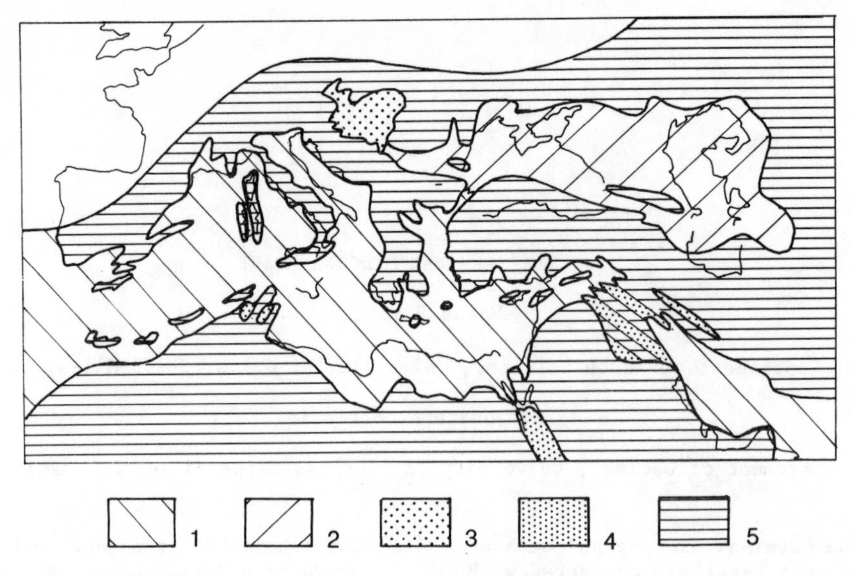

Fig. 1. Paleogeographic reconstruction of the Mediterranean and Paratethys in late Tortonian times (11.8-10.5 my BP) after Rogl and Steininger (1983). 1 = open marine facies; 2 = restricted marine facies; 3 = endemic Paratethyan facies; 4 = evaporitic facies; 5 = continental areas.

vention in Toronto. A brief summary is found in the Messinian Review Paper prepared for the Geodynamics Inter Union (Cita, 1982) and additional information may be found in Kennett (1982, 1983).

The present review of the terminal Miocene event consists of two parts: the first is devoted to the Mediterranean record, which will be outlined in its various steps and facies, and the second considers correlatable events recorded outside the Mediterranean in the open oceans or on the basin margins. We present first the Mediterranean record in order to provide the framework necessary for discussion and interpretation of the signals obtained from the deep-sea record.

Mediterranean Record

Late Miocene Scenario

After the Alpine orogeny and the termination of the Tethys seaway, the western Mediterranean basins came into existence, as a result of successive stretching phases and crustal subsidence, whereas the older eastern basins were characterized by a dominantly compressional regime. In the Late Miocene, the Mediterranean was a deep W-E trending embayment of the Atlantic. The connections with the open ocean were much wider than at present, shown by Figure 1 - a paleogeographic reconstruction of the Mediterranean and the Paratethys in Tortonian times presented by Rogl and Steininger (1983). Evaporitic basins were present in what is now the Red Sea and in Mesopotamia, and no connection existed between the Mediterranean and the Indian Ocean. The Indian connection was lost in Burdigalian times, and since then the eastern Mediterranean has suffered poor ventilation at depth, occasionally resulting in anoxic episodes. In the western basins the thermohaline circulation at depth was much more efficient.

Stratigraphic Framework

The salinity crisis occurred during the latest stage of the Miocene, called the Messinian. The neostratotype section was described in the central part of Sicily (Pasquasia-Capodarso Section, Selli, 1960). It consists essentially of evaporites (Gessoso solfifera Formation) and includes, in its lower part, laminated diatomites (Tripoli Formation) underlain by marls.

Boundary stratotypes of the stage were recently proposed in Sicily: (1) the Tortonian/Messinian boundary in the section of Falconara, 6 m beneath the base of the Tripoli diatomites (Colalongo et al., 1979) and (2) the Messinian/Zanclean boundary (=Miocene/Pliocene boundary) at the base of the Trubi Formation at Capo Rossello (Cita, 1975a,b).

The boundary stratotypes are calibrated biostratigraphically as follows (see Figure 2):

The Tortonian/Messinian boundary coincides with the first occurrence of *Globorotalia conomiozea*, postdates the first occurrence of *Globigeronoides obliquus extremus* and *Globorotalia acostaensis humerosa*, predates the first occurrrence of *Globigerina multiloba*. In terms of calcareous nannoplankton, it occurs in

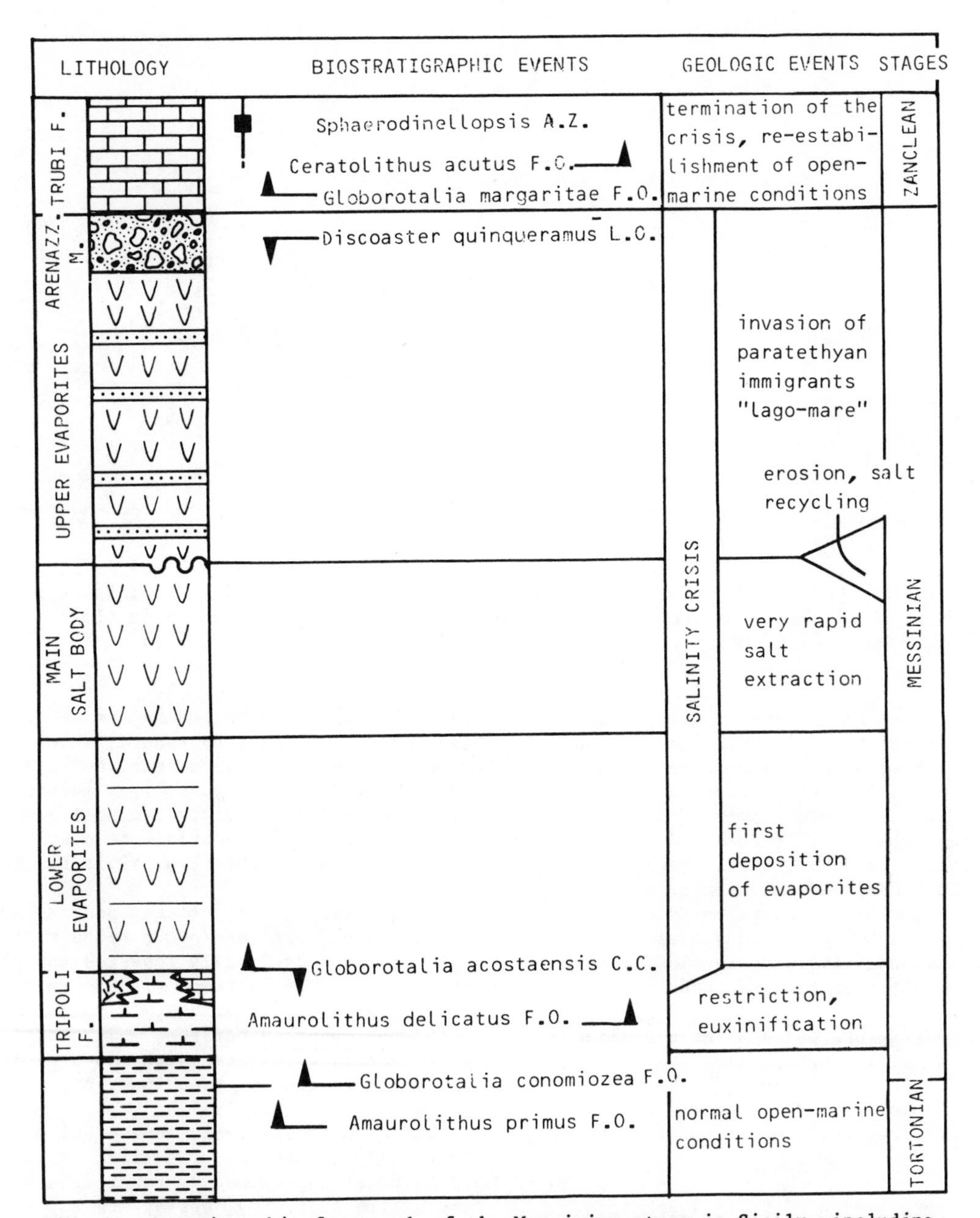

Fig. 2. Stratigraphic framework of the Messinian stage in Sicily, including lithostratigraphic succession in the type-area, biostratigraphic and geologic events. Modified after Cita (1979) and Rio et al. (1984).

between the first occurence of Amaurolithus primus and A. delicatus, within the range Discoaster berggrenii. The Messinian/Zanclean boundary coincides with the reestablishment of open marine conditions in the Mediterranean after the salinity crisis. It occurs at the base of the Sphaeroidinellopsis Acme-zone, slightly beneath the first occurrence of Globorotalia margaritae (immigration, not evolutionary appearance). In terms of calcareous nannoplankton, it occurs after the extinction of Discoaster quinqueramus, and slightly prior to the first occurrence of Ceratolithus acutus (Cita and Gartner, 1973).

Neither the recently proposed boundary stratotypes nor the Messinian type-section are calibrated paleomagnetically, notwithstanding several attempts carried out by the best specialists in the field using the most advanced equipment, including the cryogenic magnetometer. The evaporites are unfavourable lithologies for paleomagnetic studies. A

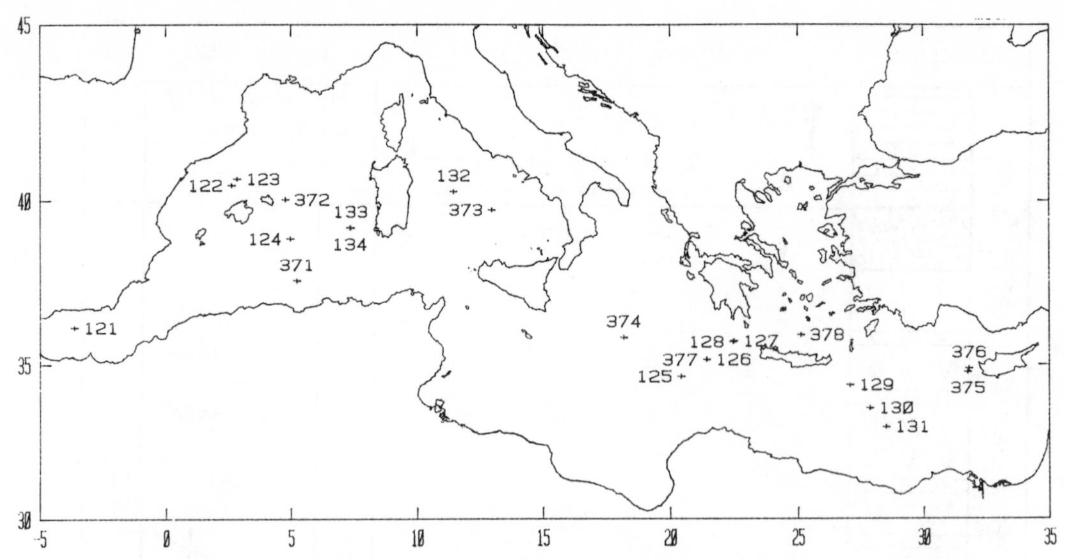

Fig. 3. Location of DSDP sites in the Mediterranean drilled in 1970 and 1975. Messinian age evaporites were recovered in the Alboran Basin (Site 121); Valencia Trough (122); Balearic Basin (124, 371, 372, 133, 134); Tyrrhenian Basin (132); Ionian Basin (374, 125); Aegean Basin (378); Levantine Basin (375, 376).

combination of an extremely weak magnetization and weathering makes the Capo Rossello section unsuitable, but preliminary paleomagnetic analysis of samples from the Tortonian/ Messinian boundary at Monte Giammoia, a section near Falconara, shows great promise (J.E. T. Channell, personal communication). Magnetic stratigraphy in marine muds of Tortonian/ Messinian age from two locations on Crete (Langreis et al., 1984) produced a clear pattern of normal and reversed signals, of which the interpretation remains controversial (Berggren et al., 1985; Hsu, 1986). Until the results from in progress paleomagnetic studies of Mediterranean sediments spanning the Tortonian/Messinian and Messinian/Zanclean boundaries are completed, the proposed paleomagnetic calibrations will remain indirect (see below).

Lithostratigraphic Record of Basinal Settings; Facies Changes; Deep-Sea Record Versus Land Record

Messinian evaporites were discovered by deep-sea drilling in all the major basins of the Mediterranean (Ryan, 1973; Hsu et al., 1973; Montadert et al., 1978). From west to east, they are Alboran Basin, Valencia Trough, Balearic Basin, Tyrrhenian Basin, Ionian Basin, Aegean Basin, and Levantine Basin (see Figure 3). Penetration into the evaporites was limited for safety reasons to a maximum of 50 meters. Fifty meters represent only a minor part of the Messinian evaporites, where this formation is fully expressed. In order to study the Messinian record in its entirety, we have to rely on the land record, which is best exposed and/or known by extensive drilling in central Sicily (see Figure 4), a former deep basin tectonically uplifted in post-Messinian times. There, it is possible to distinguish from bottom to top (Ogniben, 1957; Decima and Wezel, 1973):

(a) the Tripoli diatomites, passing laterally to coral reefs and/or to Calcare di base. Diatomites are cyclically repeated and alternate a maximum of 34 times with dolomitic claystones (McKenzie et al., 1979).;

(b) the lower evaporites, well bedded, occassionally associated with minor marls which are either barren, or yield poor and dwarf marine faunas (Cita, 1979);

(c) the main salt body, up to 600 m thick, essentially consisting of halite. Potash salts are found at approximately two thirds above the base of the salt body, not at its top as is would be expected from normal evaporation of a brine pool. An erosional unconformity is recorded in the upper part of the salt body, and geochemical evidence supports the assumption that this upper part is recycled (Decima, 1978);

(d) the upper evaporites, associated with fine clastics (marls) which contain a brackish, euryhaline fauna of mollusks and ostracods, plus reworked marine fossils. Six to seven evaporitic cycles are usually present;

(e) a coarse clastic unit (Arenazzalo), a few meters thick, with reworked fossils, closes the Messinian succession.

Facies changes are frequent on land, and

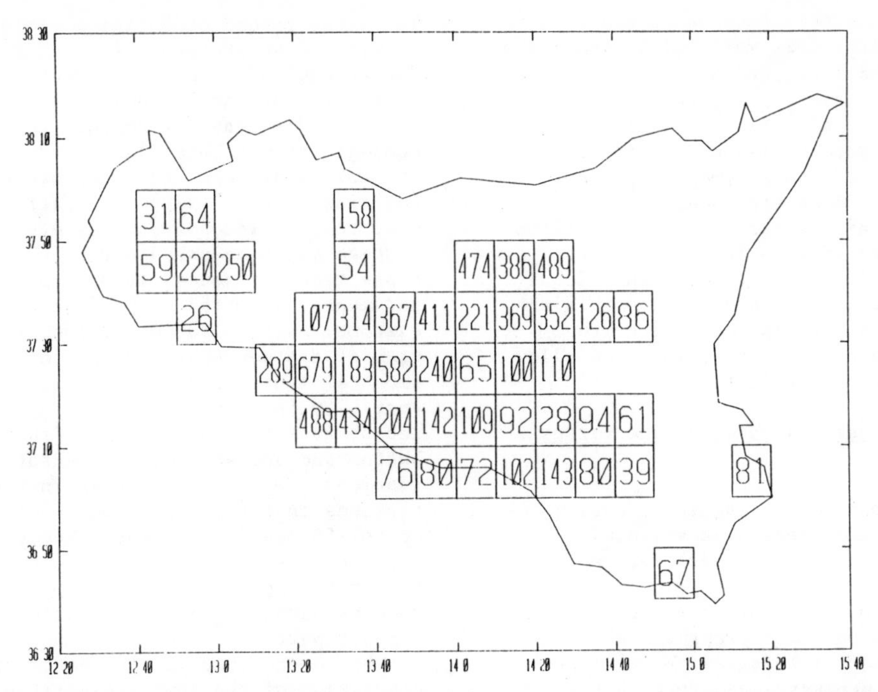

Fig. 4. Sketch of Sicily, showing the average thickness of Messinian sediments per square unit of 10' x 10' (after Camerlenghi et al., 1983). The computer map is based on 374 measured sections, most from the subsurface (329) and published. Evaporites represent the largest part of the Messinian sediments considered (see Figure 2).

record the drastic paleogeographic modifications induced by the salinity crisis. Thick non-evaporitic Messinian sequences occur in the subsurface of the Po Plain in northern Italy, and in the Laga Basin of central Apennines (Camerlenghi et al., 1983). They consist of terrigenous units, which are flysch-like and several hundred meters thick, and which accumulated in silled basins where the hydrologic budget was positive.

Messinian Erosional Surfaces and Headward Canyon Cutting

During times of evaporatic drawdown, the basin margins were exposed to subaerial erosion and denudation (Cita and Ryan, 1978a). Such erosional surfaces were clearly expressed in seismic reflection profiles, and several of them were calibrated by deep-sea drilling and commercial drilling off- and onshore (Hsu, 1978; Montadert et al., 1978; Rizzini and Dondi, 1978; Rizzini et al., 1978; Cita et al., 1979; Van Hinte et al., 1979). The sudden drop in base level compelled the major rivers to cut narrow canyons and to incise their own submarine fans.

Messinian canyons were cut subaerially by the Var, Durance and Rhone Rivers in southern France (Clauzon, 1973, 1978, 1979). The south-alpine lakes of northern Italy, previously considered as glacial in origin, are now interpreted as being related to headward erosion of the rivers Ticino (Lake Maggiore), Adda (Lake Como), Oglio (Lake Iseo), Mincio (Lake Garda) during the Messinian regression (Bini et al., 1978; Finckh, 1978). The glacial morphology is superimposed on pre-existing depressions which are too deep, as shown by seismic reflection and refraction profiles, to have been glacially formed, as for example were the north-alpine lakes (Finckh, 1978). Extended exploration for oil and natural gas in the Po basin substantiated the existence of Messinian erosional surfaces and of subaerial fans on the south-alpine margin (Rizzini and Dondi, 1978, 1979).

The best studied submarine fans are those of the Rhone in the western Mediterranean (Genesseaux and Lefebre, 1980) and of the Nile in the eastern Mediterranean (Barber, 1980). In both instances it was possible to reconstruct Messinian paleocanyons, incised for several hundred meters into pre-Messinian terranes, and subsequently filled by Plio-Quaternary sediments.

Over-incision of the Nile as far as Aswan, some 1200 km from the rivermouth, was known since the construction of the dam (Choumakov, 1967). More recently, Barber (1981) was able to reconstruct the entrenchment of the Nile cone by the Proto-Nile during the evaporitic drawdown. Barber's data base consisted of a network of seismic lines, most multichannel and up to some 8800 km long, and a number of wells which serve to calibrate the lines. The Nile paleocanyon reaches a width of some 12 km and a depth of 1.5 km beneath Alexandria; it is filled by post-Messinian sediments and cuts into terranes as old as the Eocene.

Study of Messinian erosional surfaces (Ryan and Cita, 1978) suggests paleorelief of the Mediterranean at 2 to 3 km beneath the global sea level. This approach is of interest, since the study of evaporitic facies (which are restricted to the basin floors) does not allow quantification of paleorelief.

Intra-Messinian Fossil Record; Marine Versus Non-marine Biota

Various steps can be distinguished during what is known as the "Messinian salinity crisis." The first step is characterized by restriction and euxinification prior to the complete cut-off of the Mediterranean and its eventual desiccation. The progressive reduction of an efficient interchange between the Atlantic and the Mediterranean through the Iberian portal (Benson, 1976) resulted in basin restriction, accompanied by the development of unusual coral reefs on the basin margin and diatom blooms in the basin centers. Detailed investigations by Esteban (1979) revealed a progressive downslope migration of the coral reefs, accompanied by a decrease in diversity and a strong dominance of Porites. In the latest phases Porites buildups alternated with gigantic stromatolites developed in hypersaline waters. The occurrence of late Miocene diatomites, usually laminated, non-bioturbated and occassionally impregnated with hydrocarbons, is wide-spread at the base of the Messinian evaporites in the satellite basins all around the Mediterranean.

Planktic foraminfers which were abundant, well diversified and rapidly evolving in the latest Tortonian and earliest Messinian (d'Onofrio et al., 1975) suddenly became highly specialized and dwarf. They lost most of the less tolerant species, including practically all the Globorotalia, before disappearing altogether. Globigerina multiloba, a small-sized species apparently well suited to stress conditions, is the zonal marker of the youngest biozone (subzone) of the Messinian. This taxon is accompanied by a few cosmopolitan, long-range species, the only survivors of a previously well balanced assemblage.

Benthic foraminifers also record a deteriorating environment (Orzag-Sperber et al., 1980; Van der Zwaan and den Hartog Jager, 1983) namely a progressive oxygen-deficiency. Buliminids, which are known to tolerate low oxygen levels, replace other taxa in their ecologic niches, and dominate the assemblages. The first record of Bulimina echinata is considered an isochronous event (d'Onofrio and Iaccarino, 1978) occurring just prior to the onset of evaporitic conditions.

If the response of benthic foraminifers to ecologic stress heralding the Messinian salinity crisis predates the response of diatoms (Martina et al., 1979), the latter provide the best tool to reconstruct paleoceanographic changes which occurred during the Tripoli deposition (Gersonde, 1978). Diatomaceous sediments below the evaporites are very rich in diatoms: average 30-60 million frustules per gram of dry sediment. To preserve such high amounts of diatoms in the sediment, high productivity is required (upwelling) accompanied by fast transport of the frustules to the bottom and undisturbed sedimentation (lamination). Diversity is high in the Sicilian sections investigated by Gersonde (1980) with up to 150 species present. Towards the top of the sections, oceanic species and most of the biostratigraphic markers disappear, whereas species diversity decreases drastically in the topmost part.

The second step is represented by the deposition of the lower evaporites (see above section "Lithostratigraphic record of basinal settings; facies changes; deep-sea record versus land record"). Evaporites are the sedimentary expression of an essentially abiotic environment. Since the two drilling campaigns of the Glomar Challenger demonstrated beyond any doubt that Messinian evaporites are ubiquitous in the Mediterranean (see Figure 3) the conclusion is drawn that the marine fauna which used to populate the Mediterranean waters were basically destroyed (Cita, 1976, 1979), but short-lived marine incursions are documented by a sparse fossil record.

Fossils both marine and non-marine have been found associated with the evaporites. Diatoms identified in marls interbedded with dolomitic muds underlying a massive anhydrite unit in the Balearic Basin (DSDP Site 124) and in the Messinian abyssal plain (DSDP Site 374) indicate a shallow, brackish environment with salinities ranging from 17 to 40 per mil in the western basin and from 0 to 17 per mil in the eastern ones (Schrader and Gersonde, 1978). The brackish marine influx at the Balearic Rise (DSDP Site 124) could have resulted from periodic input of Atlantic waters. The presence of 34 m of marine diatomites intercalated with marls and gypsum horizons within the Messinian evaporites of Fortuna Basin, Spain, as well as reefs growing on gypsum substrate, (Mueller, 1986) attest to renewed marine conditions in the Betic seaway during the salinity crisis. Likewise, autochthonous reefs found directly overlying gypsum in Cyprus suggest the reestablishment of marine conditions in the eastern Mediterranean during the crisis (Orszag-Sperber and Rouchy, 1979). The presence of marine microfossils, such as oligotypic planktonic foraminifers and cysts of marine planktonic algae, in several horizons of the Lago Mare sediments on the Florence Rise (DSDP Site 375/376) also indicate occasional influxes of marine waters into the eastern basins (Hsu, Montadert et al., 1978). The foraminiferal faunas in the marls associated with the evaporites are often dwarfed and poorly diversified suggesting stressed or abnormal marine conditions. In general, the inventory of fossils contained within Messinian evaporite sediments from land sections are non-marine including a number of fishes, dragonfly larvae, leaves, and even birds (Sturani, 1973, 1978, Sorbini and Rancan Tirapelle, 1979; Landini and Sorbini, 1980).

The third and final step of the salinity crisis is marked by the invasion of Paratethyan immigrants brought about by a hypothetical capture of highstanding Parathethyan lakes caused by headward erosion and canyon cutting during times of evaporitic drawdown (Cita and Ryan, 1973; Hsu et al., 1977). The term "lago-mare" is frequently used with reference to this very special biotope, which was subaqueous but essentially non-marine. Paratethyan immigrants include ostracods, pelecypods, gastropods (Decima, 1964; Ruggieri, 1967, Sissingh, 1976; Archambault-Guezu, 1976; Bossio et al., 1978). These fossils are contained in marls or clays associated with the upper evaporites or above this unit. The first Paratethyan immigrants are recorded from immediately above the salt layer (Martina et al., 1979).

Characterization and Significance of the Pliocene Transgression

The lago-mare biotope was destroyed by the basal Pliocene transgression, which re-introduced the marine fauna from the Atlantic into the Mediterranean. The Miocene/Pliocene boundary, as defined at the base of the Trubi Formation at Capo Rossello (Cita, 1975a), records the reestablishment of open-marine conditions in the Mediterranean after the Messinian salinity crisis. The change was drastic, and no survivors of the Paratethyan fauna has been reported anywhere in the Mediterranean realm (Ruggieri and Sprovieri, 1974).

The Pliocene transgression is a pan-Mediterranean event and is usually marked by an abrupt lithologic change, both in the deep-sea record and in the record from land. Sedimentation rates, which were usually high during the Messinian with up to several millimeters a year for the evaporitic rocks (Ogniben, 1957; Hsu, 1986), drop to a few centimeters per thousand years in the early Pliocene. The anomalously low rate of sediment supply to settings within the Pliocene Mediterranean, which are now basinal, is interpreted as the result of a rapid eustatic sea-level rise which accompanied the marine flooding of a depressed water body at the close of the late Miocene salinity crisis (Cita et al., 1978). Due to the continuous climatic deterioration in the North Atlantic, the marine fauna which re-populated the Mediterranean was definitely cooler than that which lived in the Mediterranean prior to the salinity crisis. Colonial corals disappeared forever from the area.

Re-colonization of the bottom started soon after the immigration of the planktonic fauna. At the beginning, benthic foraminifers belong to half a dozen species which have a deep habitat, but a fairly wide bathymetric range (Wright, 1978, 1979); an upwards increase in diversity is noticed both in the deep-sea record, and in the land record (Cita, 1972; Sprovieri, 1974; Wright, 1979). The occurrence of psychrosphric ostracods in the Early Pliocene indicates not only that the Mediterranean basins were deep, but also that the Gibralter Strait had to be deeper than at present (Benson, 1973).

The basal part of the Pliocene is referred to as *Sphaeroidinellopsis* subzone (Cati et al., 1968) or *Sphaeroidinellopsis* Acme-zone (Cita, 1973, 1975b). This short-duration biostratigraphic unit, only a few meters thick, is characterized by a peak abundance of the nominal taxon, a thick-walled planktonic form whose extant descendants live in the mesopelagic zone. Their relative abundance at the base of the transgressive deep-sea deposits of the basal Pliocene is interpreted as the result of strong vertical mixing prior to the re-establishment of a permanent thermocline (Cita, 1976). *Globorotalia margaritae*, which evolved in the North Atlantic in the latest Miocene, when the Mediterranean was isolated, first appeared in the Early Pliocene after a permanent thermocline was re-established. Thus, its first appearance is not an evolutionary but an immigration event.

Budgeting the Mediterranean Evaporites

The present day Mediterranean has a surface of 2.5 million km , a volume of 3.7 million km, a mean depth of approximately 1.5 km. The present day hydrologic budget is passive, and the estimated annual evaporative loss is 3310 km (4690 km evaporation; 1380 km precipitation plus river runoff). The evaporative loss is compensated by excess inflow from Gibralter (Hsu et al., 1973).

The present day Mediterranean has an anti-estuarine type circulation with less dense Atlantic water entering at the surface and highly saline Levantine water existing at intermediate depths. As the surface waters of the Mediterranean flow from west to east their salinity steadily increases. The highest salinities in excess of 39 per mil are found in

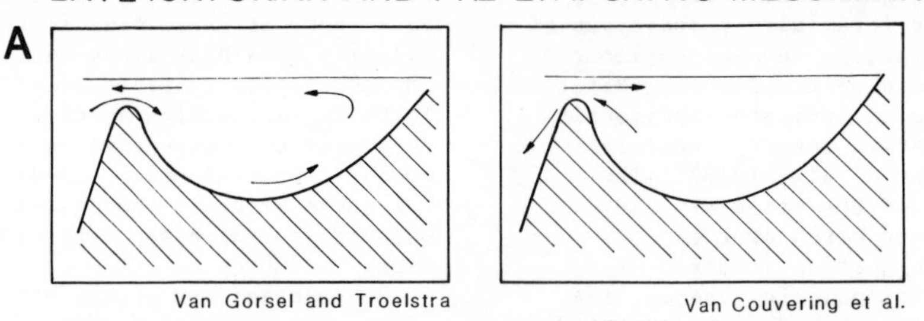

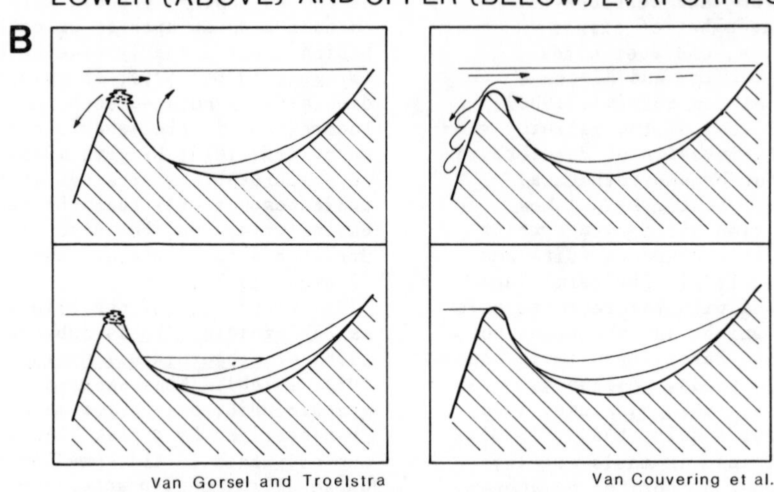

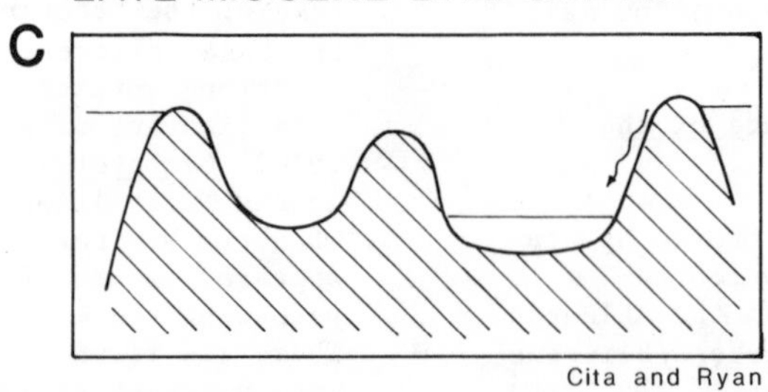

Fig. 5. Comparison of various models proposed for the deposition of Messinian evaporites in the Mediterranean by Cita and Ryan (1973), van Couvering et al. (1976), van Gorsel and Troelstra (1980). For discussion see text.

the Levantine Basin, which is surrounded by very hot and dry lands. If today we were to build a dam connecting Gibralter with Ceuta, the inevitable consequence of this artificial barrier, due to the passive hydrologic budget, would be the sudden lowering of the Mediterranean sea level on the order of one meter a year. In approximately 1500 years, geologically instantaneous, the Mediterranean would be on the average desiccated with occasional deep-seated brine pools, depending upon basin morphology.

According to the above estimates, evaporation of the entire Mediterranean would produce a volume of evaporites that is one order of magnitude less than the volume calculated on the basis of seismic reflection profiles, i.e. one million (Ryan, 1973) or one and half million cubic kilometers (Gvirtzman and Buchbinder, 1978). Repetitive marine incursions or a steady inflow from the Atlantic are required to account for the actual thickness of Messinian evaporites in a deep-basin desiccation model. Other authors, (i.e. van Couvering et al., 1976; van Gorsel and Troelstra, 1980; Sonnenfeld, 1974; Debenedetti, 1976) hypothesize a two-way connection which enhances the anti-estuarine type circulation of the Mediterranean since the outflow comprises very dense waters (see Figure 5).

We prefer the former or one-way alternative because it is compatible with the concept of an early desiccation and consistent with the record of a dissolution peak in the North Atlantic in the terminal Miocene (Thunell, 1987). Evidences for an early desiccation are twofold: (a) the deep-sea record provides evidence that Messinian erosional surfaces are more than one. The deepest one may be traced to the basin floors beneath the lower evaporites; (b) the land record provides evidence that desiccation occurred during the deposition of Calcare di base, the lowest evaporitic unit (Sturani, 1978; Ryan and Cita, 1978) and that the lower evaporites contain algal mats, indicative of a very shallow water depth (Vai and Ricci Lucchi, 1978). The regression (evaporative drawdown) reached its apex during the deposition of the salt body. At that time, certainly no outflow of saline water could be expected but a continuous inflow of marine water would be required to produce the massive salt body.

In summary, erosional surfaces could not be created if the Mediterranean were not desiccated and it is inconceivable to have an outflow from a desiccated Mediterranean. The huge amount of salts trapped in the Mediterranean were subtracted from the world's oceans. Using the more conservative one million cubic kilometers estimate, which corresponds to 6% of the mean salinity of the world's oceans, the Messinian salinity crisis must have decreased the ocean salinity by 2 per mil. This salt extraction together with the cessation of the return flow to the Atlantic of the dense Mediterranean water is undoubtedly reflected in the terminal Miocene record in the open ocean.

The Terminal Miocene Event in the Open Ocean

Cooling and Glacial Expansion in the Southern Hemisphere

The late Miocene was, in general, a time of global cooling with major glacial expansions occurring in the southern hemisphere (Kennett, 1977). A terminal Miocene regression is well expressed in the generalized eustatic curve of Vail et al. (1977). and Vail and Hardenbol (1979), and evidence for late Miocene glaciation of the Ross Sea region was revealed by seismic investigation during DSDP Leg 28 deep-sea drilling (Hayes, Frakes et al., 1975). From their oxygen-isotope curves of deep-sea sediments cored at drill sites in the Southern Ocean, Shackleton and Kennett (1975a,b) surmised a rapid build-up of a permanent ice cap in East Antarctica between 14 and 10 Ma which was followed in the latest Miocene by an even sharper deterioration of climatic conditions. They suggested that this latter period of severe continental glaciation may have exceeded all other Neogene glaciations in its intensity. Further, biostratigraphic studies of carbonate versus siliceous ooze distributions in sediments from DSDP sites in the Southern Ocean and South Atlantic (summarized in Wise, 1981) indicate a sudden northward shift of the Antarctic Convergence in the latest Miocene. In addition to this biogenic transition, the presence of ice-rafted debris and erosional episodes related to intensification of bottom-water currents could, in paleomagnetically dated piston cores from the Maurice Ewing Bank, indicate the formation of a floating but partially grounded West Antarctic ice sheet in the latest Miocene and the production of the first true Antarctic Bottom Water (AABW) (Ciesielski et al., 1982). During the latest Miocene, two periods of widespread deep-sea hiatuses, which apparently correspond to times of increased Antarctic glaciation and increased volumes of AABW production resulting in extensive erosion and dissolution of deep-sea sediments, occurred and are biostratigraphically dated between 7.5 and 6.2 Ma and 5.2 to 4.7 Ma (Barron and Keller, 1982).

Furthermore, these marine interpretations for a latest Miocene glacial expansion are supported by land geologists working on glacial deposits in the southern hemisphere, as reviewed by Mercer (1983). In particular, Rutford et al., (1972) proposed the establishment of an ice sheet in West Antarctica by around 7 Ma, while latest Miocene or earliest Pliocene glaciation in southern Argentina indicates climatic conditions more extreme than those prevailing today (Mercer and Sutter, 1982). Recent findings of glacial tills latest Miocene in age (dated at 6.75 - 5 m.y.) in southern Argentina (Mercer and Sutter, 1982; Mercer, 1983) strengthen the issue. The tills are extended beyond the glaciated areas of the Andes and are correlated with the initial growth of the west Antarctic ice sheet.

In the northern hemisphere, investigations of marine sections on land likewise show a latest Miocene regression, as evidenced by the interruption of carbonate buildups in tropical areas and extended phosporite deposits on the basin margins. Adams et al. (1979) estimate a global sea-level fall of 50 to 70 m after reviewing a number of land sections. Among the best dated sequences recording a latest Miocene regression within a continuous marine succession are those which were located on the Atlantic side of the former seaways connecting the Atlantic with the Mediterranean in pre-Messinian times, the Carmona section in southern Spain (Berggren and Haq, 1976) and the Bou Regreg section in Morroco (Cita and Ryan, 1978b).

Overall, the evidence from open-marine sediments for a major regression in the latest Miocene is overwhelming. The contribution of this paleoceanographic event to the Messinian desiccation of the Meditteranean can be delineated from a combination of high-resolution isotope-, magneto- and biostratigraphic studies of sediment cores taken from a deep-sea environment. It is critical to evaluate the timing of the regression versus the salinity crisis in order to evaluate the effect of both events upon the paleoceanography and paleoclimatology around the Miocene/Pliocene boundary.

Carbon-isotope Shift: Paleoceanographic Implications for the Non-Evaporative Messinian

A distinct decrease in the carbon-isotopic values recorded in benthic and frequently in planktic foramminifers at about 6 Ma has proven to be a significant stratigraphic marker in deep-sea sediments from the Indo-Pacific region, as well as from the North and South Atlantic (Keigwin, 1979; Loutit and Kennett, 1979; Keigwin and Shackleton, 1980; Vincent et al., 1980; Haq et al., 1980; Loutit and Keigwin, 1982; and Keigwin et al., in press). The "carbon shift" is recorded near the first occurrence of Amaurolithus primus in the upper reversed event of Epoch 6 which has a duration between 6.22 and 5.77 Ma. From a compilation of many studies a mean magnetostratigraphic age of 6.00 Ma (6.10-5.90) has been determined for the "carbon shift" (Haq et al., 1980). Further, Loutit and Kennett (1979) were able to calibrate a sequence of events, occurring in the upper half of the upper reversed event of Epoch 6 in latest Miocene sediments at Blind River, New Zealand: (1) the "carbon shift" between 6.2 and 6.3 Ma, (2) the evolutionary FAD of Globorotalia conomiozea at 6.1 Ma, and (3) a distinct cooling and regression between 6.2 and 6.1 Ma. Loutit and Keigwin (1982) studied the late Miocene Carmona-Dos Hermanos section in southwest Spain, a continuous marine section which was situated at the Atlantic entrance of the Betic Strait leading into the Mediterranean. They recorded a carbon-shift and sea-level fall at approximately 6.2 Ma, which was correlated to the lowermost Tripoli Formation (Sicily). The FAD of G. conomiozea occurs in sediments 6 m below the Tripoli Formation at Falconara, Sicily (Colalongo et al., 1979), i.e. at approximately 6.1 Ma according the Loutit and Kennett (1979).

From the stratigraphic evidence a causal relationship apparently exists for the late Miocene "carbon-shift", marine regression and paleoceanographic changes within the Mediterranean. The change from homogeneous pelagic marl sedimentation during the Tortonian to the lower Messinian deposition of cyclic laminated diatomites and dolomitic claystones of the Tripoli Formation (Sicily) represents a more restricted circulation pattern in the Mediterranean, possibly controlled by eustatic sea-level changes as reflected by the cyclic sedimentation (McKenzie et al., 1979). Various proposals for the origin of the "carbon-shift" include changes in global circulation patterns due to decreased influx of dense Mediterranean waters into the Atlantic intermediate waters (Bender and Keigwin, 1979) or increased rate of supply of organic matter to deep sea from coastal lowlands and shelves exposed during the regression (Vincent et al., 1980). The event is most likely a combination of both sea-water compositional effects and circulation because, although it is global, its magnitude varies being greater in the Indo-Pacific region (Keigwin et al., in press).

The apparent synchroneity of the paleoceanographic events occurring near or shortly after the Tortonian/Messinian boundary, designated by the FAD of G. conomiozea at approximately 6.1 Ma (Loutit and Kennett, 1979) suggest that a major regression resulted in restricted circulation in the Mediterranean as well as a reduced input of dense Mediterranean water into the Atlantic, which, in turn, lead to changes in the global circulation pattern and the "carbon-shift". The influx into the Atlantic of dense Mediterranean Sea Outflow Water (MSOW) contributes significantly to the formation of North Atlantic Deep Water (NADW). The diminuation or cessation of NADW formation as a result of decreased influx of MSOW with a major regression would have affected the Atlantic deep-water circulation pattern. Using carbon-isotope stratigraphy in benthic foraminifera from Neogene deep-sea sediments cored in both the Atlantic and Pacific Oceans, Blanc and Duplessy (1982) demonstrated the impact of the Messinian salinity crisis upon the global deep-water circulation. They correlate the Epoch 6 "carbon-shift" with the closure of the Mediterranean and propose that the subsequent deficiency of denser water in the Atlantic Ocean caused all deep-water formation in the Northern Hemisphere to cease until after the Mediterranean was reopened in the earliest Pliocene.

Apparently from the sedimentary record of the early or non-evaporative phases of the Messinian, circulation between the Mediterranean and Atlantic continued, although altered by the regression, and was sufficient to maintain fully marine conditions within the Mediterranean, as typified by the diatom sedimentation in the deeper basins and flourishing reef growth in the margins.

Oxygen-isotope Fluctuations: Paleoceanographic Implications for the Evaporative Messinian

In Sicily, the non-evaporative Messinian extended from the Tortonian/Messinian boundary recognized by the FAD of Globoratalia conomiozea (Colalongo et al., 1979) at approximately 6.1 Ma (Loutit and Kennett, 1979) through the Tripoli Formation from its base correlated with the upper half of Epoch 6 to

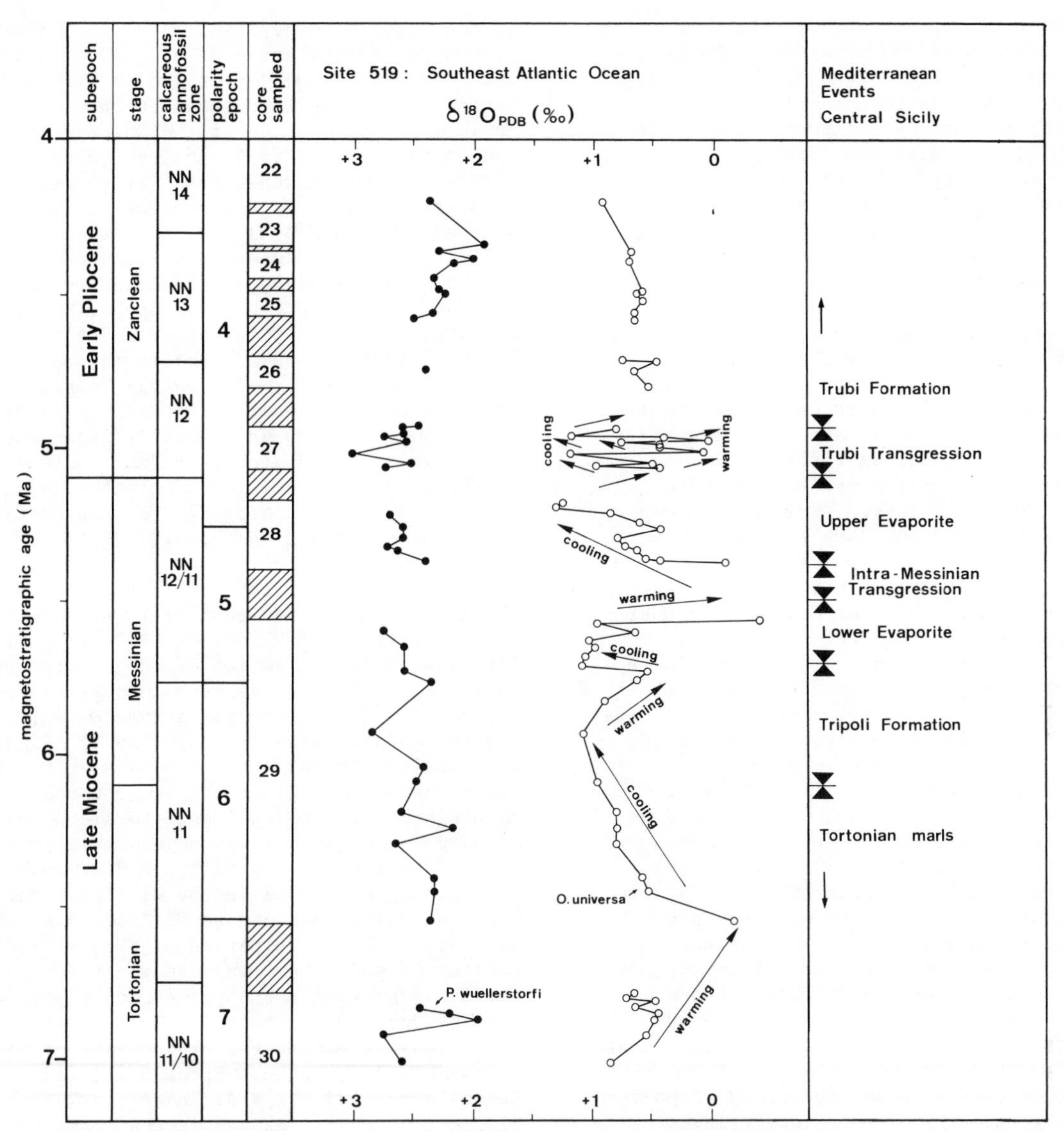

Fig. 6. Oxygen-isotope data for planktic (Orbulina universa) and benthic (Planulina wuellerstorfi) foraminifers from DSDP Site 519, plotted against time as determined by magnetostratigraphy (Hsu, LaBrecque et al., 1984) and correlated with Mediterranean geologic events (see also Figure 2). Modified after McKenzie and Oberhansli (1985).

its top corresponding to the lower normal polarity of magnetic Epoch 5 (5.6 to 5.9 ma) (Gersonde, 1980; Gersonde and Schrader, 1984). Directly above the Tripoli Formation, evaporite deposition began and continued until open marine sedimentation was re-established at the Miocene/Pliocene boundary, which has been biostratigraphically linked via FAD of Sphaeroidinella dehiscens and the FAD of Globorotalia aumida with the reversed polarity at the base of the Gilbert Epoch (Berggren, 1973; Saito et al., 1975). Therefore, the evaporative Messinian spans most of Epoch 5 and into the lower Gilbert Epoch.

During the evaporative Messinian, the outflow of MSOW ceased and the connection between the Atlantic and Mediterranean became at most a one-way street with restricted water flow from the former to the latter. A cessation of MSOW near the base of Epoch 5 was recognized by Hooper and Weaver (in press) based on their study of changes in planktic/benthic ratios related to dissolution by aggressive bottom waters in paleomagnetically dated cores from the North Atlantic (DSDP Site 609). They associated the change to aggressive bottom waters with a northward expansion of corrosive Antarctic Bottom Water (AABW) due to the termination or sharp decrease in the production of NADW. Also, based on changes in the coiling direction ratio in Neogloboquadrina pachyderma related to changes in surface-water temperature, Hooper and Weaver (in press) documented the beginning of a distinct cooling interval beginning at the top of Epoch 6 which was followed by climatic fluctuations with a periodicity of about 125,000 years, possibly indicative of glacial cycles, throughout Epoch 5 and into the base of the Gilbert. They recorded intensive dissolution on either side of the Miocene/Pliocene boundary inferred to be due to the presence of AABW with a renewed influx of less corrosive NADW on the actual boundary. Hooper and Weaver (in press) tentatively link these climatic and oceanographic changes with the Messinian salinity crisis.

Oxygen-isotope stratigraphies for paleomagnetically dated deep-sea sediments across the Miocene/Pliocene boundary likewise show climatic fluctuations, which can be related to eustatic sea-level changes and salinity events occurring within the Mediterranean throughout Epoch 5 and the lower part of the Gilbert Epoch. In an extremely detailed stable-isotope study of upper Miocene sediments from North Atlantic DSDP Site 552A, including isotope data from Sites 334, 116, 408, 410, 610E and 611C, Keigwin et al. (in press) were able to delineate two short-term glacial maxima at approximately 5.0 and 5.5 Ma, which lasted no more than 20,000 years. Covariance in the oxygen-isotope record of benthic and planktic foraminifera indicates continental ice volume increases but the record does not show prolonged glaciation during the late Miocene in sediments younger than 7 Ma. Unfortunately, it was not possible to precisely locate the Miocene/Pliocene boundary at this site, and, based on best estimates, it lies between 5.0 and 5.3 Ma.

In another extensive isotope study of paleomagnetically dated uppermost Miocene/lower Pliocene sediments from the Southwest Pacific, DSDP Site 588, Hodell et al. (1986) document a series of high frequency changes in oxygen-isotope ratios that resemble Quaternary glacial/interglacial cycles but with an amplitude of about one-third and a wavelength of 400,000 years. Glacial maxima occur at the base of Epoch 5 (5.5-5.6 Ma) and in the lower Gilbert Epoch (5.1-4.8 Ma). A short-term trangression is recorded between 5.2 and 5.1 Ma. The biostratigraphic placement of the Miocene/Pliocene boundary at Site 588 occurs in the lower part of the Gilbert Epoch at about 5.0 Ma, within a glacial maxima period.

In the South Atlantic at DSDP Site 519 the Miocene/Pliocene boundary was likewise found to occur in sediments near the base of the Gilbert Epoch with a magnetostratigraphic age of 5.1 Ma (Hsu, LaBrecque et al., 1984). As shown in Figure 6, detailed oxygen-isotope stratigraphy of the upper Miocene/lower Pliocene sediments from Site 519 (McKenzie et al., 1984; McKenzie and Oberhansli, 1985) revealed an extended period of cooling throughout Epoch 6 with a maximum at 5.9 Ma. As in the North Atlantic and Southwest Pacific Oceans, short-term glacial maxima, in this case three, were identified; (1) near the base of Epoch 5 between 5.7 and 5.6 Ma, (2) near the base of Gilbert Epoch in Miocene sediments peaking at 5.2 Ma, and (3) also, in the lower reversed polarity zone of the Gilbert Epoch, but in Pliocene sediments between 5.1 and 4.9 Ma. Site 519 also showed a dramatic decrease in the oxygen-isotope value of planktic foraminifera in the middle of Epoch 5 covering an interval between 5.6 and 5.4 Ma. This decrease was tentatively interpreted at a "melt-water" spike produced by a climatic warming which caused a reduction in the size of the West Antarctic ice sheet (McKenzie and Oberhansli, 1985).

A fourth, detailed Atlantic correlation of isotope-, magneto-, and biostratigraphy across the Miocene/Pliocene boundary in deep-sea sediments from off NW Africa (DSDP Site 397) likewise reveals a series of oxygen isotope fluctuations which have been interpreted as continental ice-volume variations or warming/cooling events (Cita and Ryan, 1979) (Figure 7). Unlike Site 552A and 519, a warming trend was recorded throughout the lower normal magnetic zone of Epoch 5 and the trend towards an oxygen-isotope maximum began first in the reversed magnetic zone of Epoch 5 at about 5.57 Ma reaching a maximum in the middle of the upper normal magnetic zone of Epoch 5 at about 5.3 Ma. Within this period of progressive cooling, a short term warming was recorded in the lower part of the upper normal magnetic zone of Epoch 5 at about 5.4 Ma. A warming trend is apparent in the oxygen-isotope record at the boundary between Epoch 5 and Gilbert Epoch (about 5.3-5.2 Ma). In the earlier Pliocene, a sharp, oxygen isotope maximum, followed by a minimum, occurred. The Epoch 5/Gilbert Epoch boundary corresponds to the Miocene/Pliocene boundary (5.26 Ma) at this site.

In summary, the oxygen-isotope record for the open-marine environment in the North and South Atlantic and Southwest Pacific Oceans (Cita and Ryan, 1979; McKenzie and Oberhansli, 1985; Keigwin et al., in press; Hodell et al., 1986) contains significant fluctuations throughout Epoch 5 and the earliest Gilbert, which can be interpreted as changes in continental ice-volume and, thus, eustatic sea level fluctuations. Further, the biostratigraphic record (Hooper and Weaver, in press) for the same time period in the North Atlantic, likewise suggests that cooling and warming cycles occurred as well as changes in bottom-water characteristics with AABW become more dominant over NADW. Although it is not pos-

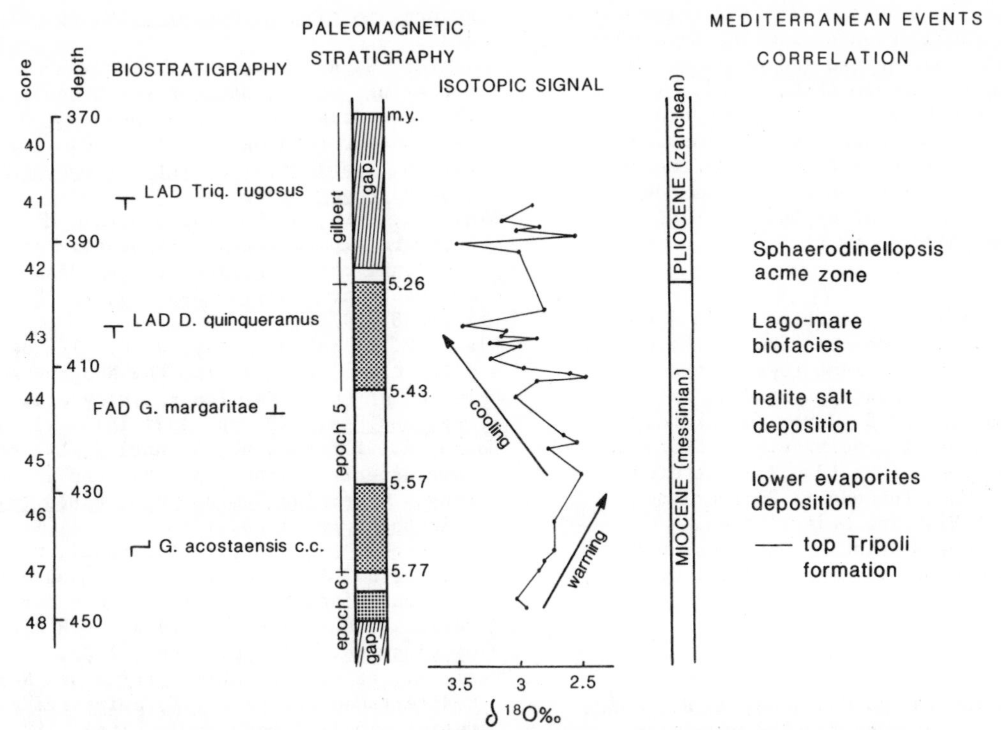

Fig. 7. Biostratigraphic events recorded at DSDP Site 397, Cape Bojador (Salvatorini and Cita, 1979), paleomagnetic record (Hamilton, 1979), and isotopic record (Shackleton and Cita, 1979) correlated with Mediterranean geologic events (see also Figure 2). Modified after Cita and Ryan (1979).

sible, as yet, to exactly correlate the various isotope events among the DSDP Sites, the marine record strongly implies that the amount of Atlantic water reaching an already restricted Mediterranean during the evaporative Messinian period could have been controlled by paleoclimate; i.e. with lowered sea level less water entered the Mediterranean across the connecting seaways resulting in evaporite deposition, while, on the contrary, rising sea level resulting in in an increased influx of marine water possibly re-establishing temporary marine conditions within the Mediterranean. This eustatic sea-level control of the Mediterranean salinity ceased once the breach at the Strait of Gibralter was achieved and the terminal marine flooding at the Miocene/Pliocene boundary occurred. It is interesting to note that the oxygen-isotope records for Sites 397, 519, 522A, and 588 all show maximum values at the boundary or into the earliest Pliocene, which can be interpreted as a regression. Does this regression correspond to the end of the evaporative Messinian period in the Mediterranean preceeding the earliest Pliocene transgression? If so, the Miocene/Pliocene boundary, as defined by the boundary stratotype in Sicily, may be more exactly pin-pointed in the marine record by oxygen-isotope stratigraphy.

Initial attempts to use strontium-isotope stratigraphy to correlate the boundary stratotype with Site 519 show that, indeed, the earliest Pliocene of Sicily (lowermost 11 m of Trubi Formation) corresponds to the earliest Pliocene sediments of Site 519, which record the lowest magnetic reversal of the Gilbert Epoch (McKenzie et al., 1985). The Sr-isotope stratigraphy indicates an age between 5.1 and 4.8 Ma for the Sicilian sediments. This age would then correlate with a period of glacial maxima recorded at the deep-sea sites. This suggests that the renewal of the connection between the Atlantic and the Mediterranean was a tectonic event, and after the Pliocene transgression, the control of glacio-eustatic sea-level changes on the salinity of the Mediterranean became insignificant compared to during the late Messinian.

To date, two attempts to correlate the isotopic record from the Atlantic with stratigraphic events in the Messinian Mediterranean have been made (Cita and Ryan, 1979; McKenzie and Oberhansli, 1985) (see Figures 6 and 7), but with varying results. The conclusions from such correlations will remain tentative awaiting more and better resolution from future studies of the marine record as well as better paleomagnetic control on sediments from the Mediterranean region. General acceptance of a

Messinian salinity crisis in the Mediterranean came with the publication of results from DSDP Legs 13 and 42A, but unresolved problems concerning the crisis remain (Hsu, 1986). In the near future new data from in-progress bio-, isotope-, litho-, and magnetostratigraphic studies of latest Miocene/Pliocene sediments from inside the Mediterranean and from the deep sea should produce better global correlations of the terminal Miocene event.

Acknowledgments. The authors would like to recognize K. J. Hsu, who contributed thoughtful discussion and encouragement towards the completion of this manuscript, and B. Das Gupta, K. Jennings and S. Joyce, who kindly provided the technical assistance to bring it to completion. One of us (J.A.M.) gratefully acknowledges EAWAG, Dubendorf, Switzerland, where she was a Visiting Scientist while working on this manuscript. Contribution No. 275 of the Laboratory for Experimental Geology, ETH Zurich.

References

Adams, C. G., Benson, R. H., Kidd, R. B., Ryan, W. B. F. and Wright, R. H., The Messinian salinity crisis and evidence of Late Miocene eustatic changes in the world ocean, Nature, 269, pp. 383-386, 1977.

Archambault-Geuzou, J., Presence de Dreissenidae euxiniques dans les depots a Congeries de la Vallee du Rhone et sur le pourtour mediterranean. Implications biogeographieques. Bull. Soc. Geol. France, 7, n.8, pp. 1267-1276, 1976.

Barber, P. M., Paleogeographic evolution of the Proto-Nile during the Messinian salinity crisis. Geol. Mediterr., 7, pp. 13-18, 1980.

Barron, J. A. and Keller, G., Widespread Miocene deep-sea hiatuses: coincidence with periods of global cooling. Geology, 10, pp. 577-581, 1982.

Bender, M. L. and Keigwin, L. D., Speculations about the Upper Miocene change in abyssal dissolved bicarbonate C. Earth Planet. Sco. Lett., 45, pp. 383-393, 1979.

Benson, R. H., Psychorospheric and continental ostracoda from ancient sediments in the floor of the Mediterranean. Init. Depts. DSDP, 13, pp. 1002-1008, 1973.

Benson, R. H., Miocene deep sea ostracods of the Iberian Portal and the Balearic Basin. Marine Micropal., 1, pp. 249-262, 1976.

Berggren, W. A., The Pliocene time scale: calibration of planktic foraminifera and calcareous nannoplankton zones. Nature, 243, pp. 392-397, 1973.

Berggren, W. A., and Haq, B., The Andalusian Stage (Late Miocene): Biostratigraphy, Biochronology, and Paleoecology. Paleogeogr., Paleoclimatol., Paleoecol., 20, pp. 67-129, 1976.

Berggren, W. A., Kent, D. V. and Van Couvering, J. A., Neogene geochronology and chronostratigraphy. In: Snelling, N. J., (ed.), Geochronology and the Geological Record. Geological Society of London Special Paper, 1985.

Bini, A., Cita, M. B., and Gaetani, M., Southern Alpine Lakes. Hypothesis of an erosional origin related to the Messinian entrenchment. Marine Geology, 27, 3/4, pp. 271-288, 1978.

Blanc, P. L. and Duplessy, J. C., The deep-water circulation during the Neogene and the impact of the Messinian salinity crisis. Deep Sea Res., 29, pp. 1391-1414, 1982.

Bosso, A., Esteban, M., Giannelli, L., et al., Some aspects of the Upper Miocene in Tuscany. Messinian Seminar n.4, Field Trip Guidebook, pp. 1-88, 1978.

Camerlenghi, A., Cita, M. B., Leoni, C., Malinverno, A., Malinverno, P., and Miranda, P., Messiniano-banca dati, logs e carte varie sul Mesiniano d'Italia. C.N.R.P.F. Geodinamica, 514, pp. 1-467, 1983.

Cati, F., et al., Biostratigrafia del Neogene Mediterraneo basata sui foraminiferi planctonici. Boll. Soc. Geol. Ital., 87, pp. 491-503, 1968.

Choumakov, I. S., Pliocene and Pleistocene deposits of the Nile Valley in Nubia and Upper Egypt. Acad. Sci. USSR, Moscow, 170 (in Russian), pp. 5-115, 1967.

Ciesielski, P. F., Ledbetter, M. T. and Ellwood, B. B., The development of Antarctic glaciation and the Neogene paleoenvironment of the Maurice Ewing Bank. Marine Geology, 46, pp. 1-51, 1982.

Cita, M. B., Il significato della transgressione pliocenica alla luce delle nuove scoperte nel Mediterraneo, Riv. Ital. Paleont., 78, pp. 527-594, 1972.

Cita, M. B., The Pliocene Record in deep sea Mediteranean sediments. I. Pliocene biostratigraphy and chronostratigraphy. Init. Repts. DSDP, 13, 2, pp. 1343-1379, 1973.

Cita, M. B., The Miocene/Pliocene boundary. History and Definition. In: T. Saito and L. H. Burckle (eds.), Late Neogene Epoch Boundaries. Micropaleontology Press, Spec. Publ. 1, pp. 1-30, 1975a.

Cita, M. B., Planktonic foraminiferal biozonation of the Mediterranean Pliocene deep-sea record. A revision. Riv. Ital. Paleont., 81, 4, pp. 527-544, 1975b.

Cita, M. B., Biodynamic effects of the Messinian salinity crisis on the evolution of planktonic foraminifers in the Mediterranean. Paleogeogr., Paleoclimatol., Paleoecol., 20, pp. 23-42, 1976.

Cita, M. B., Lacustrine and hypersaline deposits in the deep sea and their bearing on paleoenvironment and paleoecology. The

Second Maurice Ewing Symposium, AGU, pp. 402-419, 1979.
Cita, M. B., Late Neogene Environmental Evolution. Init. Repts. DSDP, 47, 1, pp. 447-460, 1979.
Cita, M. B., The Messinian Salinity Crisis in the Mediterranean: a review. Alpine Med. Geodynamics, Geodynamics Series, 7, pp. 114-140, 1982.
Cita, M. B. and Gartner, S., The stratotype Zanclean: foraminiferal and nannofossil biostratigraphy. Riv. Ital. Paleont., 79, pp. 503-588, 1973.
Cita, M. B., and Ryan, W. B. F., The Pliocene Record in deep sea Mediterranean sediments. V. Time-scale and general synthesis. Init. Repts. DSDP, 13, 2, p. 1405-1416, 1973.
Cita, M. B., and Ryan, W. B. F. (eds.), Messinian Erosional Surfaces in the Mediterranean. Marine Geology, 27, 3/4, pp. 193-363, 1978a.
Cita, M. B., and Ryan, W. B. F. (eds.), The Late Neogene section of Bou Regreg, NW Morocco: Evidence, timing and significance of a Late Miocene regressive phase. Riv. Ital. Paleont., 84, 4, pp. 1051-1082, 1978b.
Cita, M. B., Follieri, M., Longinelli, A., Revisione di alcuni pozzi profondi della Pianura Padana nel quadro del significato geodinamico della crisi di salinita del Messiniano. Bol. Soc. Geol. Ital., 97, pp. 297-316, 1979.
Cita, M. B., Ryan, W. B. F. and Kidd, R. B., Sedimentation rates in Neogene deep-sea sediments from the Mediterranean and geodynamic implications of their changes. Init. Repts. DSDP, 42A, pp. 991-1002, 1978.
Clauzon, G., The eustatic hypothesis and the pre-Pliocene cutting of the Rhone Valley. Init. Repts. DSDP 13, 2, pp. 1251-1256, 1973.
Clauzon, G., The Messinian Var Canyon (Provence, southern France), Paleogeographic implications. Marine Geology, 27, 3/4, pp. 231-246, 1978.
Clauzon, G., Le canyon messinien de la Durance (Provence, France): une preuve paleogeographique du bassin profond de dessication. Paleogeogr. Paleolclimatol. Paleoecol., 29, pp. 15-40, 1979.
Colalongo, M. L., et al., A proposal for the Tortonian/Messinian boundary. Ann. Geol. Pays Hellen., T. Horsser., I, pp. 285-294, 1979.
Debenedetti, A., Messinian salt deposits in the Mediterranean: evaporites or precipitates? Boll. Soc. Geol. Ital., 95, pp. 941-950, 1976.
Decima, A., Ostracodi del gen. Cyprideos nel Neogene e nel Quaternario italiani. Paleontogr. Italica, 57, pp. 81-127, 1964.
Decima, A., Initial data on the bromine distribution in the Miocene salt formation of Southern Sicily. Mem. Soc. Geol. It., 16, pp. 39-43, 1978.
Decima, A. and Wezel, F. C., Late Miocene Evaporites in the Central Sicilian Basin. Init. Repts. DSDP, 13, pp. 1234-1240, 1973.
D'Onofrio, S., and Iaccarino, S., Benthonic foraminifera from Italian Tortonian-Messinian sections. Int. Meeting of Geodynamic and Biodynamic Effects of the Messinian Salinity Crisis. Messinian Seminar, 4, Abstracts, 1978.
D'Oniofrio, S., Giannelli, L., Iaccarino, S. et al., Planktonic foraminifera of the Upper Miocene from some Italian sections and the problem of the lower boundary of the Messinian. Boll. Soc. Paleont. Ital., 14, p. 177-196, 1975.
Esteban, M., Significance of the Upper Miocene reefs in the western Mediterranean. Paleogeogr. Paleoclimatol. Paleoecol., 29, 1/2, pp. 169-187, 1979.
Finckh, P. G., Are southern Alpine lakes former Messinian canyons? Geophysical evidence for preglacial erosion in the southern Alpine lakes. Marine Geology, 27, 3/4, pp. 289-302, 1978.
Gennesseaux, M. and Lefebvre, D., Le palecours inferieur du Rhone au Messinian. Geol. Mediterr., 7, 1, pp. 71-80, 1980.
Gersonde, R., Diatoms paleocology in the Mediterranean, Messinian. Int. Meeting on Geodynamic and Biodynamic Effects of the Messinian Salinity Crisis. Messinian Seminar, 4, Abstracts, 1978.
Gersonde, R., Palaeoekologische und biostratigraphische Auswertung von Diatomeenassoziationen aus dem Messinium des Caltanissetta-Beckens (Sizilien) und einiger Vergleis-Profile in Southwest Spanien, northwest Algerien und auf Kreeta. Diss. Univ. Kiel, pp. 1-393, 1980.
Gersonde, R. and Schrader, H., Marine planktic diatom correlation of lower Messinian deposits in the Western Mediterranean. Marine Micropaleontology, 9, pp. 93-110, 1984.
Gvirtzman, G., Buchbinder, B., Miocene desiccations of the Tethys-Mediterranean Ocean, X. Int. Congress on Sedimentology, Jerusaleum. Abstracts, pp. 280-281, 1978.
Hamilton, N., A paleomagnetic study of sediments from Site 397, northwest African continental margin. Init. Repts. DSDP, 47, 1, p. 463-478, 1979.
Haq, B. U., Worsley, T. R., Burckle, I. H., Douglas, R. G., Keigwin, I. D., Jr., Opdyke, N. D., Savin, S. M., Sommer, M. A., Vincent, E., and Woodruff, F., Late Miocene marine carbon-isotopic shift and synchroneity of some phytoplanktonic biostratigraphic events. Geology, 8, pp. 427-431, 1980.
Hayes, D. E., Frakes, L. A., Barret, P. J., et al., Init. Repts. DSDP, 28, pp.1-1017,1975.

Hodell, D. A., Elstrom, K. M., and Kennett, J. P., Latest Miocene benthic O changes, global ice volume, sea level, and the "Messinian Salinity Crisis". Nature, 320, pp. 411-414, 1986.
Hooper, P. W. P. and Weaver, P. P. E., Paleoceanographic significance of late Miocene to early Pliocene planktonic foraminifers at DSDP Site 609. Init. Repts. DSDP, 94, in press.
Hsu, K. J., Unresolved problems concerning the Messinian Salinity Crisis, in press.
Hsu, K. J., Cita, M. B., Ryan, W. B. F., The origin of the Mediterranean evaporites. In: Ryan, W. B. F., et al. (eds.), Init. Repts. DSDP, 13, pp. 1203-1231, 1973.
Hsu, K. J., La Brecque, J. I. et al., Init. Repts. DSDP, 73, pp. 1-798, 1984.
Hsu, K. J., Montadert, L. et al., Init. Repts. DSDP, 42A, pp. 1-1249, 1978.
Hsu, K. J., Montadert, L., Bernoulli, D., Cita, M. B., Erickson, A., Garrison, R. E., Kidd, R. B., Meliers, F., Muller, C. and Wright, R. H., History of the Mediterranean salinity crisis. Nature, 267, 5610, pp. 399-402, 1977.
Keigwin, L. D., Late Cenozoic stable isotope stratigraphy and paleoceanography of Deep Sea Drilling Project Sites from the East Equatorial and Central North Pacific Oceans. Earth Planet. Sci. Lett., 45, pp. 361-382, 1979.
Keigwin, L. D., Aubry, M. P. and Kent, D. V., North Atlantic Late Miocene stable isotope stratigraphy, biostratigraphy, and magnetostratigraphy. Init. Repts. DSDP, 94, in press.
Keigwin, L. D., Jr. and Shackleton, N. J., Uppermost Miocene carbon isotope stratigraphy of a piston core in the equatorial Pacific. Nature, 284, p. 613, 1980.
Kennett, J. P., Marine Geology, Prentice-Hall, Englewood Cliffs, N.J., 1982.
Kennett, J. P., Paleo-oceanography: global ocean evolution. Review of Geophysics and Space Physics, 21, 5, pp. 1258-1274, 1977.
Landini, W. and Sorbini, L., Pesci del Miocene superiore. In: I Vertebrati fossili italiani. Catalogo della Mostra. p. 189-193, 1980.
Langereis, C. G., Zachariasse, W. J. and Zijderveld, J. D. A., Late Miocene magnetobiostratigraphy of Crete. Marine Micropaleontology, 8, pp. 261-281, 1983/84.
Loutit, T., Keigwin, L. D., Stable isotope evidence for latest Miocene sea-level fall in the Mediterranean region. Nature, 300, pp. 163-166, 1982.
Loutit, T. S., and Kennett, J. P., Application of carbon isotope stratigraphy to Late Miocene shallow marine sediments, New Zealand. Science, 204, pp. 1196-1199, 1979.
Martina, E., Casati, P., Cita, M. B. et al., Notes on the Messinian stratigraphy of the Crotone Basin, Calabria. Ann. Geol. Pays Hellen., Hors Ser., 2, pp. 75-765, 1979.
Mazzei, R., Raffi, I., Rio, D., Hamilton, N. and Cita, M. B., Calibration of Late Neogene Calcareous Plankton Datum Planes with the Paleomagnetic Record of Site 397 and Correlation with Moroccan and Mediterranean Sections, Init. Repts. DSDP, 47A, 1979.
McKenzie, J. A. and Oberhansli, H., Paleoceanographic expression of the Messinian Salinity Crisis. In: Hsu, K. J. and Weissert, H. J. (eds.), South Atlantic Paleoceanography. Cambridge University Press, p. 99-123, 1985.
McKenzie, J. A., Jenkyns, H. C. and Bennett, G. C., Stable Isotope Study of the cyclic diatomite-claystones from the Tripoli Formation, Sicily: a prelude to the Messinian salinity crisis. Paleogeogr. Paleoclimatol. Paleoecol., 29, 1/2, pp. 125-142, 1979.
McKenzie, J. A., Mueller, P. and Mueller, D., The Miocene/Pliocene boundry and the terminal flood. GSA Abstracts with Programs, 17, 7, p. 659, 1985.
McKenzie, J. A., Weissert, H., Poore, R. Z., Wright, R. C., Percival, S. F., Oberhansli, H. and Casey, M., Init. Repts. DSDP, 73, pp. 717-724, 1984.
Mercer, J. H., Cenozoic glaciation in the Southern Hemisphere. Ann. Rev. Earth Planet. Sci., 11, pp. 99-132, 1983.
Mercer, J. H. and Sutter, J. F., Late Miocene-Earliest Pliocene glaciation in southern Argentina: Implications for global ice-sheet history. Paleogeogr. Paleoclimatol. Paleoecol., 38, pp. 185-206, 1982.
Mueller, D., Palaozeanographische und sedimentologisches untersuchungen, in den Becker von Fortuna and Sorbas (Sudost-Spanien): Korrelation stratigraphisches und isotopengeochemischer Ereignisse in der Mediterranean Salinitatskrise im Messinian. Diss. ETH-Zurich, 1986.
Ogniben, L., Petrografia della Serie Solfifera siciliana e considerazoni geologiche relative. Mem. Descr. Carta Geol. Ital., 33, pp. 1-275, 1957.
Orszag-Sperber, F. and Rouchy, J. M., Le Miocene terminal et le Pliocene inferieur au sud de Chyper,Livert Guide de l'excursion, v seminaire sur le Messinian, Chypre, 1979.
Orszag-Sperber, F., Rouchy, J. M., Bizon, G. et al., La sedimentation messinienne dans le basin de Polemi (Chypre). Geol. Mediter., 7, 1, pp. 91-102, 1980.
Rizzini, A., Vezzani, F., Cococcetta, V. and Miland, G., Stratigraphy and sedimentation of a Neogene-Quaternary section of the Nile Delta area (A.R.E.). Marine Geology, 27, 3/4, pp. 327-348, 1978.
Rizzini, A., and Dondi, L., Erosional surface of Messinian age in the sub-surface of the

Lombardian Plain (Italy). Marine Geology, 27, 3/4, pp. 304-325, 1978.
Rizzini, A. and Dondi, L., Messinian evolution of the Po Basin and its economic implications (hydrocarbons). Paleogeogr. Paleoclimatol. Paleoecol., 29, 1/2, pp. 41-74, 1979.
Rogl, F. and Steininger, F. F., Vom Zerfall der Tethys zu Mediterran und Paratethys. Die neogene Palaeogeograhie und Palinspastik des zirkum-mediterranen Raumes, Ann. Naturhist. Mus. Wien, 85/A, pp. 135-163, 1983.
Ruggieri, G., The Miocene and later evolution of the Mediterranean Sea. In: Adams, C. G. and Ager, D. V. (eds.). Aspects of Tethyan Biogeography. Syst. Assoc. Publ., 7, pp. 283-290, 1967.
Ruggieri, G. and Sprovieri, R., The lacustrine faunas in Sicily and the desiccation theory of Messinian salinity crisis. Lavori Ist. Geol. Univ. Palermo, 13, pp. 1-6, 1974.
Rutford, R. H., Craddock, C., White, C. M. and Armstrong, R. L., Tertiary glaciation in the Jones Mountains. In: Adie, R. J. (ed.) Antarctic Geology and Geophysics. Oslo Universitesforlaget, pp. 239-250, 1972.
Ryan, W. B. F., Geodynamic implication of the Messinian crisis of salinity. In: Drogger, C. W. (ed.) Messinian Events in the Mediterranean. K. ned. Akad. Wetensch., pp. 26-38, 1973.
Ryan, W. B. F. and Cita, M. B., The nature and distribution of Messinian erosional surfaces. Indicators of a several kilometers deep Mediterranean in the Miocene. Marine Geology, 27, 3/4, pp. 193-230, 1978.
Ryan, W. B. F., Hsu, K. J. et al., Init. Repts. DSDP, 13, pp. 1-1447, 1973.
Saito, T., Burckle, L. H. and Hays, J. D., Late Miocene to Pleistocene biostratigraphy of equatorial Pacific sediments. In: Saito, T. and Burckle, L. H. (eds.) Late Neogene Epoch Boundaries. Micropaleontology Press, New York, pp. 226-244, 1975.
Salvatorini, G. and Cita, M. B., Miocene foraminiferal stratigraphy DSDP Site 397 (Cape Bojador, North Atlantic). Init. Repts. DSDP, 47, 1, p. 317-374, 1979.
Schrader, H. J. and Gersonde, R., The Late Miocene brackish to fresh water environment, diatom floral evidence. Init. Repts. DSDP, 42A, pp. 741-755, 1978.
Selli, R., Il Messiniano Mayer-Eymar. Proposta di un neostratotipo. Giorn. Geol., 28, pp. 1-33.
Shackleton, N. and Kennett, J. P., Paleotemperature history of the Cenozoic and the initiation of Antarctic glaciation: oxygen and carbon isotope analyses in DSDP Sites 277, 279 and 281. Init. Repts. DSDP, 29, pp. 743-755, 1975.
Shackleton, N. J. and Cita, M. B., Oxygen and Carbon Isotope stratigraphy of benthonic foraminifera in Site 397: fine-structure of climatic change in the Late Neogene. Init. Repts. DSDP, 47A, pp. 433-446, 1979.
Shackleton, N. J. and Kennett, J. P., Late Cenozoic oxygen and carbon isotopic changes at DSDP Site 284: implications for glacial history of the northern hemisphere and Antarctica. Init. Repts. DSDP, 71, 1, pp. 307-316, 1975a.
Shackleton, N. J. and Kennett, J. P., Paleotemperature history of the Cenozoic and the initiation of Antarctic glaciation: oxygen and carbonisotope analyses in DSDP Sites 277, 279, and 281. Init. Repts. DSDP, 29, pp. 743-755, 1975b.
Sissingh, W., Aspects of the Late Cenzoic Evolution of the South Aegean Ostracode Fauna. Paleogeogr. Paleoclimatol. Paleoecol. 20, pp. 131-145, 1976.
Sonnenfeld, P., The Upper Miocene evaporite basins in the Mediterranean region - a study in paleoceanography. Geol. Rundschau, 63, 3, pp. 1133-1172, 1974.
Sorbini, L. and Tirapelle Rancan, R., Messinian fossil fishes of the Mediterranean. Paleogeogr., Paleoclimatol., Paleoecol., 29, 1/2, pp. 143-154, 1979.
Sprovieri, R., La sezione infrapliocenica di Ricera: considerzioni stratigrafiche e paleoambientali sui Trubi siciliani. Boll. Soc. Geol. Ital., 93, pp. 181-214, 1974.
Sturani, C., A fossil eel (Anguilla sp.) from the Messinian of paleogeographic implications. In: Drogger, C. W. (ed.) Messinian Events in the Mediterranean. K. Ned. Akad. Wetensch., pp. 243-255, 1973.
Sturani, C., Messinian facies in the Piedmont Basin. Mem. Soc. Geol. Ital. 16, pp. 11-25, 1978.
Thunell, R. C., Carbonate dissolution and abyssal hydrography in the Atlantic Ocean. Marine Geology, 47, pp. 165-180, 1982.
Vai, G. B., and Ricci Lucchi F., The Vena del Gesso in Northern Apennines: growth and mechanical break-down of gypsified algal crusts. Mem. Soc. Geol. Ital., 16, pp. 217-250, 1978.
Vail, P. R., and Hardenbol, J., Sea level changes during the Tertiary: Oceanus, 22, 3, pp. 71-80, 1979.
Vail, P. R., Mitchum, R. M., and Thompson, S., Global cycles of relative changes of sea level. AAPG Mem. 26, pp. 83-97, 1977.
Van Couvering, J. A., Berggren, W. A., Drake, R. E., Aguirre, E., Curtis, G. H., The terminal Miocene event. Marine Micropaleont., 1, pp. 262-286, 1976.
Van der Zwaan, G. J. and Den Hartog Jager, D., Paleoecology of Late Miocene Sicilian benthic foraminifera. Proc. Kon. Nederl. Akad. Wetensch., 82, 86(2), pp. 211-223, 1983.
Van Gorsel, J. and Troelstra, S., Late Neogene climate change and the Messinian salinity

crisis. Geol. Mediterr., 7, 1, pp. 127-134, 1980.

Van Hinte, J. O., Colin, J. P., and Lehman, R., Micropaleontological record of the Messinian event in Esso Libya Inc. 31-NC 35A on the Pelagian Platform. Second Symposium on the Geology of Libya, 1979.

Vincent, E., Killingsley, J. S. and Berger, W. H., The magnetic Epoch 6 carbon shift: a change in the oceans C/ C ratio 6.2 million years ago. Marine Micropal., 5, pp. 185-203, 1980.

Wise, S. W., Jr., Deep sea drilling in the Antarctic: focus on late Miocene glaciation and applications of smear-slide biostratigraphy. In: Warme, J. E., Douglas, R. G., and Winterer, E. L. (eds.) The Deep Sea Drilling Project: A decade of progress. Soc. of Econ. Paleont. and Mineral., Spec. Publ., 32, pp. 471-487, 1981.

Wright, R., Neogene benthic Foraminifera DSDP, leg 42A, Mediterranean sea. Init. Repts. DSDP, 42A, 1, pp. 709-726, 1978.

Wright, R. H., Benthic foraminiferal repopulation of the Mediterranean after the Messinian (late Miocene) salinity crisis. Paleogeogr. Paleoclimatol. Paleoecol., 29, 1/2, pp. 189-214, 1979.

THE PLIO-PLEISTOCENE OCEAN: STABLE ISOTOPE HISTORY

N. J. Shackleton

University of Cambridge, Godwin Laboratory for Quaternary Research
Free School Lane, Cambridge CB2 3RS, England

Introduction

It is now becoming well known that the environmental changes that occurred during the Late Pliocene and Pleistocene had a major impact on the surface of the oceans (CLIMAP, 1976; 1981). The change was particularly striking in the North Atlantic, where during each glacial the Gulf Stream flowed almost due East across the ocean towards Spain instead of warming the northerly coasts of Europe (Ruddiman and McIntyre, 1976). The history of movements of this current is important because they provide the means by which oceanic climatic change may be correlated with events on the European continent where the stratigraphic record of glacial-interglacial events was first described. While the northern Atlantic Ocean was in a glacial mode, the adjacent continent must also have been. DSDP Site 552A, drilled with the Hydraulic Piston Corer on Rockall Bank to the West of Ireland, shows a dramatic sequence of glacial-interglacial fluctuations which started abruptly at 2.37 million years ago and continued up to the present (Shackleton et al., 1984). It will never be possible to recover such a detailed record on the adjacent continent, but this deep sea record will enormously aid our understanding of those fragments of the record that are available in Europe.

Alterations to ocean deep water circulation may have been as important for our understanding of climatic variation as were the changes at the surface. During the glacials the deep water that formed in the North Atlantic made a smaller contribution to the ocean deep waters than it does today (Duplessy et al., 1980). The contrast in dissolved oxygen content between the depths of the Atlantic and of the Pacific scarcely existed at the glacial maximum (Shackleton, Imbrie and Hall, 1983). Changes in the chemistry of the ocean caused the atmospheric carbon dioxide content to be lower during the glacial than it is now (Delmas et al., 1977; Broecker, 1982; Shackleton et al., 1983), whereas during the last interglacial it was higher than it was during most of the Holocene (although it was not as high as it will become as a result of man's combustion of fossil fuels).

Oxygen Isotope Stratigraphy

The use of the oxygen isotope record in benthic or planktic foraminifera as a stratigraphic correlation tool for deep sea sediments has become almost routine. This practice stemmed from the realization that for almost any oxygen isotope record generated using oceanic microfossils, changes in ocean-water isotopic composition contribute the greater part of the signal, with the effects of temperature being subordinate. The average glacial-interglacial amplitude of published Pleistocene oxygen isotope records is about 1.5 per mil. The first estimate for the change in ocean-water isotopic composition caused by the abstraction of isotopically light water to the continental ice sheets was about 0.5 per mil (Emiliani, 1955), but subsequent realistic models of the ice sheets involved led to larger estimates of 0.9 to 1.4 per mil for this effect and to a realisation that this was the major contributor to the isotopic record observed in Quaternary deep-sea sediment sequences (Olausson, 1965; Shackleton, 1967a; Craig, 1968; Dansgaard and Tauber, 1969). This led to the suggestion that the isotope record in marine sediments could be used for stratigraphic correlation (Shackleton, 1967b) and to its widespread use. CLIMAP (1976) used the oxygen isotope record as the primary tool for global correlation, and several biostratigraphic studies used the oxygen isotopic record as a means of assessing the synchroneity of each biostratigraphic datum evaluated (Hays and Shackleton, 1976; Thierstein et al, 1977; Morley and Shackleton, 1978; Burckle et al., 1978; Berggren et al., 1980). Today a large number of records are available for comparison and synthesis. I first wish to consider the reproducibility of these oxygen isotope records, and the question as to whether there exists a "standard" record.

Similarity Between Benthic Oxygen Isotope Records

Since horizontal and vertical gradients in most physical and chemical properties in the deep ocean are small by comparison with the variability in their geological estimators through time, it is to be expected that the differences between the stable isotope records in two cores taken close to one another will become less as the quality of the two records is improved. Unfortunately it is not immediately obvious how best to compare two records. The chief problem is that although gradients in water column properties may be small, sediment accumulation rate may vary over quite small distances. To compare two records requires that they are expressed on a common scale, preferably a timescale. This is achieved using the stable isotope data themselves as a basis for correlation, so that the exercise of expressing two records on the same timescale is itself an attempt at improving the similarity between the two. Nevertheless, the fact that the more detailed oxygen isotope records are those that are most easily correlated is an indication that this is in fact a valid procedure. Good records seldom contain features that hinder attempts at correlation, although the detailed discussion of this problem by Pisias et al (1984) does reveal one such occurrence.

Figure 1 compares the oxygen isotope records of cores V19-29 and V19-30, on their respective depth scales. These two cores were taken about 50km apart on the Carnegie Ridge on the south side of the Panama Basin and they also show very similar records of carbonate content (Ninkovich and Shackleton, 1975). Examination of Figure 1 shows that the two cores preserve very similar oxygen isotope records: analogous events occur at similar depths in both cores and the isotopic values are similar although the more detailed sampling of V19-30 provides a more realistic impression of the analytical noise level in such measurments. This is the most detailed published record of this time interval.

Figure 2 compares the record of V19-30 with that of core RC13-228 from the South Atlantic plotted on a common timescale. Figure 2 suggests that the similarity between the benthic record of any other core and that of V19-30 is more a function of the sampling density and detail of the individual record than of the difference between the water mass at the core site and that at the site of V19-30. However, careful examination shows that there are detectable and real differences between the deep water masses sampled by at least some pairs of cores. For example, glacial ^{18}O values deviate from Holocene values by a greater amount in the North Atlantic than in the Pacific (Duplessy et al., 1980; Shackleton et al, 1983), suggesting a cooling in the deep North Atlantic during glacial times. Thus the concept of a "standard" benthic oxygen isotope record is not entirely appropriate. Shackleton et al. (1983) worked on the hypothesis that a deep Pacific record can usefully be regarded as a reference. They compared deep water records from the north Atlantic and the equatorial Pacific by estimating the inter-oceanic oxygen isotope difference as a function of age. This record showed that the deep water of the north Atlantic was relatively colder in glacial than interglacial stages.

So far, the most detailed deep Pacific records have been obtained by analysing *Uvigerina* species, but it would be desirable to obtain equally detailed records from other species to ensure that there are no time-dependent or watermass-dependent departures from isotopic equilibrium in any species. When this has been done, it should be possible to create a standard deep Pacific oxygen isotope record with better than lka resolution and better than 0.05 per mil accuracy.

Similarity Between Planktic Oxygen Isotope Records

Emiliani (1955) was the first to compare oxygen isotope records for planktic foraminifera from nearby cores. He clearly recognised the importance of high-resolution cores for obtaining the best record. Thus when he constructed a 'generalized record' he put considerable weight on the detailed records of Atlantic cores 234, 246 and 280 which were obviously superior to his stratigraphically longer Caribbean records. There are many differences between the records that Emiliani (1955, 1964, 1966) obtained from the Caribbean. The datasets illustrated by Imbrie et al (1984) are based on planktic foraminifera and include the Atlantic, the Pacific and the Caribbean. Inspection of these records suggests the existence of systematic and not fully understood differences between records from different regions; for example, the section immediately overlying the Brunhes-Matuyama boundary seems to be different in Atlantic and Pacific cores. Since the records are from low latitudes where CLIMAP (1976) suggests that temperature variability was slight, the explanation for the differences probably lies in the interaction of varying sedimentation rate, dissolution rate and intensity of bioturbation.

Oxygen isotope records from the northern part of the North Atlantic and Norwegian Sea (Kellogg et al., 1978) do show considerably larger isotopic excursions, which clearly include a substantial temperature component. The Mediterranean Sea is another area where quite large oxygen isotope ranges are observed (Vergnaud-Grazzini et al., 1977). Since when averaged over the entire ocean, the sea surface temperature difference between the last glacial maximum and today was less than 2degC (CLIMAP, 1981), it is to be expected that for most areas temperature change has made only a small contribution to the oxygen isotope signal in planktic Foraminifera.

Generalized or Idealized Oxygen Isotope Records

Emiliani (1966) derived a 'generalized oxygen isotope record' in an informal manner, by

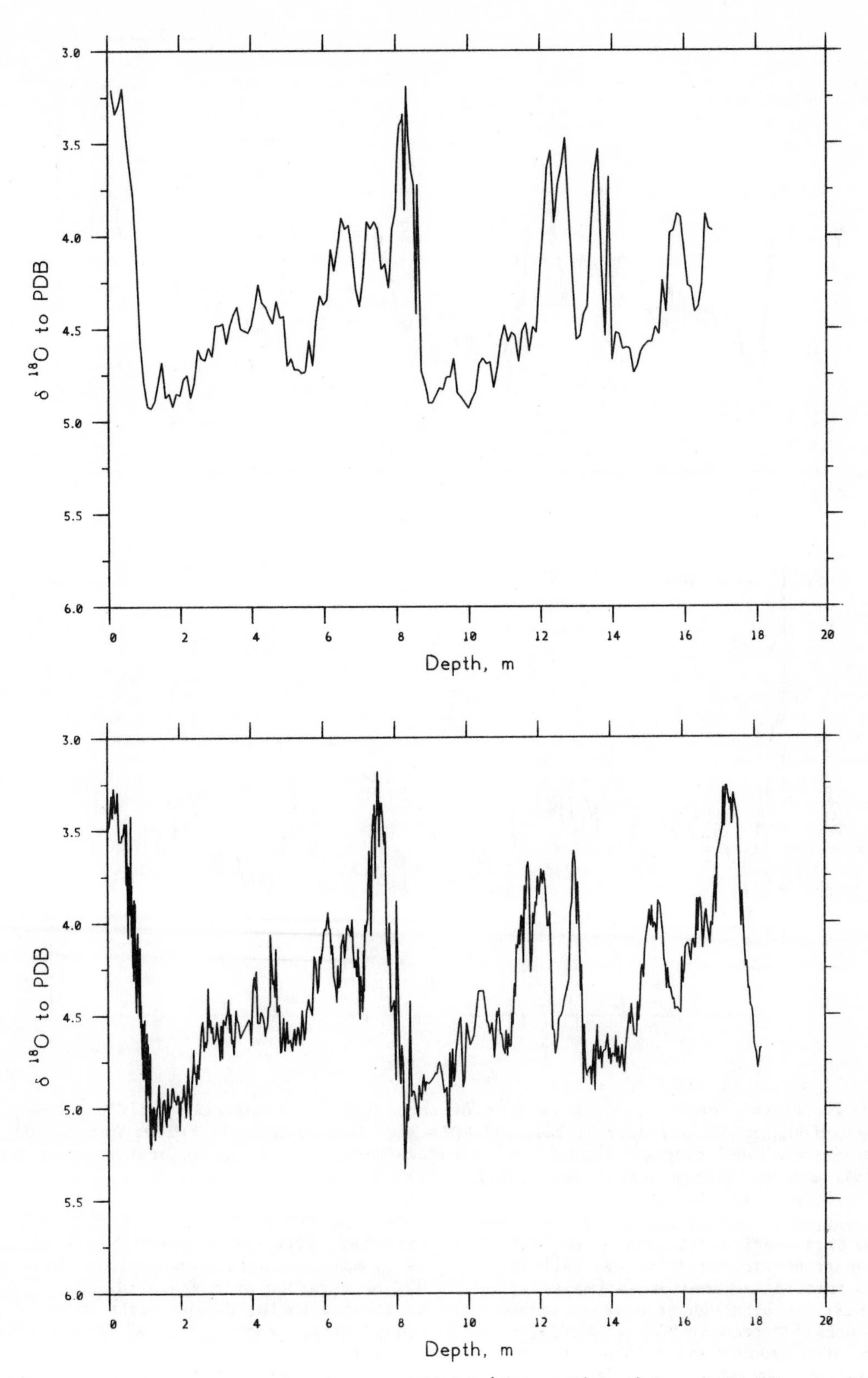

Fig. 1. Oxygen isotope records of cores V19-29 (above; Ninkovich and Shackleton, 1975) and V19-30 (below; Shackleton, Imbrie and Hall, 1983) plotted against depth in sediment. For both records the benthic genus Uvigerina was analysed.

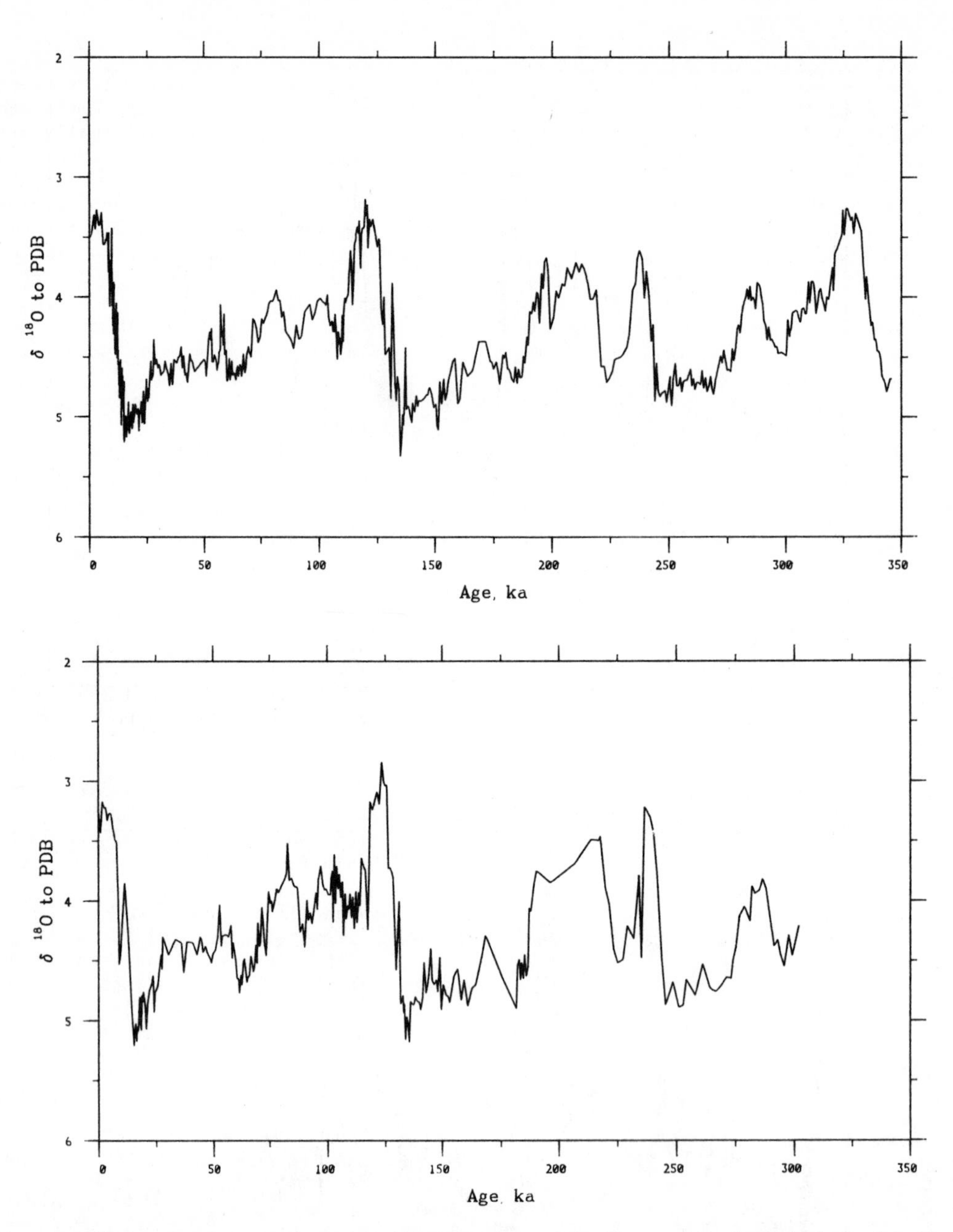

Fig. 2. Benthic oxygen isotope records of cores V19-30, equatorial Pacific (above; Shackleton, Imbrie and Hall, 1983), and RC13-228, South Atlantic (below; Morley and Shackleton, 1984) plotted against estimated age using the timescale of Imbrie et al., (1984) and Shackleton and Pisias (1985).

utilizing the best sections in each of the cores that he analysed. He expressed his generalized oxygen isotope record in terms of estimated temperature history, although of course one can re-scale it as an ^{18}O record. More recently, attempts have been made to generate generalized records objectively, by stacking different data sets (Imbrie et al., 1984; Pisias et al., 1984; Berger et al., 1985; Prell et al. MS 1985). Two somewhat different approaches have been taken. One is to make an initial assumption that one record may be regarded as a standard, and other records correlated to the depths scale of the standard record using "Shaw Diagrams" (Shaw, 1964). The stacked 300,000 year benthic record of Pisias et al. (1984) was obtained by correlating several records with that of core V19-29. An alternative method is to obtain a timescale first, so that the final stack is already a time series. This has the advantage that the result is not needlessly biassed by anomalies in the record chosen as a standard. A methodology for doing this would be to use an isotope record modelled from the calculated astronomical record (e.g. that of Imbrie and Imbrie, 1980) as a starting point; the advantages and disadvantages of this approach are laid out by Martinson et al. (1986).

The stacked record discussed by Imbrie et al. (1984) was generated from quite varied isotope records; each record was reduced to zero mean and unit variance so as to avoid biassing the record to a particular area. Thus the true isotope scale is not preserved. It is interesting to notice, however, that the stacked record bears greater similarities to the benthic isotope record than one might expect on the basis of the original data. Differences between planktic and benthic records are discussed further below.

The recent interest in generating idealized or stacked records raises a fundamental question regarding the formal status of the Oxygen Isotope Stages as stratigraphic units. If it is accepted that the oxygen isotope record of a single core is less perfect than a record obtained by averaging data from more than one core, can one sustain the principle advocated by Shackleton and Opdyke (1973), that Stage boundaries ought to be defined according to the "Golden Spike" principle, at specified depths in a particular core? Or should boundaries be defined on an idealized curve? I believe that it is preferable to stick to the conceptual definition in terms of a single core record (although as with any geological boundary, stratotypes can be re-designated). In an actual core, many alternative stratigraphic correlation methods can be applied. For example, in the case of core V28-238, amino-acid stratigraphy (King and Neville, 1977); biostratigraphy of foraminifera (Thompson, 1976; Thompson et al., 1979); biostratigraphy of radiolaria (Hays and Shackleton, 1976); biostratigraphy of coccoliths (Thierstein et al., 1977) and several chemical approaches (Kominz et al, 1979) have all been applied. Although it is true that in principal all these methods may be used to correlate the various records that are combined in a generalized record, it seems preferable to use a physical sediment sequence as a primary reference.

The fact remains that core V28-238 is not perfect; for the past 300,000 years the oxygen isotope record of core V19-30 is considerably superior, primarily because it is from a region where the sediment accumulation rate is higher. If a stratigraphically longer benthic record were obtained from this same region, it would probably be appropriate to redefine stage (and substage) boundaries in that region. This would have the additional merit that long-distance correlation is more accurate between benthic than between planktic oxygen isotope records.

Pisias et al. (1984) and Prell et al. (MS 1985) used distinctive "features" of the oxygen isotope record, with a hierarchical terminology, in their efforts to obtain an accurate inter-core correlation prior to stacking. Their terminology uses the stage numbers as originally proposed by Emiliani (1955) but represents an alternative to the sub-stages defined by Shackleton (1969) in Stage 5. Although their approach is clearly valuable as a means of formalizing the recognition of fine-structure in the oxygen isotope record (and, presumably, fine structure in the underlying climatic record) I consider that in the long term, a subdivision based on boundaries between contrasting episodes is preferable to one based on the acmes of those episodes. It is only if a record is under-sampled with respect to the scale of a particular feature, that the acme of that feature can be located more accurately than its boundaries.

The Oxygen Isotope and Sea Level Relationship

The rigorous use of the oxygen isotope record as a tool for stratigraphic correlation hinges on the assumption that the underlying cause of the observed variation is changes in the isotopic composition of the ocean. This immediately implies that there should be a relationship between oxygen isotope value and glacio-eustatic sea level, and indeed one major attraction of the marine oxygen isotope record is its intimate association with both ice volume and sea level. Shackleton and Opdyke (1973) suggested that a figure of 0.1 per mil per 10m sea level lowering was an appropriate relationship for practical use. This is equivalent to a mean oxygen isotope ratio for the accumulated ice of about -33 per mil. Fairbanks and Matthews (1978) have obtained a field calibration by analysing the isotopic composition of coral that they collected at measured vertical positions with reference to the uplifted interglacial terraces on the island of Barbados; they obtained independent estimates of 0.10 and 0.12 per mil per 10m for two transitions.

Mix and Ruddiman (1984) have made valuable numerical models that simulate the extent to which the oceanic isotope record may have lagged the true ice volume record as a result of the quite significant residence time of ice within an ice sheet. These authors also discussed the effects of geographical movements of the ice sheet centres. They showed that a lag of 1ka to 3ka in the oxygen isotope record with reference to true ice volume may be realistic. One way to constrain these models will be by extending the approach of Fairbanks and Matthews (1978) so as to obtain an accurate expression of the isotope-altitude relationship for ice growth and ice melting. However, it must be remembered that there is also a very significant phase difference between ice volume and sea level records resulting from the slow isostatic response to continental and oceanic loading and unloading as water is transferred to and fro between ice sheets and ocean (Walcott, 1972).

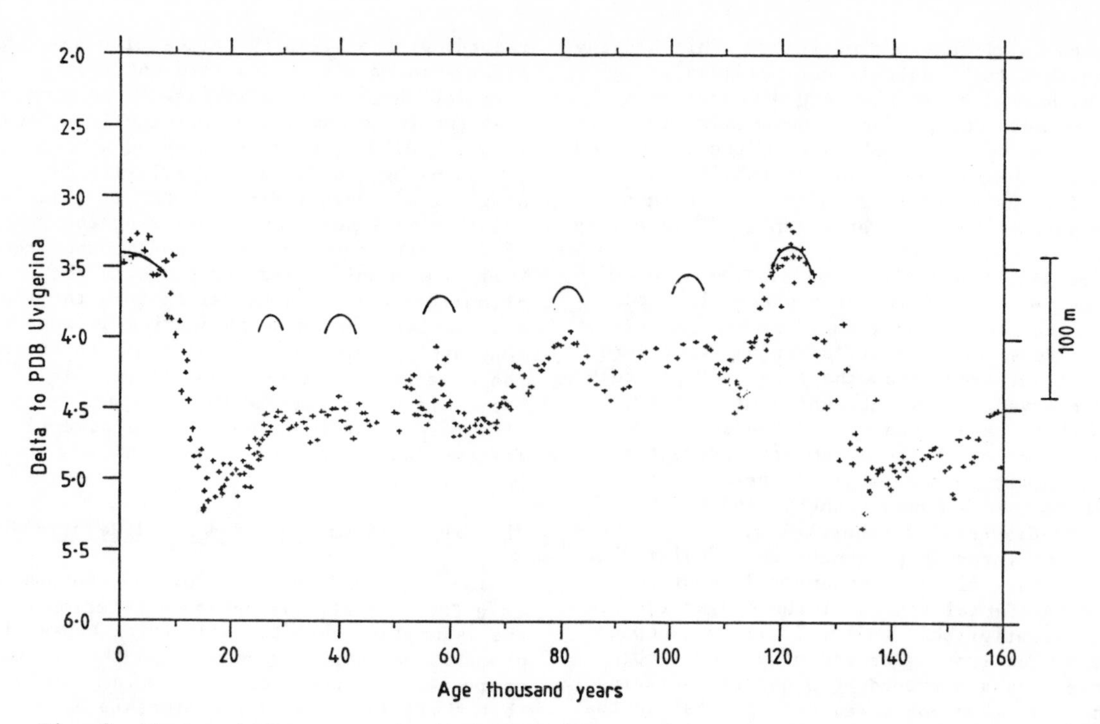

Fig. 3. Oxygen isotope record of core V19-30 for the past 150ka, and estimated sealevels based on data from Barbados, New Guinea and other areas as summarised by Moore (1982).

A serious discrepancy between the isotope record and sea level arises when we come to a comparison between the inferred sealevel of Substages 5a and 5e with the corresponding isotope values. It has been established from several studies, especially on the islands of Barbados (Mesolella et al., 1969), New Guinea (Bloom et al., 1974) and Haiti (Dodge et al., 1983) that the sea was within about 15m of its present level both at about 104ka and at about 82ka, i.e. during substages 5c and 5a. However, although this appeared to be consistent with the low resolution record of Shackleton and Opdyke (1973), more detailed records (e.g. V19-30 in Figure 2) show ^{18}O values that are consistently 0.6 per mil more positive at those times compared with today, implying sealevel 60m below present. Figure 3 illustrates the problem by comparing the sea level high-stands inferred from Barbados, New Guinea and elsewhere (as summarised by Moore, 1982) with the detailed oxygen isotope record of core V19-30. To facilitate comparison, the sea-level positions are plotted at a scale 10m for 0.1 per mil ^{18}O change. Thus if this simple relationship adequately explained the ocean isotope record, each indicated sea-level peak would overlie the equivalent peak in the isotope record. Clearly this is not the case. Williams et al., (1981) make a similar comparison to that shown in Figure 3.

Figure 4 (Chappell and Shackleton, MS 1986) compares the oxygen isotope record of core V19-30 point by point with the sea level record for New Guinea published by Chappell (1983). Figure 4 shows a remarkably good linear relationship between ^{18}O and sea level for all points except the holocene and substage 5e. It has been proposed in the past that the discrepancy between the oxygen isotope record and the sea level history for Stage 5 could be explained by assuming that there was a major ice sheet in the Arctic which stored isotopically light ice, but displaced ocean water and thus did not alter sea level (Mercer, 1970; Broecker, 1975; Fillon and Williams, 1983). This explanation would suffice to explain Figure 5. However the recent publication of a detailed oxygen isotope record from the eastern Arctic Basin (Markussen et al., 1985) finally eliminates the possibility that a volumetrically important amount of ice was stored in the Arctic during the last glacial. I consider it more likely that the true explanation is that the temperature of the Pacific deep water was not constant, but that it was significantly lower continuously from substage 5d to about 11ka, which would explain the data shown in Figure 5. If the shortfall is taken as 0.4 per mil, this would represent of the order 1.5 degC of cooling.

The question of the sea level record for the Pleistocene will have to be reconsidered in some detail following the above demonstration that the benthic record contains a 0.4 per mil temperature component. In particular, it may prove to be the case that there were some episodes that were "interglacial" in the sense of having as little

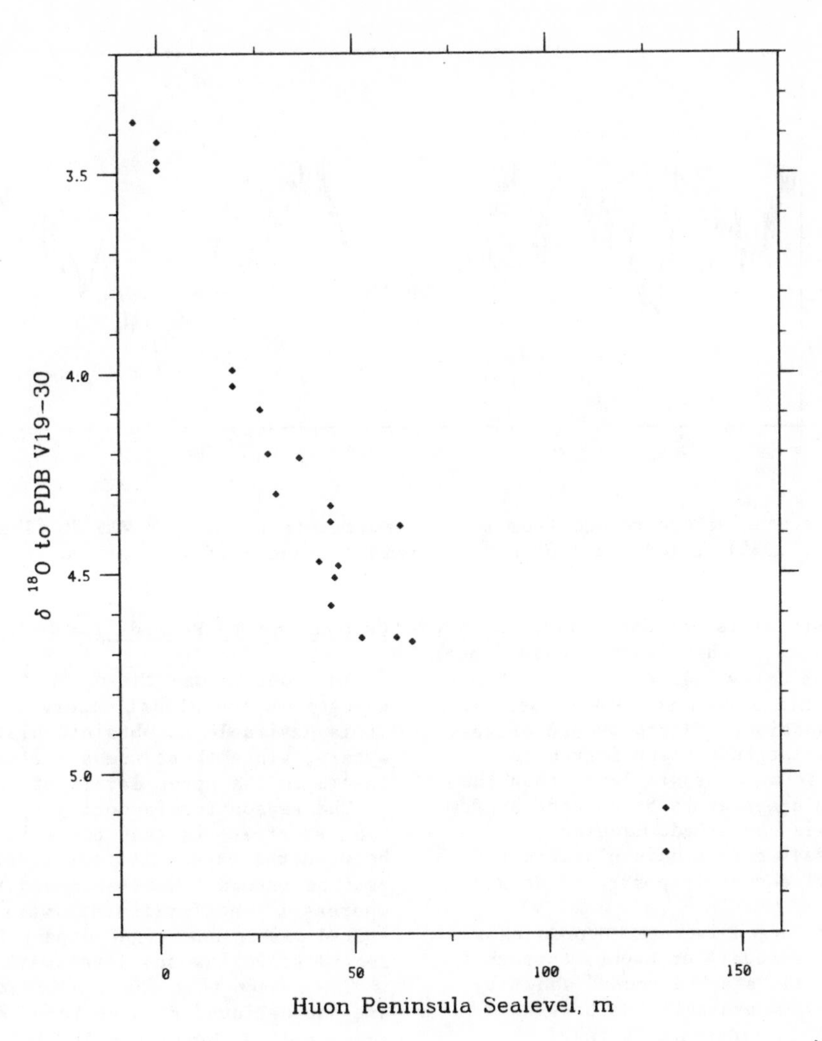

Fig. 4. Estimates of maxima and minima in sealevel as estimated from New Guinea coral terraces compared with oxygen isotope values for stratigraphically equivalent levels in core V19-30 (Chappell and Shackleton MS 1986).

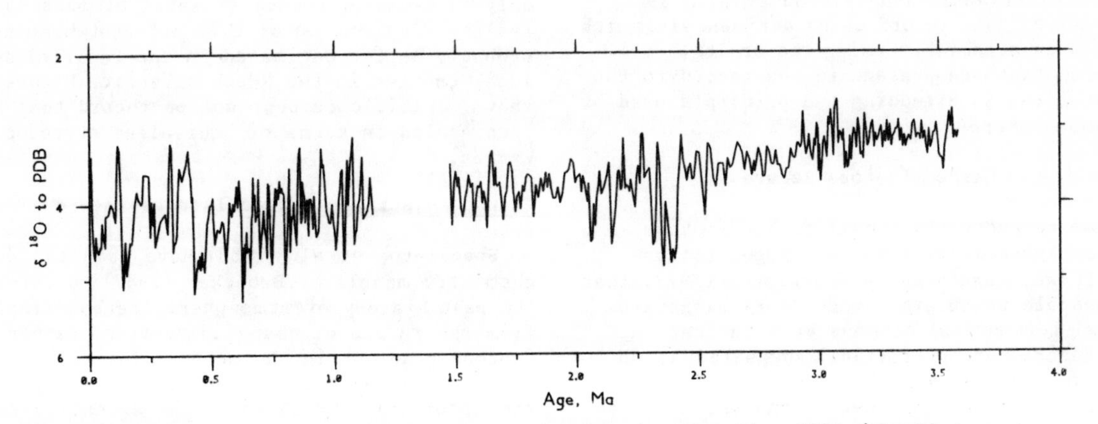

Fig. 5. Oxygen isotope record for the past 3.4Ma in DSDP Site 552A (Shackleton et al., 1984).

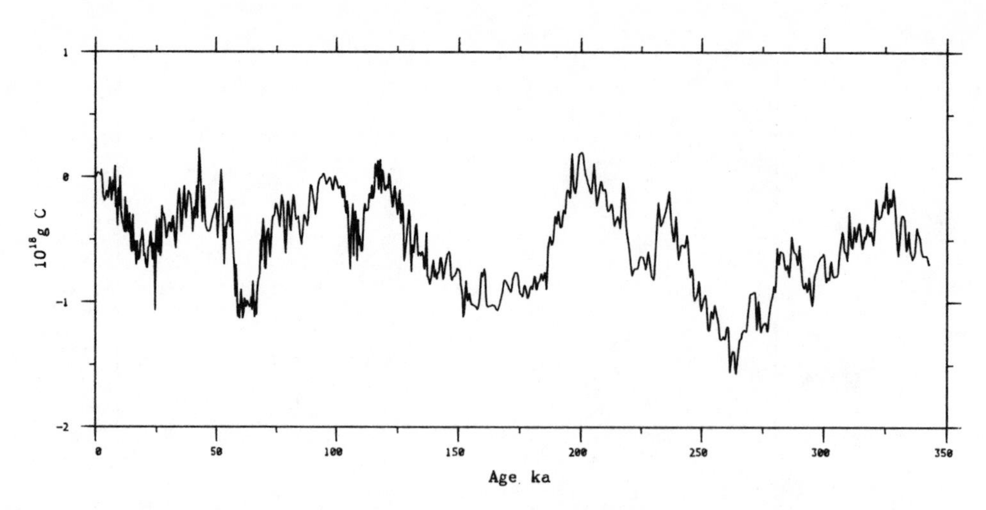

Fig. 6. Carbon isotope record from benthic Foraminifera in core V19-30 (Shackleton and Pisias, 1985) scaled for continental biomass changes (grams carbon).

ice on the continents as is present today, and yet retain cold deepwaters. Such episodes would appear only slightly isotopically lighter than substages 5a and 5c in a benthic oxygen isotope record. In principle, the suggestion of Matthews and Poore (1980) that the low latitude planktic record provides a better estimate of sea level than the benthic record, would appear to be correct so far as the Pleistocene is concerned. However, in practice it is no easier to find a planktic than a benthic record that adequately portrays the sea level record.

The 350,000 year oxygen isotope records shown in Figure 2 provide a model for events through the past 900,000 years; the stacked record shown by Imbrie et al. (1984) is probably the best depiction of the whole sequence. The earlier record has a somewhat different character (Shackleton and Opdyke, 1976 and 1977). The most detailed benthic oxygen isotope record available for the whole Pleistocene and late Pliocene is from DSDP site 552A on the flank of Rockall Bank in the North Atlantic (Shackleton et al., 1984). Figure 5 shows this record using a timescale which represents an attempt at using the orbital frequencies that are present in the record to tune it astronomically, extending the principle used by Imbrie et al. (1984).

The Carbon Isotope Record

In some respects the global carbon isotope record is as fascinating as the oxygen isotope record. It was suggested by Shackleton (1977) that it is possible to obtain a record of changes in the global terrestrial biomass from the carbon isotope record. Broecker (1982) suggested as an alternative that changes in the amount of organic carbon stored on the continental shelves might be a more important effect.

Carbon Isotope Records for Deep Water

In order to use the ocean ^{13}C record to monitor changes in the global reservoir of organic carbon, it is desirable to obtain a history for ocean deep waters, since there are significant vertical gradient in ^{13}C in the upper layers of the ocean.

The reason for favouring the continental biomass effect is that there is no phase lag between the oxygen isotope record and the carbon isotope record (Shackleton and Pisias, 1985), whereas if shelf sediments were the locus of the stored carbon one might expect the carbon isotope record to follow sea level with a lag.

The reason that the continental biomass changes are not believed to have been so large as was suggested by Shackleton (1977b) is that it is now known that there have been substantial changes in inter-oceanic carbon isotope gradients during the Pleistocene (Curry and Lohmann, 1982, 1983; Shackleton, Imbrie and Hall, 1983) so that the record that Shackleton (1977) first interpreted only in terms of the continental biomass is in reality a summation of this effect (which is probably better estimated in the Pacific) and of local changes in the North Atlantic. Figure 6 shows a Pacific carbon isotope record that has been scaled in terms of equivalent stored organic carbon.

The Ocean Surface Carbon Isotope Record

Shackleton et al. (1983) have used the ocean chemistry models of Broecker (1982) to interpret the past history of atmospheric carbon dioxide from the record of changing vertical carbon isotope gradient in the oceans. This is possible because the level of carbon dioxide in the atmosphere is controlled by the extent to which photosynthesis removes carbon from the ocean

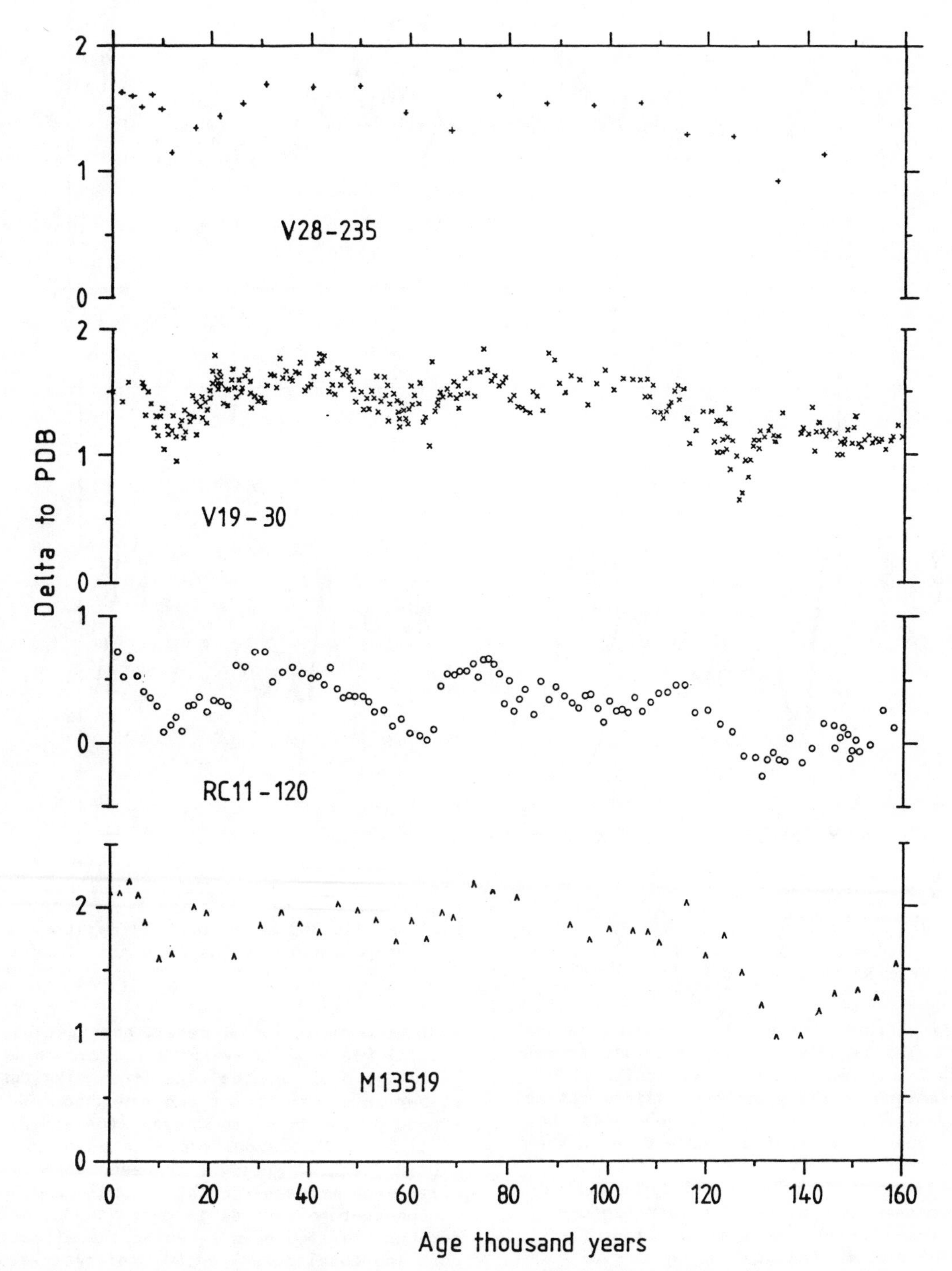

Fig. 7. Carbon isotope records for planktic Foraminifera in cores V19-30 and RC11-120 (Shackleton and Pisias, 1985), M13519 (Sarnthein et al, 1984) and V28-235 (Shackleton, unpublished).

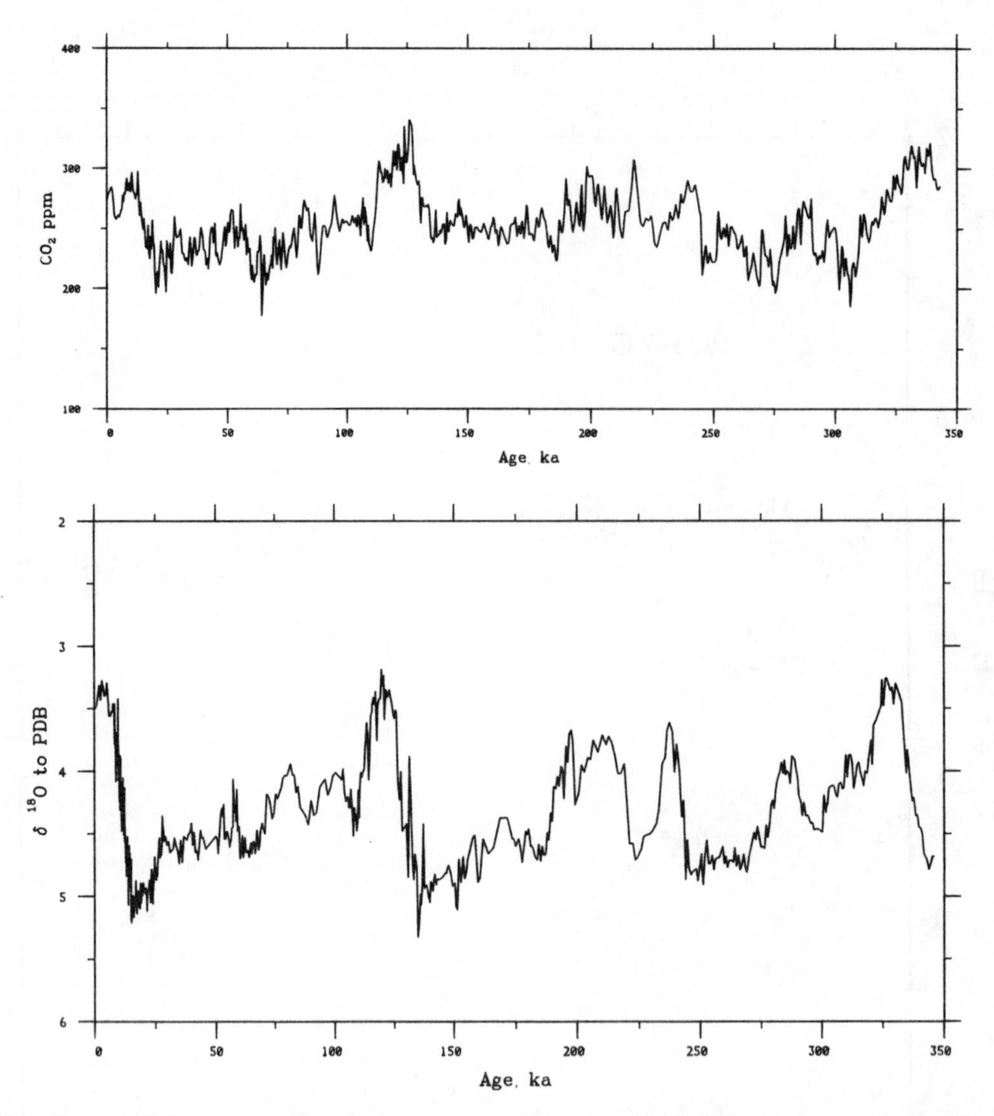

Fig. 8. Atmospheric carbon dioxide record for the past 340 ka estimated from the carbon isotope data from core V19-30, and the oxygen isotope record in the same core (modified from Shackleton and Pisias, 1985)

surface. Since photosynthetically reduced carbon is isotopically light with respect to its source, this mechanism gives rise to a vertical carbon isotope gradient in the upper few hundred metres of the ocean. Shackleton et al. (1983) used the carbon isotope record from a single core on the assumption that cores from other regions would yield similar records if analysed appropriately. Figure 7 compares the planktic carbon isotope record of core V19-30 with that of three other cores: V28-235 from the western equatorial Pacific, M13519 from the tropical Atlantic (Sarnthein et al., 1984 and Sarnthein, personal comm.) and RC11-120 from the subantarctic. The broad similarity between the records shown in Figure 7 supports the notion that there have indeed been global changes in ocean surface carbon isotope composition. However, considerably more work is required to evaluate the record more reliably, because the signal-to-noise ratio is rather unfavourable for the planktic ^{13}C record. Figure 8 (modified from Shackleton and Pisias, 1985) shows the reconstructed atmospheric carbon dioxide record. Figure 8 is based chiefly on the difference between the the planktic and benthic carbon isotope records in core V19-30. However, account has also been taken of the effect of changing total dissolved CO_2, of ocean salinity, and of temperature, along the lines suggested by Broecker (1982).

The surface carbon isotope records discussed are all from low or mid-latitudes. Mix and Fairbanks (1985) and Labeyrie et al. (1985) have shown that at high latitudes a significantly different carbon isotope record is preserved by planktonic foraminifera. The implication of this finding for the attempt to reconstruct the history of atmospheric carbon dioxide concentration is not well understood at present.

Conclusions

After three decades of activity, stable isotope measurements in deep sea sediments continue to provide new information on the history of the Quaternary oceans. It seems likely both that routine stable isotope stratigraphy will continue to be one of the primary records to be sought in the course of any new investigation based on deep-sea sediment cores, and that further insights into the operation of the global climate system will be possible by making novel uses of the stable isotope data.

References

Berger, W. H., J. S. Killingley, C. V. Metzler, and E. Vincent, Two-step deglaciation: ^{14}C dated high-resolution $\delta^{18}O$ records from the tropical Atlantic Ocean. Quaternary Res. 23, 258-271, 1985.

Berggren, W. A., L. H. Burckle, M. B. Cita, H. B. S. Cooke, B. M. Funnell, S. Gartner, J. D. Hays, J. P. Kennett, N. D. Opdyke, L. Pastouret, N. J. Shackleton, and Y. Takayanagi, Towards a Quaternary time scale. Quaternary Res. 13, 277-302, 1980.

Bloom, A. L., W. S. Broecker, J. M. A. Chappell, R. K. Matthews, and K. J. Mesolella, Quaternary sea level fluctuations on a tectonic coast: new $^{230}Th/^{234}U$ dates from the Huon Peninsula, New Guinea. Quaternary Res. 4, 185-205, 1974.

Broecker, W. S., Floating glacial ice caps in the Arctic Ocean. Science, 188, 1116-1118, 1975.

Broecker, W. S., Glacial to interglacial chages in ocean chemistry. Progr. Oceanogr. 11, 151-197, 1982.

Burckle, L. D., D. B. Clarke, and N. J. Shackleton, Isochronous last-abundant-appearance datum (LAAD) of the diatom Hemidiscus karstenii in the sub-Antarctic. Geology 6, 243-246, 1978.

Chappell, J. A revised sea-level record for the last 300,000 years from Papua New Guinea. Search 14, 99-101, 1983.

Chappell, J. and H. H. Veeh, $^{230}Th/^{234}U$ age support of an interstadial sea level of -40m at 30,000 yr. BP. Nature 276, 602-603, 1978.

CLIMAP Project Members, The surface of the ice-age earth. Science 191, 1131-1137, 1976.

CLIMAP Project Members. Seasonal reconstructions of the earth's surface at the last glacial maximum. Geol. Soc. Amer. Map and Chart Series MC-36, 1981.

Craig, H. The measurement of oxygen isotope paleotemperatures. in Procedings of the Third Spoleto Conference, ed. E. Tongiorgi, p. 161-182. Pisa: Lischi and Figli, 1968.

Curry, W. B. and G. P. Lohmann, Carbon isotopic changes in benthic foraminifera from the western south Atlantic: reconstruction of glacial abyssal circulation patterns. Quaternary Res. 18, 218-235, 1982.

Curry, W. B. and G. P. Lohmann, Reduced advection into Atlantic Ocean deep eastern basins during last glaciation maximum. Nature 306, 577-580, 1983.

Dansgaard, W. and H. Tauber, Glacier oxygen-18 content and Pleistocene ocean temperatures. Science 166, 499-502, 1969.

Delmas, R. J., J.-M. Ascensio, and M. Legrand, Polar ice evidence that atmosphere CO_2 20,000 yr. B.P. was 59% of present. Nature 284, 155-157, 1980.

Dodge, R. E., R. G. Fairbanks, L. K. Benninger, and F. Maurrasse, F., Pleistocene sea levels from raised coral reefs of Haiti. Science 219, 1423-1425, 1983.

Duplessy, J.-C., J. Moyes, and C. Pujol, Deep water formation in the North Atlantic Ocean during the last ice age. Nature 286, 479-481, 1980.

Duplessy, J.-C. and N. J. Shackleton, Carbon-13 in the World Ocean during the last interglaciation and the penultimate glacial maximum. Progr. Biometeorology 3, 48-54, 1984.

Emiliani, C., Pleistocene temperatures. Jour. Geol. 63, 538-578, 1955.

Emiliani, C., Palaeotemperature analysis of the Caribbean cores A 254-BR-C and CP-28. Geol. Soc. Amer. Bull. 75, 129-144, 1964.

Emiliani, C., Paleotemperature analysis of Caribbean cores P6304-8 and P6304-9 and a generalised temperature curve for the last 425,000 years. Jour. Geol. 74, 109-126, 1966.

Fairbanks, R. G. and R. K. Matthews, The marine oxygen isotope record in Pleistocene coral, Barbados, West Indies. Quaternary Res. 10, 181-196, 1978.

Fillon, R. H. and D. F. Williams, Glacial evolution of the Plio-Pleistocene: role of continental and arctic ocean ice sheets. Palaeogeogr., Palaeoclimatol., Palaeoecol., 42, 7-33, 1983.

Hays, J. D. and N. J. Shackleton, Globally synchronous extinction of the radiolarian Stylatractus universus. Geology 4, 649-652, 1976.

Imbrie, J., J. Hays, D. G. Martinson, A. McIntyre, A. C. Mix, J. J. Morley, N. G. Pisias, W. L. Prell, and N. J. Shackleton, The orbital theory of Pleistocene climate: suport from a revised chronology of the marine $o^{18}O$ record. in Milankovitch and Climate, edit. A. Berger et al. pp. 269-306. Dordrecht, Holland: D. Reidel, 1984.

Imbrie, J. and J. Z. Imbrie, Modelling the climatic response to orbital variations. Science 207, 943-954, 1980.

Kellogg, T. B., J.-C. Duplessy, and N. J. Shackleton, Planktonic foraminiferal and oxygen isotope stratigraphy and paleoclimatology of Norwegian Sea deep-sea cores. Boreas 7, 61-73, 1978.

King, K. and C. Neville, Isoleucine epimerization for dating marine sediments: importance of analysing monospecific foraminiferal samples. Science 195, 1333-1335, 1977.

Kominz, M. A., G. R. Heath, T.-L. Ku, and N. G. Pisias, Brunhes time scales and the interpretation of climatic change. Earth Planet. Sci. Lett. 45, 394-410, 1979.

Labeyrie, L. D., and J.-C. Duplessy, Changes in the oceanic $^{13}C/^{12}C$ ratio during the last 140000 years: high-latitude surface water records. Palaeogeogr., Palaeoclimatol. Palaeoecol., 50, 217-240, 1985.

Markussen,B., Zahn, R. and Thiede, J. Late Quaternary sedimentation in the eastern Arctic Basin: stratigraphy and depositional environment. Palaeogeogr., Palaeoclimatol., Palaeoecol., 50, 271-284, 1985.

Matthews, R. K. and R. Z. Poore, Tertiary $\delta^{18}O$ record and glacio-eustatic sea-level fluctuations. Geology 8, 501-504, 1980.

Mercer, J. H., A former ice sheet in the Arctic Ocean. Palaeogeogr., Palaeoclimatol., Palaeoecol. 8, 19-25, 1970.

Mesolella, K. J., R. K. Matthews, W. S. Broecker, and D. L. Thurber, The astronomical theory of climatic change: Barbados data. Jour. Geol. 77, 250-274, 1969.

Mix, A. C. and R. G. Fairbanks, North Atlantic surface-ocean control of Pleistocene deep-ocean circulation. Earth Planet. Sci. Lett. 73, 231-243, 1985.

Mix, A. C. and W. R. Ruddiman, Oxygen-isotope analyses and Pleistocene ice volumes. Quaternary Res. 21, 1-20, 1984.

Moore, W. S. Late Pleistocene Sea-Level History. In Uranium Series Disequilibrium: Applications to Environmental Problems, edit. M. Ivanovich and R. S. Harmon, pp. 481-496. Clarendon Press, Oxford, 1982.

Morley, J. J. and N. J. Shackleton, Extension of the radiolarian Stylatractus universus as a biostratigraphic datum to the Atlantic Ocean. Geology 6, 309-311, 1978.

Morley, J. J. and N. J. Shackleton, The effect of accumulation rate on the spectrum of geologic time series: evidence from two south Atlantic sediment cores. in Milankovitch and Climate, edit. A. Berger et al., pp. 467-480. Dordrecht, Holland: D. Reidel, 1984.

Ninkovich, D. and N. J. Shackleton, Distribution, stratigraphic position and age of ash layer "L" in the Panama Basin region. Earth Planet. Sci. Lett. 27, 20-34, 1975.

Olausson, E., Evidence of climatic changes in North Atlantic deep-sea cores. Progr. Oceanogr. 3, 221-252, 1965.

Pisias, N. G., D. G. Martinson, T. C. Moore, N. J. Shackleton, W. Prell, J. D. Hays, and G. Boden, High resolution stratigraphic correlation of benthic oxygen isotope records spanning the last 300,000 years. Marine Geol. 56, 119-136, 1984.

Ruddiman, W. R. and A. McIntyre, A., Northeast Atlantic paleoclimatic changes over the past 600,000 years. In Investigation of Late Quaternary Paleoceanography and Paleoclimatology, ed. R. M. Cline and J. D. Hays, pp. 111-146. Geol. Soc. Amer. Memoir 145, 1976.

Sarnthein, M., H. Erlenkeuser, R. von Grafenstein, and C. Schroeder, Stable-isotope stratigraphy for the last 750,000 years: "Meteor" core 13519 from the eastern equatorial Atlantic. "Meteor" Forsch.-Ergebnisse C, 38, 9-24, 1984.

Shackleton, N. J., The measurement of palaeotemperatures in the Quaternary Era. PhD Thesis, Univ. Cambridge, 1967.

Shackleton, N. J., Oxygen isotope analyses and Pleistocene temperatures re-assessed. Nature 215, 15-17, 1967.

Shackleton, N. J. The last interglacial in the marine and terrestrial records. Proc. Roy. Soc. (London) B 174, 135-154, 1969.

Shackleton, N. J., Carbon-13 in Uvigerina: tropical rainforest history and the Equatorial Pacific carbonate dissolution cycles. In The fate of fossil fuel CO_2 in the oceans, ed. N. R. Andersen and A. Malahoff, pp. 401-427. New York: Plenum Press, 1977.

Shackleton, N. J., J. Backman, H. Zimmerman, D. V. Kent, M. A. Hall, D. G. Roberts, D. Schnitker, J. G. Baldauf, A. Desprairies, R. Homrighousen, P. Huddlestun, J. B. Keene, A. J. Kaltenback, K. A. O. Krumsiek, A. C. Morton, J. W. Murray, and J. Westberg-Smith, Oxygen isotope calibration of the onset of ice-rafting and history of glaciation in the North Atlantic region. Nature 307, 620-623, 1984.

Shackleton, N. J., M. A. Hall, J. Line, and Cang Shuxi, Carbon isotope data in core V19-30 confirm reduced carbon dioxide concentration in the ice age atmosphere. Nature 306, 319-322, 1983.

Shackleton, N. J., J. Imbrie, and M. A. Hall, Oxygen and carbon isotope record of East Pacific core V19-30: implications for the formation of deep water in the North Atlantic. Earth Planet. Sci. Lett. 65, 233-244, 1983.

Shackleton, N. J., and N. D. Opdyke, Oxygen isotope and palaeomagnetic stratigraphy of equatorial Pacific core V28-238: oxygen isotope temperatures and ice volumes on a 10^5 year and 10^6 year scale. Quaternary Res. 3, 39-55, 1973.

Shackleton, N. J., and N. D. Opdyke, Oxygen isotope and palaeomagnetic stratigraphy of equatorial Pacific core V28-239, Late Pliocene to latest Pleistocene. In Investigation of Late Quaternary Paleoceanography and Paleoclimatology, ed. R. M. Cline and J. D. Hays, pp. 449-464. Geol. Soc. Amer. Memoir 145, 1976.

Shackleton, N. J., and N. D. Opdyke, Oxygen isotope and palaeomagnetic evidence for early Northern Hemisphere glaciation. Nature 270, 216-219, 1977.

Shackleton, N. J. and N. G. Pisias, Atmospheric Carbon Dioxide, Orbital Forcing and Climate, in The Carbon Cycle and Atmospheric CO_2: Natural

Variations Archean to Present, Geophysical Monograph 32, edit. E. T. Sundquist and W. S. Broecker, pp. 303-317, AGU Washington D.C., 1985.

Shaw, A. B. Time in Stratigraphy. New York: McGraw-Hill, 365pp, 1964.

Thierstein, H. R., K. Geitzenauer, B. Molfino, and N. J. Shackleton, Global synchroneity of Late Quaternary coccolith datums: validation by oxygen isotopes. Geology 5, 400-404, 1977.

Thompson, P. R., Planktonic foraminiferal dissolution and the progress towards a Pleistocene equatorial Pacific transfer function. Jour. Foraminiferal Res., 6, 208-227, 1976.

Thompson, P. R., A. W. H. Be, J.-C. Duplessy, and N. J. Shackleton, Disappearance of pink-pigmented Globigerinoides ruber at 120,000 yr BP in the Indian and Pacific Oceans. Nature 280, 554-557, 1979.

Vergnaud-Grazzini, C., W. B. F. Ryan, and M. B. Cita, Stable isotopic fractionation, climatic change and episodic stagnation in the eastern Mediterranean during the Late Quaternary. Marine Micropaleontol. 2, 353-370, 1977.

Walcott, R. I., Past sea levels, eustacy and deformation of the earth. Quaternary Res. 2, 1-14, 1972.

Williams, D. F., W. S. Moore, and R. H. Fillon, Role of glacial Arctic Ocean ice sheets in Pleistocene oxygen isotope and sea level records. Earth and Planetary Science Letters 56, 157-166, 1981.